Nuclear Cardiology

Cardiovascular Clinics Series

*Not Available

Nuclear Cardiology

James T. Willerson, M.D. | Editor

Professor of Medicine
The University of Texas
Southwestern Medical School at Dallas
Dallas, Texas

CARDIOVASCULAR CLINICS

Albert N. Brest, M.D. | Editor-in-Chief

James C. Wilson Professor of Medicine
Director, Division of Cardiology
Jefferson Medical College
Philadelphia, Pennsylvania

 **F. A. DAVIS COMPANY, PHILADELPHIA**

Library of Congress Cataloging in Publication Data

Main entry under title:

Nuclear cardiology.

(Cardiovascular clinics; v. 10, no. 2)
Includes bibliographical references and index.
1. Heart—Diseases—Diagnosis.   2. Radioisotopes in cardiology.   3. Radioisotope scanning.   I. Willerson, James T.   II. Series. [DNLM: 1. Heart diseases—Radionuclide imaging. W1 CA77N v. 10 no. 2 / WG141.5.R2 N964]
RC681.A1C27 vol. 10, no. 2 [RC683.5.R33] 78-31500
ISBN 0-8036-9329-X   616.1'008s   [616.1'2'07575]

Preface

Nuclear cardiology is a rapidly growing field that has made very remarkable progress during the past several years. Developments in this area now place several different, relatively noninvasive methodologies at the disposal of the cardiovascular clinician and scientist, allowing improved diagnostic capability and/or assessment of therapeutic effect in patients with many different types of cardiovascular problems. The purpose of this book is to describe and place in proper perspective these radionuclide techniques. Theory, practice, usefulness, and interpretation of these radionuclide approaches are discussed in detail with the expectation that further understanding will aid the concerned physician in selecting the intervention potentially most valuable in a particular clinical situation. It is our hope that such an approach will make this book valuable for all those interested in cardiovascular disease.

I wish to thank all of the contributing authors for their detailed and well-developed discussions of special topics in nuclear cardiology. I would also like to express my appreciation for their patience and understanding in meeting necessary deadlines and in providing manuscripts as up-to-date as possible. I acknowledge with gratitude the constructive criticism and help of my colleagues and collaborators at the University of Texas Health Science Center at Dallas, and in this regard I would like to extend special thanks to Dr. Robert Parkey, Dr. L. Maximilian Buja, Dr. Samuel Lewis, and Dr. Frederick Bonte.

James T. Willerson, M.D.
Guest Editor

Editor's Commentary

Although nuclear cardiology had its beginnings many years earlier, the broad clinical benefits of this technology have fully dawned only during the past 20 years. Interest in this discipline has burgeoned as its clinical potential unfolds. Included among its important clinical applications are recognition of myocardial ischemia and infarction, visualization of cardiac chambers, measurement of right and left ventricular function, and quantification of intracardiac shunts. This issue of *Cardiovascular Clinics* aims to explore some of the important fundamental aspects of nuclear cardiology as well as clinical utility and ongoing developments. I am extremely grateful to Dr. James T. Willerson for his invaluable guidance in the formulation of this issue, and both of us are deeply indebted to the individual contributors for their generous participation in this endeavor.

Albert N. Brest, M.D.
Editor-in-Chief

Contributors

Page A. W. Anderson, M.D.
Associate Professor of Pediatrics, Duke University Medical Center, Durham, North Carolina

George A. Beller, M.D.
Professor of Medicine, Head, Cardiology Division, University of Virginia Medical Center, Charlottesville, Virginia

Harvey J. Berger, M.D.
Research Associate in Cardiovascular Nuclear Medicine, Yale University School of Medicine, New Haven, Connecticut

Frederick J. Bonte, M.D.
Dean and Professor of Radiology, The University of Texas Southwestern Medical School at Dallas, Dallas, Texas

Thomas F. Budinger, M.D., Ph.D.
Henry Miller Professor of Medical Research, University of California, Berkeley, Berkeley, California

L. Maximilian Buja, M.D.
Associate Professor of Pathology, The University of Texas Southwestern Medical School at Dallas, Dallas, Texas

James H. Caldwell, M.D.
Staff Cardiologist, Veterans Administration Hospital; Associate Investigator, Division of Cardiology, University of Washington, Seattle, Washington

Paul J. Cannon, M.D.
Professor of Medicine, College of Physicians and Surgeons of Columbia University, New York, New York

John T. Fallon, M.D.
Assistant Professor in Pathology, Harvard Medical School, Boston, Massachusetts

William R. Gray, M.D.
Assistant Professor of Radiology, The University of Texas Southwestern Medical School at Dallas, Dallas, Texas

Glen W. Hamilton, M.D.
Associate Professor of Medicine, University of Washington; Chief, Nuclear Medicine, Veterans Administration Hospital, Seattle, Washington

Robert H. Jones, M.D.
Assistant Professor of Surgery and Radiology, Howard Hughes Medical Investigator, Department of Surgery, Duke University Medical Center, Durham, North Carolina

Milton S. Klein, M.D.
Assistant Professor of Medicine, Washington University School of Medicine, St. Louis, Missouri

Robert W. Parkey, M.D.
Professor and Chairman, Department of Radiology, The University of Texas Southwestern Medical School at Dallas, Dallas, Texas

Bertram Pitt, M.D.
Professor of Internal Medicine, Director, Cardiovascular Division, The University of Michigan Medical School, Ann Arbor, Michigan

Gerald M. Pohost, M.D.
Assistant Professor of Medicine, Harvard Medical School; Assistant in Medicine, Massachusetts General Hospital, Boston, Massachusetts

James L. Ritchie, M.D.
Staff Cardiologist, Veterans Administration Hospital; Assistant Professor of Medicine, University of Washington, Seattle, Washington

Burton E. Sobel, M.D.
Professor of Medicine, Director, Cardiovascular Division, Washington University School of Medicine, St. Louis, Missouri

Peter M. Scholz, M.D.
Scholar in Academic Surgery, Duke University Medical Center, Durham, North Carolina

H. William Strauss, M.D.
Assistant Professor of Radiology, Harvard Medical School; Assistant Radiologist, Massachusetts General Hospital, Boston, Massachusetts

Donald Twieg, Ph.D.
Assistant Professor of Radiology, The University of Texas, Southwestern Medical School at Dallas, Dallas, Texas

James T. Willerson, M.D.
Professor of Medicine, The University of Texas Southwestern Medical School at Dallas, Dallas, Texas

Barry L. Zaret, M.D.
Associate Professor of Medicine and Diagnostic Radiology, Yale University School of Medicine, New Haven, Connecticut

Past, Present and Future
of Nuclear Cardiology

Frederick J. Bonte, M.D., Robert W. Parkey, M.D., and James T. Willerson, M.D.

The field of nuclear cardiology was established with the brilliant experiment of Blumgart and Weiss[1] in 1927. These investigators utilized the radioactive tracer method, which had been described earlier by Hevesy,[2] of injecting a dose of radium C salt into an arm vein, and measuring circulation time by recording the arrival of radioactivity in the opposite arm with a Wilson cloud chamber.

This same experiment was revived by Prinzmetal and associates[3] in 1948, using the artificial radionuclide ^{24}Na, a Geiger-Muller counter, and a recorder. With the new tools of the Atomic Age these investigators repeated the work of Blumgart and Weiss on determination of circulation time, and they also recorded the passage of radioactivity through the heart and lungs. The product of this study was the "radiocardiogram," which Prinzmetal and his colleagues proposed as a means of studying the function of the left ventricle in health and disease, and detecting shunts. In 1949 Kety and coworkers[4] described a method of determining blood flow by measuring the rate of clearance of ^{24}Na from a subcutaneous or intramuscular injection site.

With their associates Veall[5] and MacIntyre[6] learned how to derive the cardiac output from the left ventricular portion of the radiocardiographic curve. Although more complex versions of detection equipment have since been devised, the radiocardiogram has survived to the present era, having been refined by Donato[7] and Van Dyke[8] and their coworkers. Single- and multiple-probe recording techniques may now also be used to estimate pulmonary blood volume and ventricular ejection fraction. Although image-based methods have largely replaced the radiocardiogram, this technique is still a valuable one, for it is noninvasive and may be performed easily with relatively inexpensive detecting and recording equipment.

Imaging of the cardiovascular system with radionuclides was first described by Rejali and his colleagues[9] in 1958. These investigators reasoned that if one injected the newly available tracer, ^{131}I-labeled human serum albumin, into a patient's vein and allowed it to come to equilibrium in the body vascular space, it might be possible to image that space with a rectilinear scanner. Resultant images were initially used to detect pericardial effusions. Other teams, such as those of Bonte,[10-12] MacIntyre,[13] Sklaroff,[14] and Wagner,[15] found that blood pool imaging could be used for four general purposes: (1) identification of pericardial effusion; (2) differential diagnosis of midline thoracoabdominal masses between aneurysms and solid tumors; (3) normal and abnormal anatomy of mediastinal arteries and large veins; and (4) detection of certain abnormalities of the heart, such as gross chamber enlargement and intracardiac tumors.

Important changes came to blood pool imaging technology with the availability of the

large-crystal scintillation camera, described by Anger[16] and ^{99m}Tc-labeled tracers, developed by Richards[17] and by Harper[18] and their colleagues. It now became possible to view the passage of a tracer bolus through the central venous and arterial circulation as a dynamic series of events in multiple, serial scintillation camera images, rather than as a summation of events occurring over a long period of time, collected in a static image by a rectilinear scanner. The test which was evolved came to be called radionuclide angiocardiography[19-21]; it is still used as a convenient, accurate means of appraising mediastinal vascular anatomy and the general state of cardiac function and lung perfusion.

The concept of storing and processing radionuclide image data using magnetic tape and specialized circuits was developed by MacIntyre[23] and by Bonte[24] and their associates. Processing was significantly advanced by Brown[25] in 1964, when he described the use of a digital computer to analyze and display images derived from a rectilinear scanner. These techniques were rapidly adapted to the processing of scintillation camera images and led to the development of sophisticated nuclear angiographic systems by the teams of Ashburn[26] and Wellman[27] in 1968, and by Kriss and coworkers[28] in 1971. By 1971 nuclear imaging equipment had reached such a degree of sophistication that it was possible to appraise volume changes in individual cardiac chambers. Using a variation of the classic roentgenographic method of Chapman and coworkers,[29] Mullins and Ashburn and their colleagues[30] employed radionuclide angiocardiography for the determination of left ventricular volume. This advance led to the rapid development of techniques for the estimation of cardiac output, and in 1971 a method for measuring left ventricular ejection fraction was devised by Strauss and his associates.[31]

With further refinement in image processing technology the cinematographic display of radionuclide tracer images of cardiac chambers could be obtained from gating mechanisms triggered by the patient's electrocardiogram. In addition to determining changes in the ventricular volume and ejection fraction, the nuclear physician could then evaluate the pump function of the left ventricle as well by observing contractility and elasticity of the chamber wall from several projections. This procedure, often termed a "wall-motion study"[32], provides useful information by identifying poorly functioning segments of ventricular wall, such as those which may exist following myocardial infarction. Motionless or paradoxically moving segments, such as those involved in ventricular aneurysm formation, may also be demonstrated. Thus, the concept of global ventricular function may now be applied to cineradionuclide imaging techniques.

The ability to derive quantitative information by processing stored image data also led to the development of procedures for identifying left-to-right and right-to-left shunts by Folse and Braunwald,[33] and by Rosenthall.[34] These investigators learned how to formulate flow curves derived from areas of interest within stored, digitized images; this provided information similar to that obtained with single-probe techniques. This finding led several groups of investigators to study the work of Kety and associates[35] who, in 1948, had utilized radionuclide tracers to determine blood flow in myocardium and other tissues. His work was developed into a procedure in which a diffusible tracer such as ^{133}Xe or ^{85}Kr is injected into a coronary artery—or into the heart muscle itself—and the disappearance, or "washout," of radioactivity was measured with a radiation detection system giving an index of tissue blood flow. At first this procedure was carried out with single-probe detectors, but later investigators learned how to utilize data derived from serially tape-recorded scintillation camera images to estimate myocardial blood flow by arbitrarily subdividing the recorded images into zones. With these techniques, several groups[36-38] have been able to obtain reproducible values for regional myocardial blood flow. The major work in this field has been done by Cannon and his colleagues;[36] they employed a multicrystal scintillation camera to calculate myocardial blood flow values for regions corresponding to each of the camera's crystals.

An alternative method for determining regional myocardial blood flow followed the

2

observation made by Sapirstein,[39] that organ blood flow could be measured by the fractional distribution of particulate tracers which were slightly larger in diameter than tissue capillaries. Quinn and coworkers[40] showed that Sapirstein's method could be utilized to make pictures of the distribution of coronary perfusion following intracoronary–arterial radioactive particle injection. This procedure was further developed by Ashburn[41] and by Jansen[42] and their colleagues; they performed particulate tracer injection studies of the distribution of myocardial perfusion in large numbers of patients, thus establishing its diagnostic value.

Both the diffusible and particulate tracer methods are invasive procedures, and although valuable as research tests, they require cardiac catheterization, in which there is a certain inherent morbidity. But nuclear physicians had long been eager to find radiotracers which would selectively label either normal or diseased myocardium. The distribution of a tracer which selectively labeled normal myocardium would be at once an index of myocardial perfusion and of myocardial cell function. Furthermore, if such a tracer were to be injected intravenously it might provide a noninvasive procedure for eliciting this valuable information.

Considerable work has been done over the years in the development of these "tracer families,"[43-45] with much of the important research attributable to Carr, Beierwaltes, and their colleagues. A number of radioisotopes of rubidium and cesium, analogs of the intracellular cation, potassium, have all been employed without a great deal of success. However, in 1971 ^{43}K became available.[46] Although the photon emissions of this radionuclide are of relatively high energy, rectilinear scans of good quality can be made with proper collimation. Strauss and his associates[47] have used ^{43}K not only to obtain diagnostic images of normal myocardium, but to demonstrate areas of transient ischemia occurring after exercise in a group of patients with coronary arterial disease.

Other agents have been proposed for the visualization of perfused, normal myocardium, some with partial success. In 1965 Evans and his group[48] showed that radioiodinated long-chain unsaturated fatty acids could be used as an imaging tracer for perfused myocardium, since the myocardium utilizes such substances as prime energy sources. Additional work by other groups has failed to develop a really satisfactory tracer of this sort for single photon imaging. However, Ter-Pogossian and coworkers[49] have used a positron-tomographic camera and the positron-emitting radionuclide ^{11}C incorporated into carbon monoxide or palmitate to image myocardium. Harper and associates[50] have used another positron-emitter, ^{13}N, in the form ammonia with interesting results.

The present tracer of choice for imaging the perfused and normally functioning myocardium is ^{201}Tl, introduced by Lebowitz and his colleagues[51] and utilized to advantage by many groups, such as that of Wackers.[52] With this agent, nuclear physicians can now demonstrate areas of transient ischemia following exercise, and can show unperfused areas representing fresh infarcts. ^{201}Tl can even be used to screen patients prior to admission to a coronary intensive care unit. Perfusion defects due to scarring by remote disease can also be demonstrated, although the ultimate cause for an area of nonlocalization of this tracer must be elicited by other means, such as history, ECG, ventriculography, enzyme tests, and infarct-avid imaging.

While the search for a means of labeling normal myocardium progressed, many of the same investigators were also looking for a compound which would selectively label a myocardial infarct. It is likely that the first successful experimental infarct scan was made in 1962 by Carr, Beierwaltes, and their associates[53] who used ^{203}Hg-chloromerodrin, an agent which was thought to label necrotic tissue. This tracer proved unsatisfactory for clinical use, however. Malek and coworkers[54] reported some success with ^{203}Hg-fluorescein, and subsequently experimented with labeled tetracycline. In 1973 Holman and his colleagues[55] developed a ^{99m}Tc-tetracycline and found it to be the best agent reported up until then for direct infarct imaging; however, optimum concentration of the tracer in infarct occurred late after administration, an unfortunate circumstance when the radionuclide label is a short-lived emitter such as [99m]Tc. In 1973 Bonte and his group[56] described the localization of ^{99m}Tc-stannous pyrophosphate in experimental infarcts in animals; later, with Parkey,[57] Willerson,[58] and others, he showed successful application of an infarct-avid imaging procedure in patients. Working with Buja[59] these investigators have shown that this radiopharmaceutical labels necrotic myocardium produced by infarct, by blunt trauma, or by the diffuse and chronic necrotizing process which is often found in patients who have unstable angina pectoris. A similar process also accompanies the development of ventricular aneurysms, so that positive ^{99m}Tc-phosphate images may be encountered with each of these entities. Widely used at the present time, infarct-avid imaging with ^{99m}Tc-phosphate compounds has also been found to be very useful in detecting infarct extensions during treatment, and in the evaluation of myocardial damage incidental to coronary jump-graft surgery.

At the present time the diagnosis and estimation of the consequences of myocardial infarction by nuclear imaging methods is usually carried out by some combination of testing with ^{201}Tl and ^{99m}Tc phosphate tracers, together with ECG-gated ventriculography performed with ^{99m}Tc-labeled human serum albumin or red cells. This combination can measure ejection fraction and evaluate wall motion. Today's nuclear physician now has at his command an impressive battery of diagnostic tests which can be done under noninvasive circumstances at the patient's bedside in the coronary care unit.

The immediate future of nuclear cardiology will see several significant advances. One of these will be the evolution of computer-assisted, three-dimensional image reconstruction, the concept originally introduced into nuclear medicine by Kuhl and Edwards in 1964.[60] Ter-Pogossian and associates[49] have been able to accomplish this objective with their positron-tomographic camera and positron-emitting tracers, as Kuhl and Edwards had done using single-photon tomographic techniques. It should be remembered, however, that positron-based imaging procedures require expensive detection equipment and the ready availability of a cyclotron: formidable requirements which, for now, place these techniques beyond the reach of the average hospital's nuclear medicine department. What is needed, therefore, in order to fully implement three-dimensional imaging in nuclear cardiology is a tomographic instrument utilizing single photon emissions from radionuclides such as ^{99m}Tc and [201]Tl.

Lewis and coworkers[61] have shown the potential of infarct-sizing inherent in three-dimensional reconstructional methods, and there is reason to believe that computerized axial tomographic systems based on single-photon imaging will soon be commercially available.[62] The impact of computerized axial tomographic scanning on diagnostic radiology has been enormous; radiologists were presented for the first time with cross-sectional views of anatomic structure rendered with exceedingly fine detail. There is no doubt that a similar impact will take place in diagnostic nuclear imaging when cross-sectional views become generally available to nuclear physicians.

Advances are being made in instrument systems. Furthermore, these systems will eventually be more routinely available at the patient's bedside in a coronary care unit, in the catheterization and exercise laboratories, and in other settings in which cardiologic testing is done. New techniques and radiopharmaceuticals[63] will become available for imaging of both the perfused and the infarcted myocardium. It seems likely, too, that nuclear cine-ventriculography and rapid, sequential derivation of ejection fraction and other similar variables will become routine procedures, and that these will be used for short-term evaluation of the effects of therapeutic interventions with drugs of various families.

Advances in circuitry and design, and their manufacture, will hopefully result in instrument systems which are more simplified and available at a lower cost. If this does not occur, nuclear cardiology will come to the unfavorable attention of health planners as an ultra–high-cost technology, which, for reasons of its cost alone, might be withheld from patients who would benefit from its application.

4

In the meantime, two formidable competing technologies have appeared. The first of these is the application of computerized axial roentgen tomography (CAT) for detecting myocardial infarcts; this has been used successfully on an experimental basis.[64] It is not impossible to imagine the ultimate evolution of a CAT unit with roentgen cinematographic capabilities, in which all aspects of myocardial structure and function might become accessible to the nuclear physician.

The second competitor is ultrasound, which has now attained cinematographic capability, permitting estimates of global ventricular function and chamber function. The development of a cinematographic ultrasound unit with capabilities similar to those of existing nuclear diagnostic units would compel the attention of the thoughtful physician who is anxious to avoid even those small risks inherent in the use of ionizing radiation for diagnostic purposes.

REFERENCES

1. BLUMGART, H. L., AND WEISS, S.: *Studies on the velocity of blood flow.* J. Clin. Invest. 4:15, 1927.

2. HEVESY, G.: *The absorption and translocation of lead by plants: A contribution to the application of the method of radioactive indicators in the investigation in the change of substance in plants.* Biochem. J. 17:439, 1923.

3. PRINZMETAL, M., CORDAY, E., BERGMAN, H. C., ET AL.: *Radiocardiography: A new method for studying the blood flow through the chambers of the heart in human beings.* Science 108:340, 1948.

4. KETY, S. S.: *Measurement of regional circulation by the local clearance of radioactive sodium.* Am. Heart J. 38:321, 1949.

5. VEALL, N., PEARSON, J. D., HENLEY, E., ET AL.: *A method for determination of cardiac output.* Radioisotopes Conference, Oxford-Medical and Physiological Applications. London, Butterworth, 1954, p. 183.

6. MACINTYRE, W. J., PRITCHARD, W. H., AND MOIR, T. W.: *The determination of cardiac output by the dilution method without arterial sampling. I: Analytic concepts.* Circulation 18:1139, 1958.

7. DONATO, L., GIUNTINI, C., LEWIS, M. L., ET AL.: *Quantitative radiocardiography. I: Theoretical considerations.* Circulation 26:174, 1962.

8. VAN DYKE, D., ANGER, H. A., SULLIVAN, R. W., ET AL.: *Cardiac evaluation from radioisotope dynamics.* J. Nucl. Med. 13:585, 1972.

9. REJALI, A. M., MACINTYRE, W. J., AND FRIEDELL, H. L.: *Radioisotope method of visualization of blood pools.* Am. J. Roentgenol. 79:129, 1958.

10. BONTE, F. J., KROHMER, J. S., TSENG, C. H., ET AL.: *Scintillation scanning in differential diagnosis: Thoracoabdominal midline masses.* JAMA 175:221, 1961.

11. BONTE, F. J., ANDREWS, G. J., ELMENDORF, E. A., ET AL.: *Radioisotope scanning in the detection of pericardial effusion.* South. Med. J. 55:577, 1962.

12. BONTE, F. J., CHRISTENSEN, E. E., AND CURRY, T. S.: *Tc-99m pertechnetate angiocardiography in the diagnosis of superior mediastinal masses and pericardial effusions.* Am. J. Roentgenol. 107:404, 1969.

13. MACINTYRE, W. J., CRESPO, G. G., AND CHRISTIE, J. H.: *The use of radioiodinated (I-131) iodipamide for cardiovascular scanning.* Am. J. Roentgenol. 89:315, 1963.

14. SKLAROFF, D. M., CHARKES, N. D., AND MORSE, D.: *Measurement of pericardial fluid: Correlation with I-131 cholografin IHSA heart scan.* J. Nucl. Med. 5:101, 1964.

15. WAGNER, H. N., JR., MCAFEE, J. G., AND MOZLEY, J. M.: *Diagnosis of pericardial effusion by radioisotope scanning.* Arch. Intern. Med. 108:679, 1961.

16. ANGER, H. O.: *Scintillation camera.* Rev. Sci. Instrum. 29:27, 1958.

17. RICHARDS, P.: *A survey of the production at Brookhaven National Laboratory of radioisotopes for medical research.* In Trans. 5th Nuclear Congress, New York, IEEE, 1960, pp. 225-244.

18. HARPER, P. V., LATHROP, K. A., JIMENEZ, F., ET AL.: *Technetium-99m as a scanning agent.* Radiology 85:101, 1965.

19. BONTE, F. J., CHRISTENSEN, E. E., CURRY, T. S., III: *Tc-99m pertechnetate angiocardiography in the diagnosis of superior mediastinal masses and pericardial effusions.* Am. J. Roentgenol. 107:404, 1969.

20. GOTTSCHALK, A.: *Radioisotope scintiphotography with technetium-99m and gamma scintillation camera.* Am. J. Roentgenol. 97:860, 1966.

21. KRISS, J. P., YEH, S. H., FARRER, P. A., ET AL.: *Radioisotope angiocardiography*. J. Nucl. Med. 7:367, 1966.

22. ROSENTHALL, L.: *Radionuclide venography using technetium-99m pertechnetate and gamma-ray scintillation camera*. Am. J. Roentgenol. 97:874, 1966.

23. MACINTYRE, W. J., FRIEDELL, H. L., CRESPO, G. G., ET AL.: *Accentuation scintillation scanning*. Radiology 73:329, 1959.

24. BONTE, F. J., KROHMER, J. S., AND ROMANS, W. E.: *Magnetic tape recording of scintillation scan data*. Int. J. Appl. Radiat. Isot. 14:273, 1963.

25. BROWN, D. W.: *Digital computer analysis display of the radioisotope scan*. J. Nucl. Med. 5:802, 1964.

26. ASHBURN, W. L., HARBERT, J. C., WHITEHOUSE, W. C., ET AL.: *A video system for recording dynamic radioisotope studies with the Anger scintillation camera*. J. Nucl. Med. 9:554, 1968.

27. WELLMAN, H. N., HUNKAR, D., KEREIAKES, J., ET AL.: *A new concept in dynamic function studies—quantitative cinescintivideography*. J. Nucl. Med. 9:420, 1968.

28. KRISS, J. P., ENRIGHT, L. P., HAYDEN, W. G., ET AL.: *Radioisotopic angiocardiography*. Circulation 43:792, 1971.

29. CHAPMAN, C. B., BAKER, O., REYNOLDS, J., ET AL.: *Use of biplane cinefluorography for measurement of ventricular volume*. Circulation 18:1105, 1958.

30. MULLINS, C. B., MASON, D. T., ASHBURN, W. L., ET AL.: *Determination of ventricular volume by radioisotope angiography*. Am. J. Cardiol. 24:72, 1969.

31. STRAUSS, H. W., ZARET, B. L., HURLEY, P. J., ET AL.: *A non-invasive scintiphotographic method for measuring left ventricular ejection fraction in man without cardiac catheterization*. Am. J. Cardiol. 28:575, 1971.

32. ZARET, B. L., STRAUSS, H. W., HURLEY, P. J., ET AL.: *A non-invasive scintiphotographic method for detecting regional ventricular dysfunction in man*. N. Engl. J. Med. 284:1165, 1971.

33. FOLSE, R., AND BRAUNWALD, E.: *Pulmonary vascular dilution curves recorded by external detection in diagnosis of left-to-right shunts*. Br. Heart J. 24:166, 1962.

34. ROSENTHALL, L.: *Nucleographic screening of patients for left-to-right cardiac shunts*. Radiology 99:601, 1971.

35. KETY, S. S., AND SCHMIDT, C. S.: *The determination of cerebral blood flow in man by use of nitrous oxide in low concentrations*. Am. J. Physiol. 144:53, 1945.

36. CANNON, E. J., DELL, R. B., AND DWYER, E. M.: *Measurement of regional myocardial perfusion in man with 133-xenon and a scintillation camera*. J. Clin. Invest. 51:964, 1972.

37. BENDER, M. A., AND BLAU, M.: *The autofluoroscope*. Nucleonics 21:52, 1963.

38. BONTE, F. J., PARKEY, R. W., STOKELY, E. M., ET AL.: *Radionuclide determination of myocardial blood flow*. Semin. Nucl. Med. 3:153, 1973.

39. SAPIRSTEIN, L. A.: *Regional blood flow by fractional distribution of indicators*. Am. J. Physiol. 193:161, 1958.

40. QUINN, J. L., III, SERRATTO, M., AND KEZDI, P.: *Coronary artery bed photoscanning using radioiodine albumin macroaggregates (RAMA)*. J. Nucl. Med. 7:107, 1966.

41. ASHBURN, W. L., BRAUNWALD, E., SIMON, A. L., ET AL.: *Myocardial perfusion imaging with radioactive-labeled particles injected directly into the coronary circulation of patients with coronary artery disease*. Circulation 44:851, 1971.

42. JANSEN, C., JUDKINS, M. D., GRAMES, G. M., ET AL.: *Myocardial perfusion color scintigraphy with MAA*. Radiology 109:369, 1973.

43. CARR, E. A., JR., BEIERWALTES, W. H., WEGST, A. V., ET AL.: *Myocardial scanning with rubidium-86*. J. Nucl. Med. 3:76, 1962.

44. NOLTING, D., MACK, R., LUTHY, E., ET AL.: *Measurement of coronary blood flow and myocardial rubidium uptake with Rb-86*. J. Clin. Invest. 37:921, 1958.

45. ROMHILT, D. W., ADOLPH, R. J., SODD, V. S., ET AL.: *Cesium-129 myocardial scintigraphy to detect myocardial infarct*. Circulation 48:1242, 1973.

46. HURLEY, P. J., COOPER, M., REBA, R. C., ET AL.: *^{43}KCl: A new radiopharmaceutical for imaging the heart*. J. Nucl. Med. 12:516, 1971.

47. STRAUSS, H. W., ZARET, B. L., MARTIN, N. D., ET AL.: *Non-invasive evaluation of regional myocardial perfusion with potassium-43: Technique in patients with exercise induced transient myocardial ischemia*. Radiology 108:85, 1973.

48. EVANS, J. R., GUNTON, R. W., BAKER, R. G., ET AL.: *Use of radioiodinated fatty acid for photoscans of the heart*. Circ. Res. 16:1, 1965.

49. TER-POGOSSIAN, M. M., PHELPS, M. E., HOFFMAN, E. J., ET AL.: *Positron-emission transaxial tomograph for nuclear imaging (PETT)*. Radiology 114:89, 1975.

50. HARPER, P. V., LATHROP, K. A., KRIZEK, H., ET AL.: *Clinical feasibility of myocardial imaging with $^{13}NH_3$*. J. Nucl. Med. 13:278, 1972.

51. LEBOWITZ, E., GREENE, M. W., BRADLEY-MOORE, P., ET AL.: *Thallium-201 for medical use*. J. Nucl. Med. 14:421, 1973.

52. WACKERS, F. J. T. H., BUSEMANN SOKOLE, E., SAMPSON, G., ET AL.: *Value and limitations of ^{201}Tl scintigraphy in acute myocardial infarction*. N. Engl. J. Med. 295:1, 1976.

53. CARR, E. A., BEIERWALTES, W. H., PATENO, M. E., ET AL.: *The detection of experimental myocardial infarcts by photoscanning*. Am. Heart J. 64:650, 1962.

54. MALEK, P., VAVREJN, B., RATUSKY, J., ET AL.: *Detection of myocardial infarction by in vivo scanning*. Cardiologia 51:22, 1967.

55. HOLMAN, B. L., DEWANJEE, M. K., IDOINE, J., ET AL.: *Detection and localization of experimental myocardial infarction ^{99m}Tc-tetracycline*. J. Nucl. Med. 14:595, 1973.

56. BONTE, F. J., PARKEY, R. W., GRAHAM, K. D., ET AL.: *A new method for radionuclide imaging of myocardial infarcts*. Radiology 110:473, 1974.

57. PARKEY, R. W., BONTE, F. J., MEYER, S. L., ET AL.: *A new method for radionuclide imaging of acute myocardial infarction in humans*. Circulation 50:540, 1974.

58. WILLERSON, J. T., PARKEY, R. W., BONTE, F. J., ET AL.: *Technetium stannous pyrophosphate myocardial scintigrams in patients with chest pain of varying etiology*. Circulation 51:1046, 1975.

59. BUJA, L. M., PARKEY, R. W., STOKELY, E. M., ET AL.: *Pathophysiology of technetium-99m stannous pyrophosphate and thallium-201 scintigraphy of acute anterior myocardial infarcts in dogs*. J. Clin. Invest. 57:1508, 1976.

60. KUHL, D. E., AND EDWARDS, R. Q.: *Cylindrical and section radioisotope scanning of the liver and brain*. Radiology 83:926, 1964.

61. LEWIS, M., BUJA, M., SAFFER, S., ET AL.: *Experimental infarct sizing utilizing computer processing and three-dimensional model*. Science 197:169, 1977.

62. PHELPS, M. E.: *Emission computed tomography*. Semin. Nucl. Med. 7:337, 1977.

63. CARIDE, V. J., AND ZARET, B. L.: *Liposome accumulation in regions of experimental myocardial infarction*. Science 198:735, 1977.

64. ADAMS, D. F., HESSEL, S. J., AND JUDY, P. F.: *Computed tomography of the normal and infarcted myocardium*. Am. J. Roentgenol. 126:786, 1976.

Physiology and Physics
of Nuclear Cardiology*

Thomas F. Budinger, M.D., Ph.D.

Radionuclide cardiology encompasses the assessment of heart mechanical function, blood flow, and myocardial muscle metabolism by measurement of the distribution of radioactivity after injection of various radiopharmaceuticals. The vast majority of clinically useful procedures involves use of external radionuclide imaging cameras of the Anger type,[1] the modified Bender-Blau multicrystal type,[2,3] or the positron emission computed tomographic devices.[4-8] The great appeal of nuclear cardiology is that both myocardial cell metabolism and function can be assessed by noninvasive procedures. The term "nuclear cardiology" encompasses the following procedures and concepts:

Heart Muscle	*Blood Volumes and Chamber Flow*
Myocardial imaging	Radioangiocardiography
Myocardial perfusion	Scintiangiography
Myocardial scintigraphy	Radionuclide blood pool imaging
Infarction imaging—	Gated blood pool imaging
(positive or negative)	Initial pass radionuclide imaging
Radiocardiography	

All of these terms and their various combinations are self-explanatory to the cardiologist with the possible exception of the prefix "scinti" which denotes scintillation or production of visible light in a crystal receiving gamma rays.

A useful division of the field of radionuclide cardiology separates procedures which involve evaluation of ventricular function from those which investigate myocardial blood flow and heart muscle metabolism. The former category encompasses ventriculography for measurement of ejection fraction, stroke volume, shunts, congenital anomalies, and interchamber transit times; whereas the second category encompasses infarct imaging (positive or negative), microsphere and xenon myocardial capillary flow measurements, and sugar, fatty acid, and amino acid metabolism assessment by emission computed tomography. Dwyer[9] reviewed some of these procedures in an earlier volume of this series.

The purpose of this chapter is to present a primer of the physics, instrumentation, and physiology which form the basis of radionuclide cardiology. In just the last few years there have been significant advances in instrumentation and radiopharmaceuticals; this chapter will attempt to give the cardiologist and cardiovascular researcher some perspective on the

*This work was supported by National Institutes of Health Grant 94-6031193 and National Cancer Institute Contract CB-50304 and the Department of Energy.

diversity of procedures, new instruments, and potentials of nuclear cardiology. We start with a basic introduction to the radionuclides involved in imaging and a description of the applicable instrumentation. Next we examine the basic physiology of flow and methods of evaluating the nutritional state of the myocyte. Finally we review the various procedures of noninvasive examination of ventricular function and heart muscle integrity. Each of these four major sections—*physics, flow and transit time principles, myocardial perfusion and metabolism,* and *angiography*—can be read of itself and does not depend on the other sections.

PHYSICS

The Radionuclide

Nuclear medicine imaging studies involve injection of an isotope (radionuclide) that emits a high energy gamma ray. Gamma rays are produced during radioactive decay when the nucleus of the isotope undergoes a transformation from one nuclear energy level to a lower one. In general, it is best to designate these radiations as photons with some stated energy. Radiowaves and visual light are also photons, distinguishable from x-ray and gamma radiations by their energy. In general, physicists refer to these radiations as photons whether they be light, x-rays, or gamma rays. Photons are characterized by their energy which is based on the electron volt or the energy gained by an electron (or other particle carrying a single charge) when it is accelerated through an electrical field with a potential difference of 1 V. An electron volt (eV) is a very small unit of energy; for example, a million billion electron volts are equivalent to the energy of one drop of water that has fallen 10 cm. The light reflecting from this page is comprised of about 10^5 photons per second and each photon has an energy of about 2 eV. The relationships between various methods of designating energy can be summarized as follows:

$$1\,eV = \begin{cases} 1.6 \times 10^{-19}\ \text{joule (or newton meter)} \\ 1.6 \times 10^{-19}\ \text{watt sec} \\ 1.6 \times 10^{-12}\ \text{erg (or dyne cm)} \\ 3.6 \times 10^{-20}\ \text{calorie (cal)} \end{cases}$$

Notice that the only difference between a calorie (as a designator of energy) and the electron volt is the conversion factor 3.6×10^{-20}; that is, we need 2.8×10^{19} electron volts to give the energy equivalent to 1 cal. Thus, to raise the temperature of 1 gm of water 1 °C we need 2.8×10^{19} eV; or, to get this amount of energy in 1 gm we need 28 billion (28×10^9) photons, each having an energy of 1 billion electron volts (1 billion electron volts = 1000 million electron volts = 1,000,000 kiloelectron volts = 10^9 electron volts). The photons emitted from radioisotopes can thus be visualized as tiny bullets traveling at the velocity of light with an energy measured in terms of kiloelectron volts (1000 eV). That is, each gamma ray has an energy more than 1000 times that of each light photon. For example, thallium-201 (^{201}Tl), an isotope used for myocardial imaging, emits x-rays whose photon energy is about 80,000 eV or 80 kiloelectron volts (80 keV). Other isotopes may give off energies of 1000 keV or 1 million electron volts (1 MeV).

The energies given off by radioisotopes which are used in nuclear cardiology are given in Table 1. These are injected into the patient; they localize in the myocardium unless they are attached to albumin or red cells, or chelated to some compound, in which case they remain in the blood pool. Thousands of photons must be detected by external devices in order to produce an image of the myocardium or the ventricular blood pools; 300,000 photons is a typical amount.

Radioisotopes used in nuclear cardiology decay in one of four ways: electron emission

Table 1. Decay characteristics of radionuclides used for cardiovascular studies

Tracer	Half-life	Decay Mode	Principal Photons (keV) and Intensity (%)*
^{38}K	7.7 min	β^+	511, 200%; 2170, 100%
^{42}K	12.4 hr	β^-	1525, 18%
^{43}K	22.2 hr	β^-	373, 85%; 618, 81%
^{81}Rb	4.58 hr	β^+, EC	446, 24%; 511, 67%; 190 (^{81m}Kr daughter)
^{82}Rb	1.25 min	β^+, EC	551, 192%; 776, 13%
^{84}Rb	33.0 days	β^+, EC	511, 38%; 882, 73%
^{86}Rb	18.7 days	β^-	1079, 9%
^{127}Cs	6.2 hr	β^+, EC	125, 18%; 411, 64%
^{130}Cs	30.0 min	EC, β^+, β^-	511, 92%; 35, 2%
^{131}Cs	9.7 days	EC	31 to 36, 88%
^{129}Cs	32.0 hr	EC	372, 32%; 411, 22%
^{134m}Cs	2.9 hr	IT	128, 14%
^{13}NH$_3$	10.0 min	β^+	511, 200%
^{201}Tl	73.0 hr	EC	69 to 83, 93%; 135, 2%; 167, 8%
^{127}Xe	36.0 days	EC	172, 21%; 203, 61%; 375, 18%
^{133}Xe	5.3 days	β^-	81, 35%
^{135}Xe	9.2 hr	β^-	250, 92%; 609, 3%
^{11}C	20.5 min	β^+	511, 200%
^{15}O	2.0 min	β^+	511, 200%
^{99m}Tc	6.0 hr	IT	140, 90%
^{113m}In	100.0 min	IT	393, 64%
^{123}I	13.0 hr	EC	160, 83%

*An intensity of 200% for positron annihilation photons corresponds to a 100% decay by positron emission since each positron produces two photons.

β^+ = positron emission; β^+ = electron emission; EC = K-electron capture; IT = isomeric transition.

(β^-), positron emission (β^+), isomeric transition, and K-electron capture (EC). For practical purposes it is only necessary to consider two types of decay: single photon or positron emission.

Single Photon Emission (EC, β^-, IT)

The term "single photon" is used to designate radionuclides whose photons arise from electron capture (EC), electron emission (β^-), or isomeric transition (IT). The single photon isotopes usually emit photons at a few different energies, but only one energy is normally used for detection. An example of EC is the decay of ^{201}Tl. Thallium is an element with 81 protons in the nucleus to balance the 81 electrons. The atomic number of stable (nonradioactive) thallium is 204; thus, there are 123 neutrons. The radionuclide ^{201}Tl has 120 neutrons (i.e., 120 neutrons + 81 protons = 201). Sometime during the life of one atom of ^{201}Tl the K-orbit electron flies close enough to a proton in the nucleus to result in a proton-electron interaction which creates another neutron. With the loss of a proton the atom becomes mercury, which has an atomic number of 80. But this atom of mercury (Hg) now has a vacant K-shell. The energy released when the K-shell vacancy is filled by a free electron is 80 keV, and that energy becomes a photon known as a mercury x-ray. In addition, about 8 percent of the time the mercury atom is in an excited state: 135 or 167 keV above its ground state. Thus, 8 percent of all ^{201}Tl decays result in a 135 keV or 167 keV photon in addition to the 80 keV photon of the Hg x-ray. Far more 80 keV Hg x-ray photons are available than are the 167 keV or 135 keV photons. Thus, since we confine our detection devices to accept only a definable band of energy, the "single photon" detected for ^{201}Tl is 80 keV. (Actually, this is a lumped number for the four characteristic x-rays of 69, 71, 80, and 83 keV.)

Radionuclides such as ^{86}Rb and ^{133}Xe decay by emission of an electron which results in the conversion of a neutron to a proton. The resulting atom will have an increase in atomic number by one.

Another isotope used commonly in radionuclear cardiology is a form of technetium, ^{99m}Tc. The "m" means that this radionuclide is in a metastable state. For most isotopes, the excited daughter nucleus decays to ground state immediately after emission of an electron; however, in some decay processes the daughter nucleus remains in an excited state for minutes or hours. Technetium-99m arises from the decay of ^{99}Mo. The intermediate nucleus ^{99m}Tc decays with a half-life of 6 hours by release of a 140-keV photon. ^{113m}In is another example of this process; this type of decay is known as isomeric transition.

More detail on types of decay is provided in Evans' 1955 text[10] and by Budinger and Rollo.[11]

Positron Emission (β^+)

Positron emission isotopes such as carbon-11, oxygen-15, nitrogen-13, potassium-38, rubidium-82, and gallium-68 undergo a transition in which the nuclear proton changes into a positive electron (positron). The positron is ejected from the nucleus and almost instantaneously combines with any nearby electron and annihilates into 1.022 MeV of "pure" energy in accordance with $E = mc^2$. This energy is equally divided between two photons which fly away from one another at a 180-degree angle. Each photon has an energy of 511 keV. The facts that these annihilation photons always escape at 180-degree angles from one another and arrive at opposing detectors in time coincidence or simultaneously are utilized for cardiovascular imaging, particularly computed tomography, which is discussed below.

Scattering and Attenuation

The basic concept of nuclear cardiology imaging is to determine how much radionuclide resides in some part of the body; this is done by detecting the number of photons which fly off in all directions through the use of an external detection instrument. Only a very few of the emitted photons ever reach the detector as there are two alternative possible fates: (1) photons may fly off in a direction other than that in which the detector is aimed, or (2) photons may be scattered by interaction with tissue atoms between the radionuclide and the detector.

The first problem is a combination of the solid angle effect and collimator complications. (The collimator is discussed below; it allows the viewer to focus only on the region responsible for the activity.) The solid angle effect is given by the ratio of the resolution area of the detector to $4\pi \times$ (distance between the source and detector).[2] The combined effect of collimator and solid angle leads to a transmission of about 1 in every 10,000 source photons from the isotopes used in nuclear cardiology.

The second problem, scatter and attenuation, results in a loss of 50 to 80 percent of the photons depending on the depth of the body and the energy of the isotope. The influence of depth on photon scatter is shown in Figure 1. Here the differential attenuation of an isotope in the posterior and anterior myocardium can be seen by noting the change in photon transmission with tissue depth or thickness. More information will reach the detector from the anterior myocardium than from the posterior myocardium. One of the reasons ^{131}Cs is still used for myocardial infarction scintigraphy is because 88 percent of its detected signal is from the anterior myocardium so that anterior defects can be seen with good contrast; with ^{43}K only 41 percent of the activity is from the anterior myocardium and 11 percent is

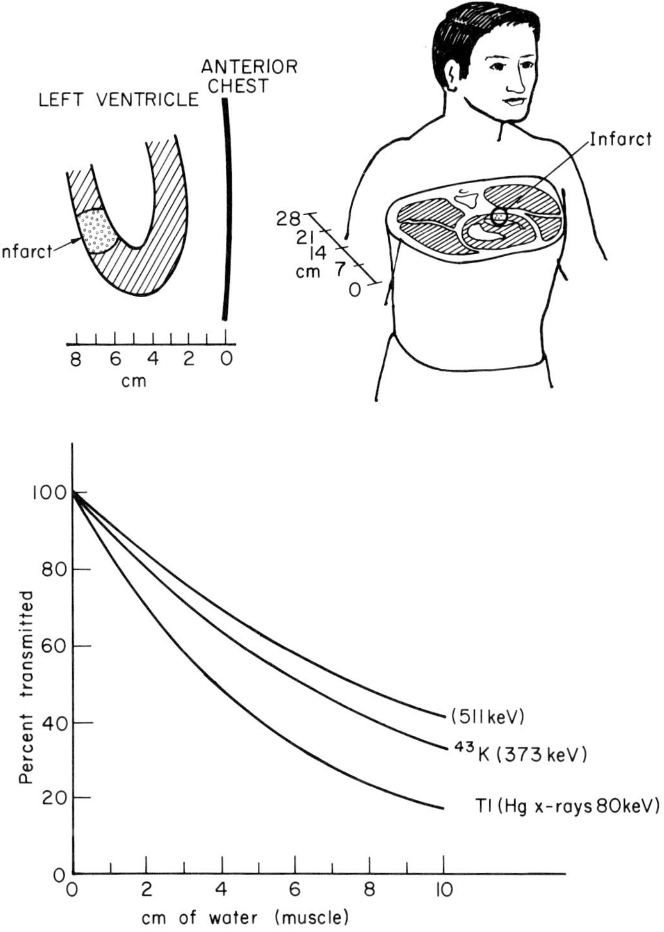

Figure 1. Demonstration of the decrease in detected photons occasioned by attenuation due to the tissue thickness between source and the body surfaces. Curves follow the equation N detector $= N$ source $e^{-\mu x}$ where x is the thickness and μ is the attenuation coefficient (e.g., μ for [201]Tl = 0.18; for [43]K (373) = 0.13; for 511 keV = 0.10).

from the posterior wall (Fig. 2). The relative contributions of radionuclides in various parts of the thorax to the projection image are due mainly to attenuation.

Detectors and Imaging Systems

The various devices used to detect the isotopes listed in Table 1 are illustrated in Figure 3. The four components of all high energy photon detection devices are collimation, photon conversion, electronic amplification, and data reduction.

Collimation

A collimator is a type of lens used to focus in on the organ or region of interest (Fig. 4). The basic purpose of a collimator is to exclude from the detector those photons which are traveling in some direction other than a straight line from the organ or region of interest. Since the photons we are dealing with are very energetic, lead is used to stop unwanted

13

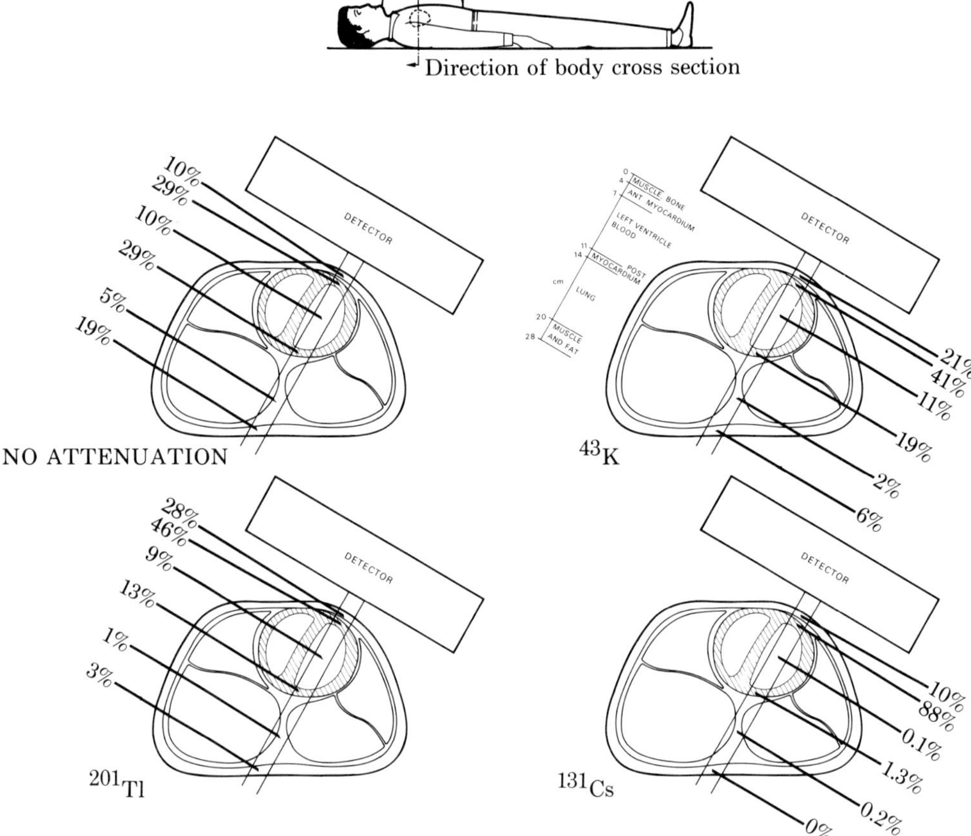

Figure 2. Demonstration of the relative amounts of activity that contribute to an image from different portions through the thorax assuming a target-to-nontarget ratio of 4:1. The contribution from the posterior myocardium is 29 percent if there is no attenuation. However, for the realistic situation of attenuation, the fractional contribution from the posterior myocardium is appreciably less, whereas the relative contribution from the anterior myocardium is greater.

photons. The simplest collimator is a block of lead with a single hole (Fig. 4A). An image of the activity distribution can be produced by sequentially moving the detector device and collimator in a rectilinear fashion and recording the number of detected photons at each position during a fixed time interval. More complicated collimators use multiple holes placed in a diverging or converging pattern such that one depth in the body will be in better focus than over- and underlying regions. Large area photon detection systems such as the Anger camera, multicrystal devices (e.g., System 77) and multiwire chambers use mainly parallel-hole multichannel collimators (Fig. 4B). It is frequently beneficial to use non-parallel hole collimators for cardiovascular imaging. For example, the 30-degree slant hole collimator consists of a multichannel lead collimator with the channels slanted at 30 degrees to accommodate a modified left anterior oblique view of the human myocardium.[12] This view gives an optimal separation of right and left ventricles for purposes of blood pool imaging when evaluating abnormal wall motion.

Other types of collimators used in cardiovascular work include converging, diverging, and pinhole collimators; there are also coded aperture collimators, such as the Fresnel zone

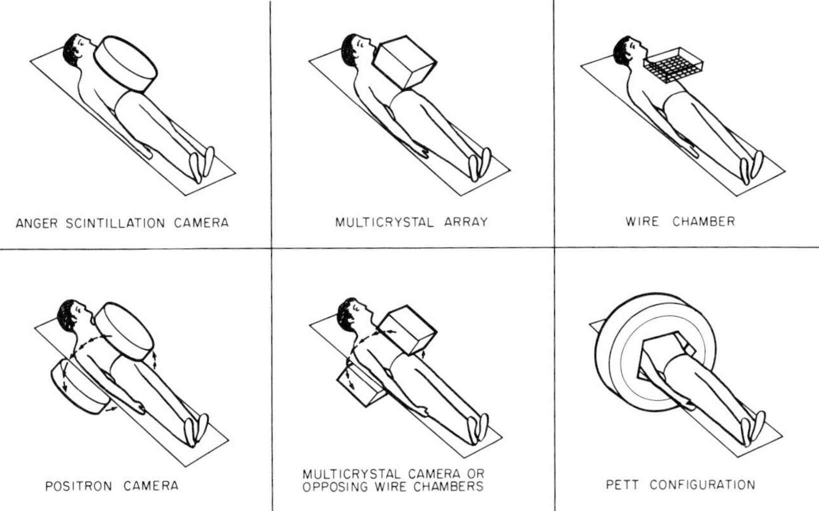

Figure 3. Six configurations of radionuclide imaging instruments for nuclear cardiology.

plate. The converging and diverging collimators are used to magnify or minify the area viewed by a detector. Collimators which provide magnification can provide an apparent improvement in spatial resolution. By accepting only converging photons from a region of interest larger than the crystal, it is possible to image lung fields larger in dimension than the available camera.

The principles of the pinhole collimator are illustrated in Figure 4C. The pinhole collimator is used where significant magnification is required (e.g., evaluation of the uptake of ^{201}Tl in a rat heart). But enlargement comes at some cost. First, the efficiency is low due to the small solid angle associated with the small (4 to 5 mm) pinhole aperture. Secondly, there is a nonuniformity in sensitivity such that the periphery of the image receives far less photons than the center. The falloff in sensitivity is in accord with $\cos^3 \theta$, where θ is the angle between the optical axis and the position of the image.

Coded apertures are comprised of a random array of holes or concentric rings. The principle of coded apertures is similar to that of hologram formation. These devices are useful for imaging small objects and have been used for cardiovascular research in animals[13] but generally have provided only limited success in clinical nuclear cardiology.

The required thickness of the collimator is determined by the energy of the photons being used. For example, to exclude all but 5 percent of the unwanted photons we need 1.1 mm of lead for the 80 keV photons of ^{201}Tl, but 20 times this thickness for the photons from ^{43}K. The resolution and efficiency or transmission of a collimator are related to the size, number, and length of holes, as well as the septa thickness between holes. Collimators used with ^{201}Tl and ^{99m}Tc have a transmission of 10^4; that is, only 1 in 10,000 emitted photons reach the camera.

Electronic Collimation and Positron Imaging

For detecting the annihilation gamma rays from positron emitting isotopes, an electronic coincidence circuit is used instead of lead. As noted already, the annihilation photons from positron emitters fly off in opposite directions with an angle between these photons of 180-degrees; there is some slight variability, about ¼ degree, due to the electron-positron momentum at the instant of annihilation. If we arrange a ring of detectors around a pa-

15

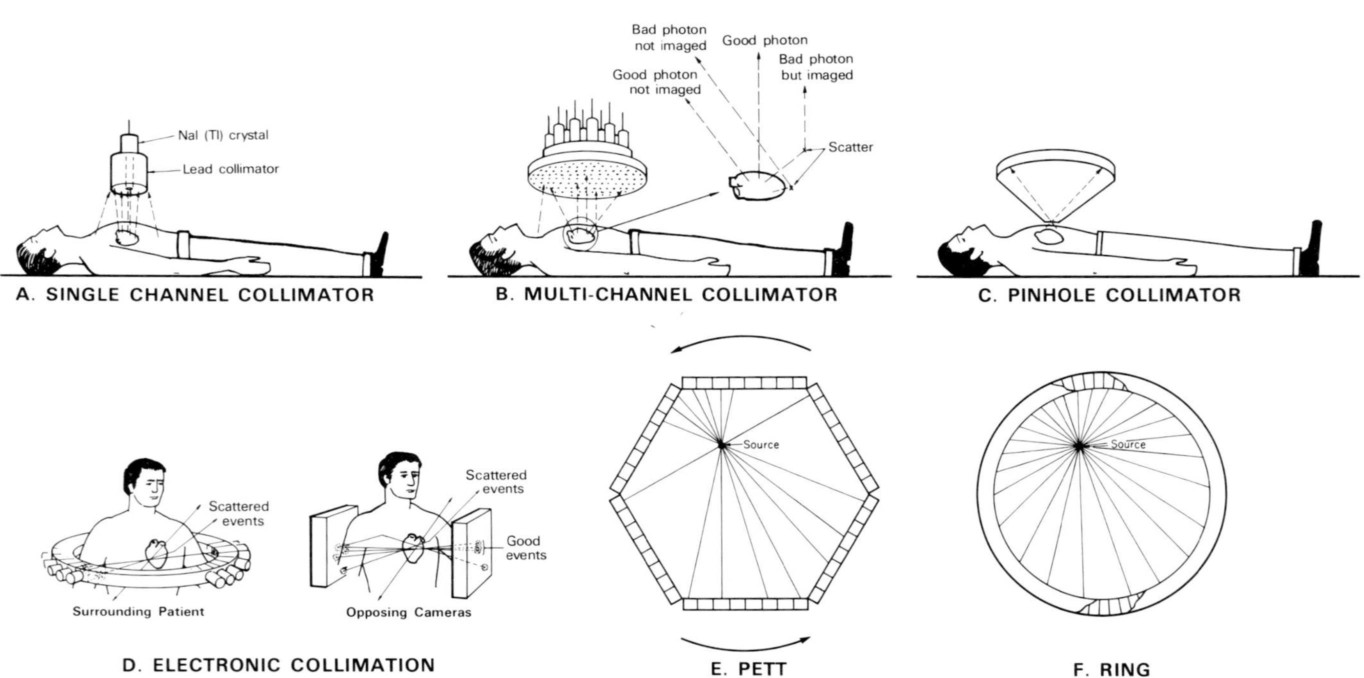

Figure 4. Various forms of photon collimation. A, B. C, Lead collimators used for single photon detection. D, E, Configurations used in positron annihilation photon coincidence detection.

tient, and two crystals were struck at the same instant, we would know that the radionuclide was somewhere on a straight line between the two crystals (Fig. 4D). If the isotope is concentrated in only one spot of the body, then the line connecting each crystal pair will pass through that one spot (Fig. 4E and F). This is the basic principle for the positron camera to be discussed later.

Photon Detection

The photons which pass through a collimator interact with a crystal of sodium iodide to produce visual light (scintillations) or with a gas, in the case of multiwire chambers, to produce electrons. Most photon detection instruments employ a sodium iodide crystal doped with thallium. The visual light produced from a high energy photon interaction in the crystal is converted to an electronic signal by a photoelectron multiplier tube (PMT). The PMT operates as follows. The light photon from the crystal hits a glass plate coated with a rare-earth material which has the property of releasing an electron when struck by a photon. Inside the PMT the released electrons are accelerated to a metal plate called the dynode. Upon striking the dynode, more electrons are released which in turn are accelerated to another dynode, where even more electrons are produced; thus the name photoelectron multiplier. As many as 10^6 electrons can be produced from one electron released from the phosphor. This current of electrons is then converted to a usable signal by electronic circuitry. The signal is proportional to the energy of the incoming photon absorbed by the crystal, since the number of electrons produced in the PMT is proportional to the amount of light produced in the crystal, which, in turn, is proportional to the energy of the incoming photon.

An important part of the detector electronics in this process involves selecting only the photoelectric (photopeak) events; other events could be caused by isotope photons which have scattered in the patient and made their way through the collimator to the detector. For high resolution imaging of the myocardium this energy selection is essential. Usually this photopeak selection is expressed as a percentage of an energy window (e.g., a 20 percent window means the investigator has set his electronics so that the spread of accepted energies around the photopeak is 20 percent of the photopeak energy). Thus, a 20 percent window for the 80-keV photon from ^{201}Tl will be 16 keV wide.

The electronic unit used to select the proper pulses from the phototube, pre-amplifier, and amplifier is called a pulse-height analyzer. This device helps to exclude pulses corresponding to Compton-scattered radiation in the patient and radiation from radioisotopes other than the one being examined. Frequently, different isotopes are used for various studies done on a patient during the examination (e.g., ^{99m}Tc-albumin for cardiac cavitary volume studies and ^{81}Rb for myocardial imaging). Use of the pulse-height analyer allows one to exclude information from the ^{99m}Tc in the subsequent ^{81}Rb study. Similarly, a ^{99m}Tc flow study can be performed after a ^{201}Tl study by changing the window setting from 80 keV to 140 keV. However, there will be some interference from the 135, 167-keV photons, but since 10 times more ^{99m}Tc can be used than ^{201}Tl, and since only one tenth of the thallium disintegrations produce 135, 167-keV photons, this interference is negligible.

The events accepted by the pulse-height analyzer are sent to a readout system that can consist of one or a combination of the following: (1) A photographic film that is exposed to a tiny light source which moves over the film in synchrony with a scanner. The intensity of the exposure is adjusted to be proportional to the intensity of photons received by the scanner at each position. (2) A cathode ray tube where each event is displayed as a flash at a position corresponding to where the event occurred in projection from the patient. This can be a persistent scope or TV type of display. (3) A videotape in which each event or the count rate at each position is stored as a signal along with its X-Y position. (4) A paper tapper which is a mechanical impact plotter that moves in synchrony with the detector while placing dots on paper in direct proportion to the observed count rate. (5) Computer memory (core, disc, or magnetic tape).

For fast studies multiple images are recorded by mechanically effecting multiple exposures at designated intervals, or by a mechanical multilens device; data can also be stored in computer format for subsequent replay. The most common method involves computer storage and computer processing of data.

Anger Camera

The Anger camera consists of a large flat NaI(Tl) crystal, and anywhere from 9 to 91 phototubes closely packed over the crystal (Fig. 5). Most cameras in current use have crystals 1.27 cm thick and 27.9 cm in diameter, but newer detectors with large fields of view have diameters up to 40.6 cm. As with the simple crystal PMT system, the photons that interact with the large camera crystal lose their energy by photoelectric or Compton interactions, or both. The process produces light photons with the total number produced being directly proportional to the total photon energy absorbed by the crystal for each photon-NaI(Tl) interaction.

The crystal is optically coupled to a series of photomultiplier tubes that are arranged in a hexagonal array. Most cameras in current use have 19 photomultiplier tubes in the array, but newer systems have as many as 37 or even 91 tubes. Each phototube absorbs the light photons produced within its field of view onto its bialkali, photosensitive surface, thereby releasing electrons by the photoelectric process. The phototube multiplies the original number of electrons by a factor of 10^6 or 10^7 to produce an output pulse that is directly proportional to the total energy absorbed within the field of view of the phototube.

A computing circuit is used in combination with the phototube array to determine the X-Y location of scintillations occurring within the crystal on the basis of the relative intensity of light seen by each of the phototubes (Fig. 5). The circuit determines X and Y deflection signals, which are applied to an oscilloscope to reproduce the scintillations as point flashes of light on the oscilloscope screen at the appropriate X-Y locations. In addition, the computing circuit sums the output from all tubes for a single event to determine the Z

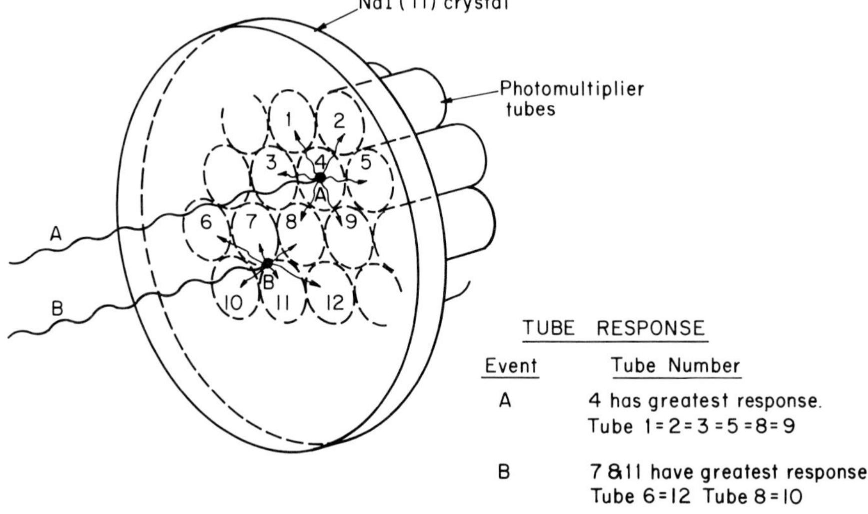

Figure 5. Basic concept behind the Anger camera. The position of a scintillation is deduced by the amount of light detected by the phototubes relative to one another.

pulse, an output signal that is proportional to the total energy absorbed by the NaI(Tl) crystal for the event. This signal is applied to the pulse-height selection circuit, which allows only those Z pulses with amplitudes falling within a preselected energy range to pass on to the data-recording portions of the imaging system. The pulse-height analyzer thereby excludes pulses from undesirable sources, such as radiation scattered within the source or pulses from photons of radioisotopes other than the one being imaged. If a Z pulse is rejected, no scintillation occurs on the oscilloscope screen. If it is accepted, a scintillation occurs on the oscilloscope screen at the X-Y position determined by the X-Y output signal.

The phosphor of the oscilloscope screen usually has low persistence, so that scintillations produced on the screen are almost instantaneous. This allows permanent images to be recorded by exposing film to the scintillations occurring on the oscilloscope screen for a particular interval of time, thereby creating a composite picture corresponding to the organ distribution of activity over that period of time. The most common recording system utilizes a Polaroid camera with a triple lens system. The three lenses are equipped with variable aperture diaphragms or three different neutral density filters. The lens system is set with lens apertures so that three pictures of the same event will occur, each leaving a different exposure. This increases the probability that at least one exposure will have the required brightness and contrast to be of diagnostic value. Other systems utilize 35-mm or 70-mm transparencies or multi-image format devices with various sizes of x-ray films to record a number of different images of the activity distribution on a single film sheet.

An important problem in evaluating a camera for cardiac dynamic work is speed. Dead time is the parameter usually given for characterizing the speed of a camera system. Usually, the smaller the dead time, the better the system, if resolution does not deteriorate at high count rates. A long dead time will cause the data of a dynamic flow study to be seriously distorted. For example, for a dead time of 20 μsec, there will be 25 percent data loss on a dynamic flow study when the true count rate should be 30,000 counts/sec. The importance of dead time is depicted in Figure 6 for theoretical curves of time versus activity in a cardiac flow study. The distortions shown here can be more serious in practical situations because dead time increases as count rate increases in some cameras. In addition to distortions in the flow curves, gamma cameras produce serious artifacts when exposed to high activity (e.g., 20 mCi); these are caused by a pile-up of events in the electronic positioning circuits.

Multicrystal Scintillation Camera

The concept of a multicrystal scintillation camera was described by Bender and Blau in 1963.[2] Their basic system, known as the autofluoroscope, was subsequently modified and is currently available commercially as the Baird Atomic System 77.[3]

The camera consists of 294 sodium iodide crystals, each measuring 8 mm square and 3.8 cm thick. The crystals are arranged in a rectangular array, 15 by 23 cm, with 14 columns (Y axis) and 21 rows (X axis). The crystals are optically coupled to 35 PMTs by means of a light-piping technique that splits the light from each NaI(Tl) crystal into two equal parts. Thus, each detector is covered by two light pipes, providing two output pulses for each event that occurs within a given crystal. One light pipe is coupled to a PMT used to identify the crystals' X coordinate, while the second light pipe is coupled to a PMT that localizes the crystals' Y coordinate. Events occurring within a given crystal produce an output pulse in each of the two PMTs coupled to the crystal. These pulses are then amplified by the preamplifier-amplifier associated with each PMT. The outputs from all the amplifiers of the 14 Y axes and 21 X axes are totaled and passed on to a single channel analyzer for pulse-height analysis.

The event detection circuits and positron information system are independent of one another in the multicrystal camera because each crystal can detect only those photons that

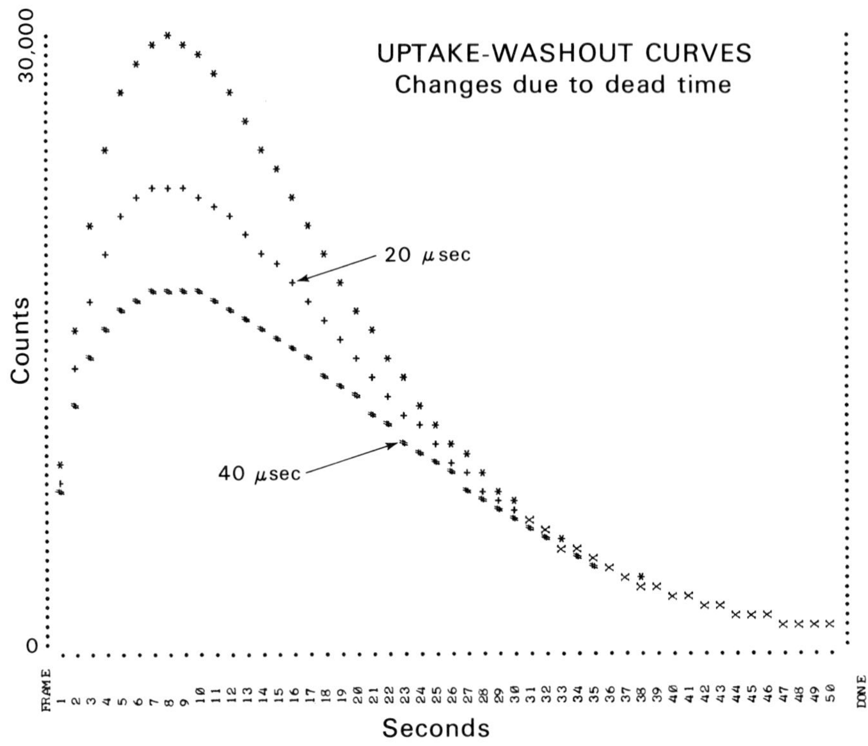

Figure 6. Uptake washout curves for isotope flowing through the left ventricle for two conditions of instrument dead time. Upper curve is the ideal data for a first pass study with 10 mCi ^{99m}Tc bolus injection.

pass through the appropriate collimator hole; thus, the collimator completely defines the spatial resolution of the system. This design allows the system to operate at counting rates of 400,000 cps without saturating the detectors or introducing errors in the positron pulses. Thus, the multicrystal camera has a high temporal resolution for cardiac flow studies. The trade-off is a loss of spatial resolution, because each crystal is 8 mm on a side and individually collimated. Only 294 resolution cells per scene are accumulated in dynamic studies. To improve resolution, the multicrystal camera is moved in small increments while obtaining data in static mode for studies using ^{201}Tl or ^{99m}Tc-pyrophosphate (PYP) for myocardial imaging.

Both the Anger camera and the multicrystal detection system, which also has computer capabilities, can be used for most nuclear cardiology studies. The latter can be used to image ^{99m}Tc-pyrophosphate for acute myocardial infarcts and ^{201}Tl for ischemic heart disease; it can perform ECG gated blood pool imaging for left ventricular function, and first pass cine mode studies to evaluate wall motion, ejection fraction, shunt flow, and transit times.

Multiwire Proportional Chambers

The multiwire proportional chamber (MWPC) uses the process of ionization to determine the positions of detected events.[14] The basic chamber consists of three planes of parallel wire grids. The spacing between the wires in a given grid is typically 1 to 3 mm, depending on the desired spatial resolution. The two outer grids are placed at ground potential, whereas the center grid has a high constant positive potential. The entire grid arrangement is housed in a chamber filled with a mixture of 90 percent xenon and 10 percent carbon dioxide at 1 to 10 atmospheres of pressure. These devices are capable of handling

20

count rates of 10^5 counts/sec while providing an intrinsic spatial resolution of 2 mm; however, they are not efficient unless low energy photons or very high pressure is used.

Photons entering the active volume of the chamber have a finite probability of undergoing a photoelectric interaction with one of the gas atoms. When the process occurs, the ejected electrons drift toward the positive central grid where they undergo an avalanche multiplication (10^5) in the immediate vicinity of the wire. This results in a current pulse on the central wire. The pulse amplitude is proportional to the energy of the photoelectrons. At the same time, an induced pulse of the same magnitude but opposite polarity is generated in the closest wires of the outside grids by the positive ions produced in the avalanche. The x-ray position of each event is ascertained by signals produced in the outer grids that are passed through individual amplifiers or electromagnetic delay lines. An oscilloscope display device is used to record a light flash at the appropriate X-Y locations. Since the total charge collected by the central grid is proportional to the energy lost in the initial photoelectric interaction, this pulse can be used in conjunction with an electronic window (pulse-height analyzer) to attain energy discrimination. Images can be obtained by photographing the events on the oscilloscope display, or the data can be digitized and stored in a computer for future image processing.

The main advantage of the MWPC is its good spatial resolution (2 mm FWHM). For myocardial imaging applications this device holds promise in ^{201}Tl (80 keV) accumulation studies. The device has good energy resolution but uniform electric fields must be maintained at the anode. For detection of photons having higher energies, special lead converters must be used to increase the detection efficiency of the system. This approach, however, has the disadvantage of precluding energy discrimination of the incident gamma rays. Thus, practical imaging with agents such as ^{99m}Tc cannot be realistically accomplished. For these reasons, multiwire proportional chambers have not yet found widespread use in cardiovascular research or clinical studies.

Image Intensifier Camera

The image intensifier uses a converter screen that transforms the photon energy to light energy which is detected by a large area image amplifier. This concept has been used recently with a mosaic of scintillation crystals that are optically coupled to the photocathode of the image intensifier. Image intensifiers are still of little use to cardiologists due to the difficulty of pulse-height analysis, high noise factor, difficulty in maintaining uniform sensitivity, and large expense of the intensifier tube. Some schemes for overcoming these limitations[15,16] have been proposed, but an adequate image intensifier camera for cardiovascular studies is not yet commercially available.

Semiconductor Detectors

A solid-state detector converts an incoming photon to an electrical signal instead of a visual photon (scintillation). High purity germanium, cadmium telluride, and mercuric iodide are candidates for this application. The major advantage of these devices is that they have very good energy resolution; thus, much of the scatter (Fig. 4) can be excluded electronically. This is an area of future development, but no device is commercially available yet.[17]

The solid-state detector schemes are potentially useful in cardiovascular research which examines circulation using microspheres to which various low energy isotopes are absorbed or chelated. In this in vitro application the use of five or more different isotopes with different photopeaks becomes feasible without recourse to special computer manipulations for "stripping" the photopeak intensity from the spectrum of overlapping photopeaks, as must be done with NaI(Tl). Because germanium has a density or photoelectric cross section that is approximately five times lower than that of sodium iodide, the sensitivity of this crystal for high energy photons is relatively low; thus, the use of germanium is restricted to either thick crystals or low energy photons.

Positron Camera

There are two general reasons for the cardiologist's interest in positron imaging. First, because of their unique method of localizing events, positron cameras do not require collimators. Therefore, these systems characteristically have relatively high detection efficiency as well as good spatial resolution. Second, since better than half of all available radionuclides are positron emitters, it should theoretically be possible to find a positron emitter for virtually any imaging problem of interest.

Over the past five years, several different types of positron cameras have been introduced. These include (1) positron cameras with two opposing Anger cameras (Fig. 5); (2) crystal mosaic planar array cameras; (3) multiwire proportional chamber detectors with lead converters; (4) hexagonal crystal arrays surrounding the patient; and (5) a ring of crystals surrounding the patient (Figs. 4, 7, and 8).

Independent of the camera type, the basic principles of detection are the same as explained above and reviewed here. An isotope such as ^{11}C, ^{113}N, ^{15}O, ^{38}K, or ^{82}Rb decays by emission of a positive electron. This positron moves a few millimeters away from the point of isotope decay while undergoing a series of collisons with surrounding matter. Through these collisions the kinetic energy is reduced to a few electron volts. At low or zero kinetic energy, the positron is captured by an electron (matter and antimatter interaction), and an annihilation process results in the emission of two 511-keV photons which fly off in an angle of 180 $\pm$ 0.25 degrees. An image of the activity distribution can be obtained by projecting the line of intersection of each set of recorded events through a given plane on the object (Fig. 7). Annihilation events which originate within the plane will appear in focus while events originating elsewhere will apear blurred. Thus, even in the simplest case, positron camera images can present tomographic information.

The first positron camera type consists of two opposing Anger scintillation cameras. The commercial version has 2.54-cm-thick NaI(Tl) crystals to increase the detection efficiency of 511-keV photons.[18] The electronics of the Anger cameras used here have been modified so that they are capable of recording a higher singles count rate than is normally possible. Clever analogue circuitry is employed to depict the activity on a plane of interest with a CRT. Computer display systems are also used for this purpose; these calculate the image in digital form from X-Y locations of the recorded events stored in the memory of a computer. The imaging techniques used vary from the simplest case of two opposing large field detectors, from which superposition (back-projection) tomographic data are obtained (Fig. 7), to multiple view computed tomography, wherein the pair of opposing view cameras is rotated around the patient (Figs. 7 and 8B).[4,6]

Multiwire proportional chambers with lead converters are also being used as positron cameras.[14] Here again two opposing detector arrays are used; the device is relatively inexpensive and has a spatial resolution that can be controlled by changing the hole size in the converter matrix. The disadvantages of this camera are low detection efficiency with current lead converters and long resolving times. Also, this system saturates with about 100 μCi in the field of view, making it difficult to obtain good statistics without inordinate patient time.

A group at Massachusetts General Hospital (MGH) has built a camera consisting of two planar crystal arrays, each having 127 NaI(Tl) crystals that provide sensitive areas of 27 by 30 cm.[4] This system is currently being marketed.* The camera achieves good resolution,

---

*The Berkeley Cyclotron Corporation, Berkeley, California.

22

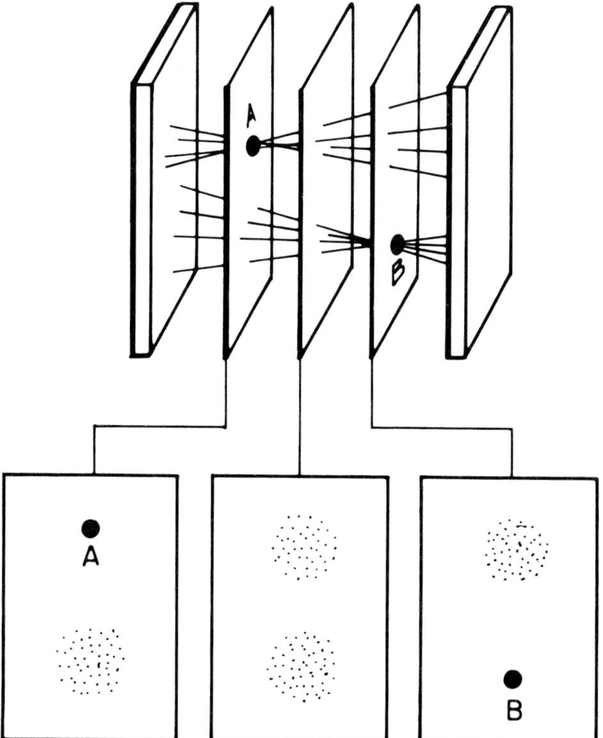

Figure 7. Concept of superposition tomography used by positron cameras. Multiple planes can be displayed using analogue methods (Anger positron camera) or digital methods available in the MGH positron device and the Searle Radiographics positron camera.

good field uniformity, and can handle 10 times the activity in the field of view as can the Anger type positron camera and the multiwire proportional chamber camera. For transverse section reconstruction, the MGH positron system of crystal arrays is rotated about the patient to provide the required angular information, as is the case for the modified Anger type positron camera. The MGH device has great potential for myocardial imaging using $^{13}NH_3$ and ^{82}Rb without rotation, since adequate data can be obtained to compute simple back-projection tomographic images (Fig. 7). Myocardium transverse section images can be obtained with these devices after bolus injection of ^{13}N (10 min)-labeled ammonia, ^{38}K (8 min), and ^{81}Rb, which has a half-life of 4.7 hr and positron emission during 67 percent of the disintegrations.

Positron Emission Transaxial Tomograph

The positron emission transaxial tomograph (PETT) is a hexagonal system (Figs. 3 and 4E). The present commercial system is built by ORTEC and has 66 crystals arranged in a hexagon.[6] This system is important to myocardial studies because of its demonstrated ability to measure the uptake of ^{13}N-ammonia, ^{11}C-palmitic acid, and ^{18}Fe or ^{11}C-deoxyglucose in the myocardium of animals and humans. (This is discussed in greater detail in the next section.) Six sets of eight detectors are mounted on platforms capable of rectilinear motion; the entire gantry is capable of rotational motion. Thus, over a period of a few minutes it is possible to record enough events to reconstruct a transverse section. The resolution of this system is approximately 1.0 cm FWHM.[6] Images of the transverse sections are produced by the convolution technique of computed tomography reconstruction. A new system, designated the PETT IV, was originated by M. Ter-Pogossian[19] at Washington University, St. Louis; it can image seven transverse sections simultaneously with a transverse section resolution of about 1.7 cm FWHM.[7]

Positron Ring Detector

The geometry embodied in a ring of detectors allows simultaneous data collection without detector or patient movement for transverse section reconstruction. NaI(Tl) crystals mounted on a ring surrounding the patient (Figs. 3 and 7) can obtain transverse sections with a resolution of 7.5 mm FWHM.[8] To achieve this, 280 crystals on a ring 80 to 90 cm in diameter are required. The Donner 280-crystal system can reconstruct images of the transverse section of viable human myocardium using ^{82}Rb (75 sec half-life). A system with 64 crystals operating at the University of Southern California has shown ^{81}Rb and ^{13}N-alanine accumulation in the myocardium.[20]

Multiple ring positron imaging devices are now under development and this area of instrumentation development is likely to offer advances of great value to cardiovascular research and nuclear cardiology.

At present, the positron emitters of importance for cardiovascular research are ^{11}C, ^{13}N, ^{15}O, ^{38}K, ^{81}Rb, ^{82}Rb and ^{38}K. The rapid development of positron camera technology, as well as the improvement of radionuclide production methods, promise to make positron imaging a vital tool for clinical and research cardiovascular medicine.

Principles of Three-Dimensional Reconstruction

An important new aspect of research and clinical cardiology involves computed tomography using x-rays, or, in the case of nuclear cardiology, emitted photons from radionuclides flowing through or accumulating in the myocardium. There are four approaches to acquisition of 3-D information:

1. Emission computed tomography using positron annihilation photons
2. Emission computed tomography using single photons
3. Longitudinal tomography (single photon or positron)
4. Three-dimensional reconstruction using limited views and a model

Emission computed tomography devices are shown in Figure 8. Longitudinal tomography is shown in Figure 7 using positron annihilation photon imaging and in Figure 9 using single photon applications. Figure 10 illustrates the general concept of transverse section reconstruction.

Single photon tomographic devices (Fig. 8A) are used at a few centers to collect data from 36 or more views around the patient. Data from each view correspond to a projection of activity for a specific angle. Serial transverse sections of the distribution of thallium or ^{99m}Tc-PYP are computed by methods similar to those used in x-ray computed tomography.[19] In nuclear cardiology some compensation for attenuation must be applied since the problem involves calculation of the amount of radionuclide in an unknown region with unknown attenuation between the region of concentration and the detector; the x-ray CT problem involves merely the calculation of the distribution of attenuation.[21] Positron annihilation photon detection systems have inherent tomographic capabilities (Figs. 4D through F and 7).[4,6,19] In the case of positron CT imaging, the data from opposing detector pairs are organized into projection information. For the PETT systems comprised of a hex-

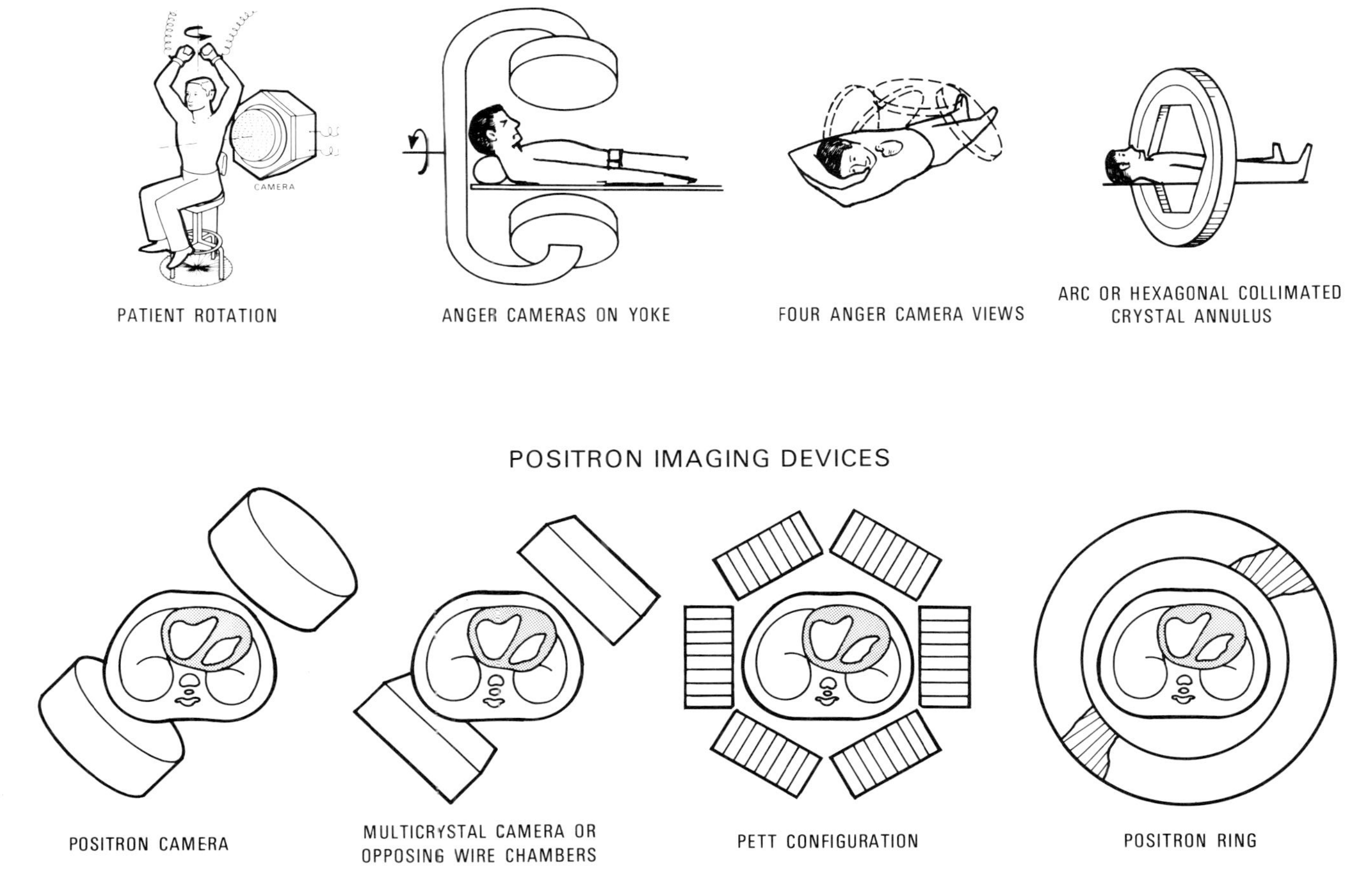

Figure 8. Eight devices for emission computed tomography.

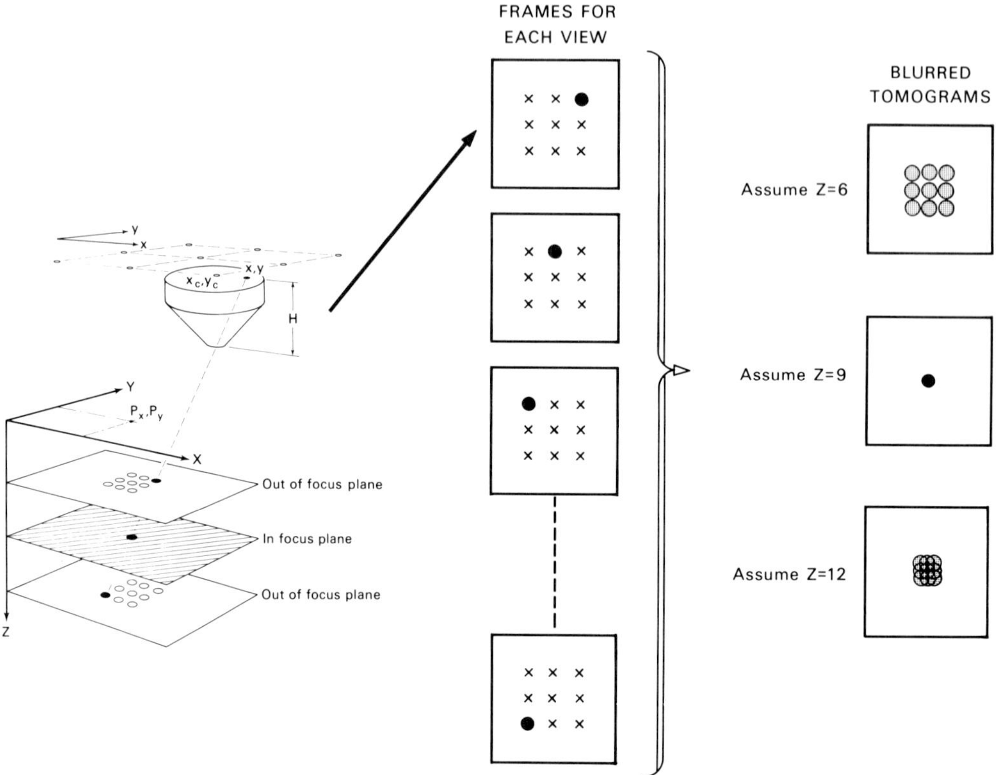

Figure 9. Longitudinal tomography can be achieved by acquiring multiple images with a pinhole collimator so that each image corresponds to a different position of the collimator. Separate images are produced using multiple positions of the camera or patient, or by a special multi-pinhole collimator which encompasses seven images simultaneously.[129]

agon of crystals, only a few projection angles are available at one instant; thus, the system must be operated from several different angles in order to acquire an adequate number of views. In the case of a ring of crystals surrounding the patient, multiple views are available instantaneously and eliminate the need for system rotation.

The number of views needed is related to the desired or obtainable resolution. As a general rule, the number of views is equal to twice the number of resolution elements across the object being examined. Thus, for 10-mm resolution across a 30-cm object we need 60 views.

Quantitative accuracy in emission computed tomography depends on the amount of activity or number of events recorded and the resolution required for the study.[22] For a uniform distribution of activity, the accuracy is given in terms of the root mean squared (rms) percent uncertainty (% coefficient of variation) as

$$\text{rms \% / resolution cell} = \frac{120 \times (\text{number of resolution cells})^{3/4}}{(\text{number of events per cell})^{1/2}}$$

This formula is applicable to uniform distributions; when applied to emission images it shows that about 10 times more data is needed than would be expected from such a naive prediction based on the denominator alone. For nonuniform distributions, such as an ^{11}C-palmitate positron emission study, the statistical relations are improved; thus,

26

$$\frac{\text{rms \% uncertainty resolution}}{\text{element of the heart}} = \frac{120\,(\text{total number of events})^{1/4}}{(\text{average number of events per heart resolution element})^{3/4}}$$

The considerations of this paragraph are of paramount importance in evaluating the future of emission computed tomography for quantitative in vivo studies. This analysis is applicable to both single photon and positron emission tomography.[23]

Positron annihilation transverse section imaging has enjoyed success with imaging of $^{13}NH_4{}^+$, ^{18}Fe-deoxyglucose, ^{11}C-palmitic acid, and ^{13}N-alanine in the myocardium as discussed later in this chapter.[4,6,19,24]

Single photon tomography is just now being applied to heart imaging.[25-27]

Longitudinal tomography involves electronic focusing, such as that used for positron

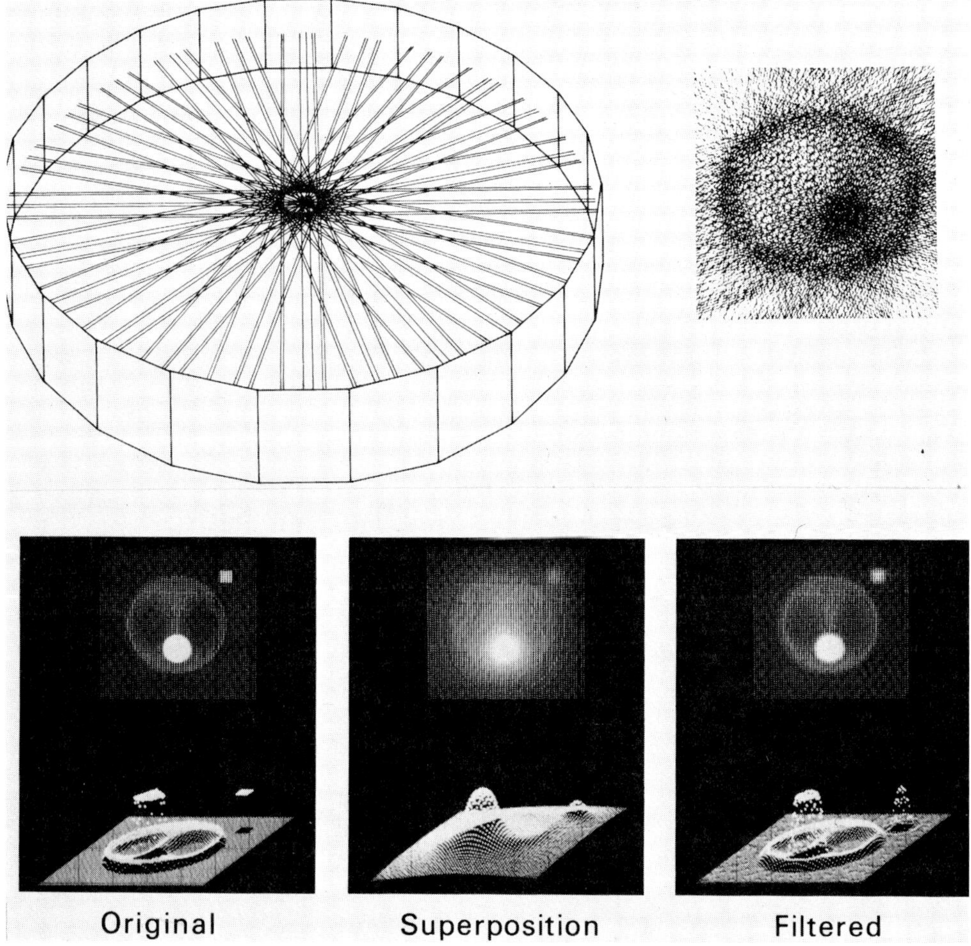

Figure 10. Transverse section reconstruction from projection data involves back-projection of the projections to form a superposition or blurred image which, when mathematically filtered, can be modified to give a near perfect reconstruction of the true distribution. The usual method of reconstruction involves modification of the projection data before back-projection by a similar mathematical filtering technique. Both methods produce the desired result. Iterative methods of finding the best match between the reconstructed distribution and the projection data are also used.[21]

emitters (Fig. 7), or computational selection of a plane parallel to the data recording plan, as with Fresnel zone plate imaging or multiple pinholes.[13,25,27]

The concept of longitudinal tomography using a pinhole or focusing collimator is similar to the concept behind the variable focus of a light microscope. The plane of interest is examined in focus while the adjacent planes are out of focus and blurred. The procedure for obtaining longitudinal tomographs of the heart is shown in Figure 9, wherein the single or multiple pinhole aperture is translated to various known positions over a plane parallel to the planes to be reconstructed. The imaging geometry can be described as two similar triangles. The relationship between a source at X, Y in the object and the image point x, y relative to the image frame center x_c, y_c is

$$\frac{X}{Y} = \frac{-(x - x_c)}{H}; \frac{Y}{Z} = \frac{-(y - y_c)}{H}$$

where Z is the distance from the pinhole aperture to the object plane, and H is the distance between the aperture and the camera. A tomographic image slice for a particular depth Z in the patient's myocardium is constructed by assuming all the activity exists at that depth for seven or more images, where each image represents the pinhole collimator view from a different position. This method is the longitudinal analogue of the transverse section back-projection or blurred tomography method common to conventional laminography techniques in radiology. In the application to nuclear cardiology the reconstruction of emission images from rubidium[25] or thallium[27] has been done by digital manipulation of the data from each separate image, such that a single image is produced for about five different depths in the patient. Each tomographic slice involves shifting all of the data points from each image into an appropriate Z depth slice, where for each depth the Z in the shifting equations is varied. The equations for this process are

$$X = -(x - x_c)\left(\frac{Z}{H}\right) - P_x; Y = -(y - y_c)\left(\frac{Z}{H}\right) - P_y$$

where P_x and P_y are the displacements of the patient images from the central axis.

Three dimensional reconstruction of the heart from a limited number of views using a model or a priori information concerning the object has some promise for clinical applications. If two or three projections are obtained of ^{99m}Tc-PYP uptake in a necrotic patch of myocardium it is possible to estimate what volume of a solid object will best correspond to these projections. Thus, if an ellipsoid or a crescent-shaped region is assumed, the most likely volume can be calculated using only the projection data.[28]

Absorbed Dose from Radionuclides

The dose absorbed by a patient from the decay of radionuclides is dependent on the energy and abundance of photons and electrons emitted by the radioactive atoms. Careful calculations involve attention to Auger electrons, x-rays, and internal conversion electrons, as well as the gamma ray photons and high energy electrons emitted as part of the primary decay event. The Society of Nuclear Medicine provides careful calculations to deal with all of these problems, and periodically publishes a pamphlet called "MIRD" (Medical Internal Radiation Dose). At present only some of the isotopes and radionuclides produced for nuclear cardiology are included in these pamphlets. For many of the radionuclides presently in use it is possible to calculate doses for the whole body with the formula given below. For a particular organ distribution the fraction of the dose deposited in that organ and the effective half-life ($T_{1/2}$) need to be taken into account.

28

The dose deposited by a radionuclide distributed in the heart is not the sum of all the energy released by the atoms because most of the energy is carried away by the gamma photons we use to detect the presence of the radioactivity. In fact, only one half to one third of the energy of most isotopes used in nuclear cardiology is absorbed by the body. This can be seen in Figure 1, where about 50 percent of the photons from ^{201}Tl are transmitted through 5 cm of tissue. It should be noted also that those organs which do not concentrate any isotope still receive energy from photons that are emitted from contiguous tissues.

The general formula for dose calculations[11] is

$$\text{body dose (rads/mCi)} = \frac{71.2 \times \text{effective } T_{1/2} \times 0.3 \times \overset{\text{gamma}}{\text{energies (MeV)}} + \overset{\text{electron}}{\text{energies}}}{\text{body weight (kg)}}$$

The coefficient 71.2 is a conversion factor relating millicuries, half-life in days, and energy in MeV, to dose in rads. The millicurie (mCi) is a basic isotope intensity unit equal to 3.7×10^7 disintegrations per second. The rad is the unit for absorbed dose and is equivalent to 100 ergs per gram of tissue.

This calculation can be easily applied to [201]Tl. The effective half-life of ^{201}Tl is 57 hours; this takes into account the biological clearance as well as the physical half-life of 73 hours. Approximately 30 percent of the photon energy from the gamma rays is absorbed in the body. Substitution of appropriate values in the formula above gives a dose of 0.11 rad/mCi. This calculation does not take into account Auger electrons, internal conversion electrons, and x-rays; when these other radiations are taken into account the whole body dose becomes 0.21 rad/mCi. The dose from 10 mCi of ^{99}Tc-albumin is one tenth that of [201]Tl. Some whole body doses for radionuclides used in cardiology are shown in Table 2. Note the low dose for [82]Rb, which has a half-life of 75 seconds and is obtained from a table top generator as the decay product from [82]Sr. For most of these isotopes the organs receiving the highest dose (typically two to three times that for the whole body) are liver, gut, kidney, and bladder.

FLOW AND TRANSIT TIME PRINCIPLES

Now that we have analyzed the physical properties of the imaging modalities and radiopharmaceuticals available to cardiology, let us turn to an analysis of the general flow

Table 2. Whole-body doses

Radionuclide	mrad/mCi
[81]Rb	100
[81]Rb + contaminants	400
[82]Rb	2
[84]Rb	16,000
[129]Cs	170
[201]Tl	210
[38]K	31
[43]K	600
[99m]Tc-albumin	18
[11]C-amino acids or fatty acids	13
[13]N-amino acids	7

properties of the cardiovascular system and present some important facts of where and how radiopharmaceuticals distribute. The techniques can be classified into those which measure flow, those which measure cell metabolism, and those which measure pure function by morphologic changes. This section presents the basic physiology of flow for diffusible tracers. Substances which are partially extracted or are metabolized are discussed in subsequent sections.

Uptake and Wash-Through Curves

Let us begin our analysis by considering the flow of a nondiffusible tracer through a circulatory network such as the 10-channel system shown in Fig. 11. Assume the diameter of all vessels is the same and the velocity is constant. When will a tracer arrive at B if a bolus is injected at A? Since the velocity is constant, the time for each channel is

$$T_i = \frac{L_i}{V_i}$$

where the index i designates each channel, L_i is the length and V_i is the volume of each channel. Thus, to determine the shape of the arrival curve we need to know only the length of each channel. A tabulation of arrival times is given in Fig. 11.

If ten units of tracer are injected at A and each channel gets one unit, then we see that the tracer from channels 4, 5, 6, and 7 will arrive at B earliest; the tracers in channels 8 and 9 will arrive next, and so forth. The curve of tracer arrival times is deduced from a histogram of the tabulation in Fig. 11. For a network of 100 channels we have a similar curve but with

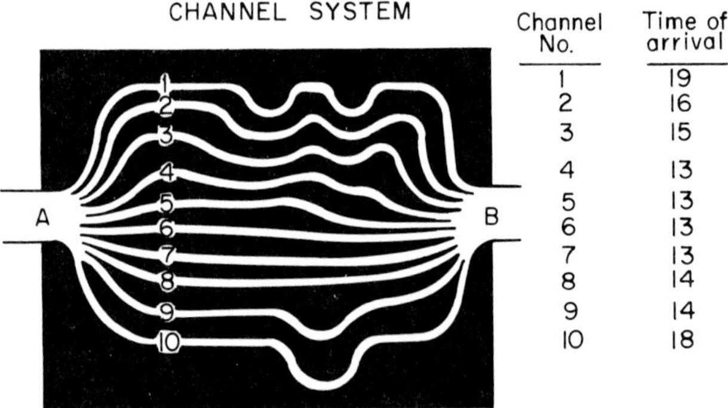

Channel No.	Time of arrival
1	19
2	16
3	15
4	13
5	13
6	13
7	13
8	14
9	14
10	18

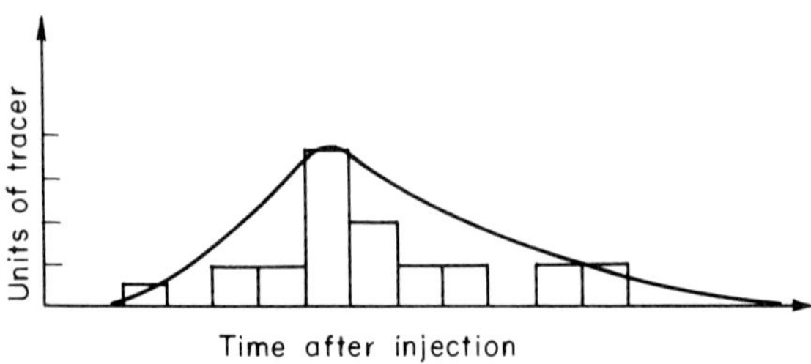

Figure 11. Simple capillary system (*top*) where vessels have different lengths but negligible resistance and similar diameters. Graph represents transit time histogram, with a curve depicting the case for many channels.

30

smoother transitions; this is justification for the solid curve of Fig. 11. Now let us change the vessel (channel) diameters. Under conditions of a constant pressure drop the velocity is inversely proportional to the area of the vessel. A model of such a network is shown in Figure 12A. The curve for the arrival of tracer in compartment B will be similar to the curve in Figure 11. Figure 12B depicts a third model, a mixture of vessel length, vessel size, and a branching network. Once again the arrival curve will be similar to the one in Figure 11. Generally, the concentration of tracer plotted against time gives a bell-shaped curve skewed to the right; that is, toward longer arrival time. This curve is referred to with several terms: the uptake-washout curve, the unnormalized probability density curve, the impulse response, the time-activity curve, or the frequency distribution curve. All these terms describe the same thing, although the last phrase is usually used in the context of frequency or time^{-1}. Usually we can measure this curve by external counting over an organ or multiple sampling, as shown in Figure 13. The shape of the curve alone tells us a great deal about the relative number of channels in each length or velocity interval. We cannot determine the number and size of each channel, but we can determine the mean transit time. And if we know the amount injected, we can evaluate flow. These simple concepts are the basis of the equations and physiologic principles associated with measurement of flow of radioactive tracers through the cardiovascular system.

The first derivative and the integral sign are used frequently in this chapter; thus, it might be helpful, especially for the busy clinical investigator, to recall some simple terms from the calculus involved:

first derivative: rate of change = velocity = change in amount with time = dq/dt where dq is a small change in the amount q and dt is a small increment of time; the purist would define

$$\frac{dq}{dt} = \text{limit} \, \frac{\Delta q}{\Delta t} \text{ as } \Delta t \to 0$$

which is essentially the same thing.

integration: act of adding the amounts = summation.
integral sign: an exaggerated "s" meaning sum.

The first derivative is important because we frequently want to know the change in the amount with time. Integration is important because we want to determine the accumulated amount, that is, the sum or integral of the quantity over time.

Basic Principles of Flow Measurement

The underlying principle of central or peripheral blood flow measurement methods is the law of conservation of mass which says that the physiologic system must be fully accountable: the amount of any substance which accumulates in the tissue under examination is always the difference between the amount carried by the flow into the tissue minus the amount which leaves, or is transported, out of the tissue. In the following section the concepts of flow and accumulation are examined using two approaches which are fundamentally identical. The first approach is sometimes called the stochastic approach; it dates back to 1897 and a classic paper by Stewart.[29] A precise physiologic exposition was provided in 1954.[30] The second approach is attributed to Kety and Schmidt.[31] Recent discussions of flow measurements by Holman and coworkers[32] and Bassingthwaighte[33] are also incorporated in the following discussion. Different viewpoints of the same concepts are presented to aid comprehension and to elucidate the important physiologic assumptions underlying most of the flow methods currently in use.

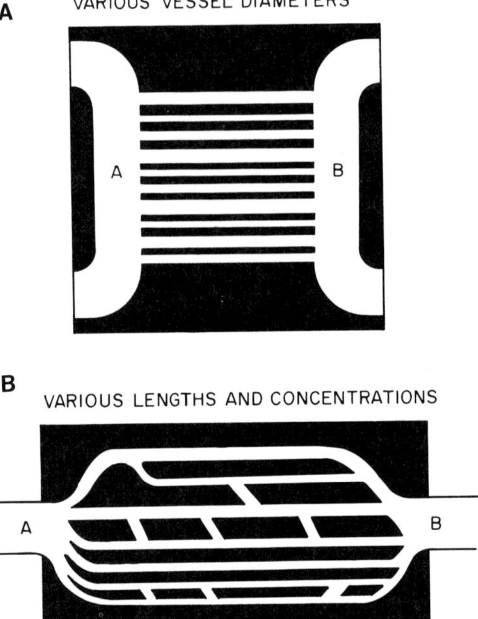

Figure 12. A, Capillary network in which the velocity is inversely proportional to the area due to a constant pressure drop. B, Branching network shows a mixture of vessel lengths and sizes. Both arrival curves will be similar to the curve in Figure 11.

Mean Transit Time and Flow Relationships

The amount of activity arriving at the detectors in Figure 13 at each moment of time is, simply, the flow times the concentration.

$$a(t) = F \cdot B(t) \tag{1}$$

where $a(t)$ is the activity per time, F is the flow in volume per time, and B is the concentration at time t. To get the fractional amount arriving at any time we simply divide the amount arriving at time t by the total amount injected. This quotient is known as the magnitude of the impulse response for time t, or the transport function. It is the fraction of the injected dose reaching the outflow each second.

$$h(t) = \frac{FB(t)}{q_0} \tag{2}$$

where q_0 is the total amount injected. Since the sum of all the fractional quantities must equal the whole we have

$$\int_0^\infty h(t)dt = 1 \tag{3}$$

At any time after a bolus injection the fractional amount remaining in the system can be determined by subtracting from 1 the amount which has accumulated or passes B; thus

32

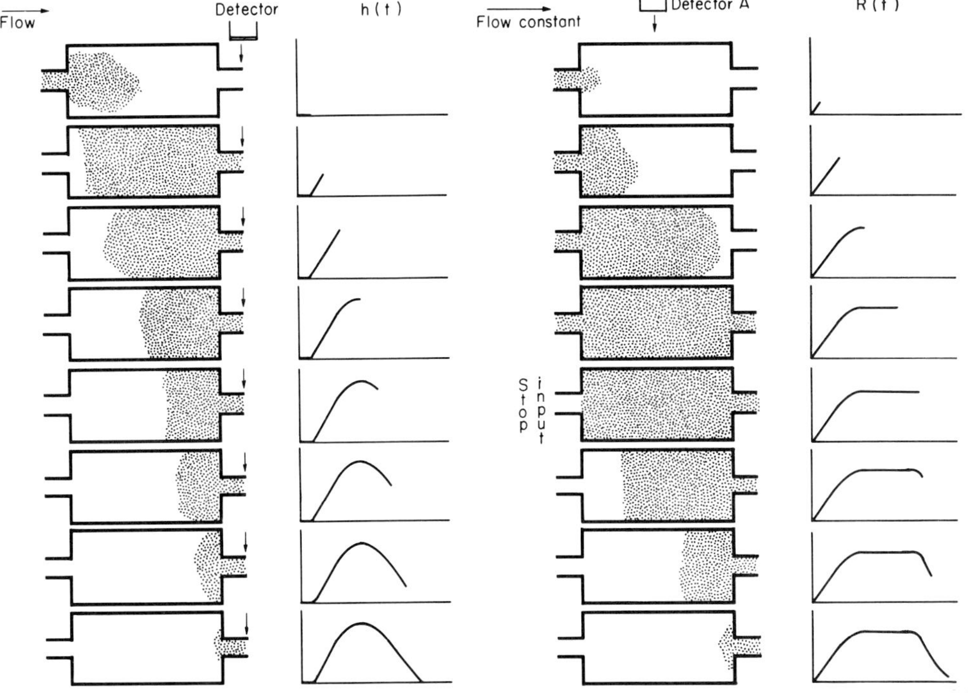

Figure 13. The moment-to-moment, activity-versus-time curve at the outlet of an organ or capillary network after a single bolus injection of activity (*left*). The amount present in the organ at any time is given by the residue curves (*right*). Note difference in the position of the detector in the two cases.

$$R(t) = 1 - \int_0^t h(\tau)d\tau \qquad\qquad 4$$

where τ is a substitute variable of integration. Note that the integration is carried to time t; $R(t)$ is the fractional amount remaining in the system. If the integration were carried to a large value of time, $R(t)$ would approach zero. This residue function $R(t)$ times the total amount is what we frequently measure in external counting over an organ into which the tracer flows (Fig. 13) or is injected (Fig. 14). The actual amount remaining in the system is

$$q(t) = q_0 \cdot R(t) = q_0 \cdot [1 - \int_0^t h(\tau)d\tau] \qquad\qquad 5$$

For example, if the detector system is over the perfusion system into which a tracer is injected, the activity versus time curve of Figure 11 will be observed.

If we carry out the integration of the residue function, we get

$$\int_0^\infty R(t)dt = \int_0^\infty \left[1 - \int_0^t h(\tau)d\tau\right]dt = \tilde{t} = \text{mean transit time} \qquad\qquad 6$$

The fact that integral of the residue function is the mean transit time was first proved by Meier and Zierler.[30] But we know that the ratio of total volume to flow is the mean transit time; thus we combine Equations 5 and 6 to get flow (F)/volume (V)

$$F/V = \tilde{t}^1 = \frac{q_0}{\displaystyle\int_0^{\infty} q(t)dt} \qquad\qquad 7$$

This equation is sometimes called the Stewart-Hamilton equation; a different approach is presented later in this section. Note that equation 7 and Figure 14 suggest we could measure specific flow from the height and area under the residue curve. This indeed is the case. If we assume the washout is monoexponential, we have

$$R(t) = e^{-kt} \qquad\qquad 8$$

Thus, from Equation 5

$$\int_0^{\infty} q(t)dt = q_0\int_0^{\infty} e^{-kt}dt = \frac{q_0[-e^{-k(\infty)} + e^{-k(0)}]}{k} = \frac{q_0}{k} \qquad\qquad 9$$

Substituting Equation 9 in Equation 7 we have

$$\frac{F}{V} = \frac{q_0}{q_0/k} = k \qquad\qquad 10$$

That is, the flow per unit volume is merely the slope of the exponential curve. The reader might recall that this is exactly the Kety-Schmidt approach to determination of flow from xenon washout after tissue injection or coronary injection of an inert tracer. This relation is derived by a more physiological approach from the Fick principle, which is discussed later.

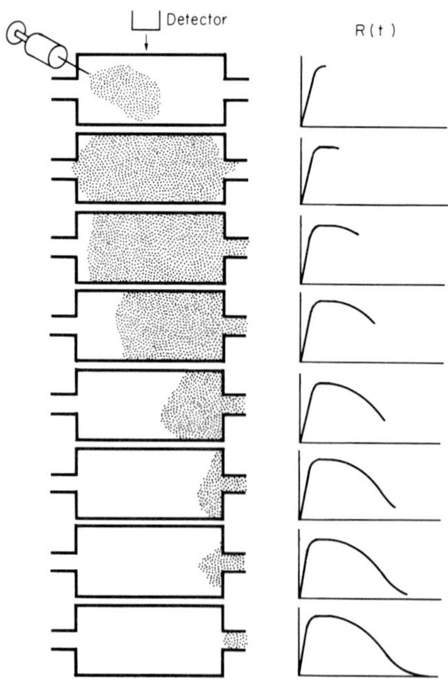

Figure 14. Moment-to-moment detected quantity, $q(t)$, in an organ or tissue space after bolus injection into that tissue.

The partition coefficient is usually introduced at this point, but we will introduce this parameter later.

Unfortunately, the amount present, $q(t)$, is usually not a simple monoexponential function. But Equation 7 says nothing about monoexponentials; it says that the flow per unit volume is merely one over the area of the residue function; that is

$$\frac{\text{flow}}{\text{volume}} = \frac{q_0}{\displaystyle\int_0^\infty q(t)dt} = \frac{q_0}{\displaystyle\int_0^\infty q_0\left[1 - \int_0^t h(\tau)d\tau\right]dt} = \frac{q_0}{\displaystyle q_0\int_0^\infty R(t)dt} = \frac{1}{\displaystyle\int_0^\infty R(t)dt}$$

Or as already stated, the flow per unit volume known as specific volume flow, is

$$\frac{F}{V} = \frac{1}{\text{mean transit time}} \qquad 12$$

Equations 7 and 11 are powerful but are seldom used. Equation 10 is frequently used but often in cases where the underlying assumptions are not valid; that is, $q(t)$ is not monoexponential. For example, suppose we are measuring the xenon washout from the heart in which there is a well known heterogeneity of flow.[34] There are three regions wherein specific volume flow differs by significant amounts. The clearance of injected xenon is described by the weighted sum of the rate constants for each of the three regions using Equation 7:

$$q(t) = \frac{q_1}{q_0}e^{-k_1 t} + \frac{q_2}{q_0}e^{-k_2 t} + \frac{q_3}{q_0}e^{-k_3 t} \qquad 13$$

where q_1/q_0, q_2/q_0, and q_3/q_0 are the fractional quantities in each region. Using the result of Equations 5 and 6 we have

$$\bar{t} = \sum_{i=1}^{3} \frac{q_i}{q_0}\bar{t} \text{ or } \frac{F}{V} = \left(\sum_{i=1}^{3} \frac{q_i}{q_0}\frac{1}{k_1}\right)^{-1} \qquad 14$$

Thus, the mean specific volume flow has some contribution from each rate constant multiplied by the amount which arrives at the region. The disappearance curve can be stripped into three components. Each intercept is the fraction of injected amount which reaches a particular region, and each slope is the specific volume flow for that region assuming a monoexponential washout for each region.

Thus, the externally detected specific volume flow after injection of xenon will mainly reflect the high flow regions which receive the highest fraction of the input. The resulting curve will be biased away from a description of the low flow information as pointed out by Adelstein and Maseri.[35]

If the tracer enters the system as some function of time other than a bolus or sudden injection (e.g., a delta function), the resulting equations which relate flow to the activity detected are similar to those for a sharp bolus injection. However, assumptions regarding simple exponentials are definitely not valid. Consider the problem where the input function looks like the output at B in Figures 11 or 13. If we convolve this input function $B(t)$ with a new transport function $h(t)$ for another capillary system we get an output function

$$\int_0^t B(t - \tau)h(\tau)d\tau = C(t) \qquad 15$$

As before, the quantity remaining in the system at time t is the difference between accumulated input and accumulated output, or

$$q(t) = F \int_0^t B(\tau)d\tau - F \int_0^t C(\tau)d\tau \qquad 16$$

These integrations, when solved, could give us a considerable knowledge of flow; however, the areas under curves, even if obtainable, are not accurate due to recirculation, as shown in Figure 15. To simplify our problem, let $F \cdot B(t) = a$; that is, the input (amount per time) is constant, which could be achieved by constant infusion. Thus, after multiplying both sides of Equation 15 by flow, we get

$$FC(t) = a \int_0^t h(\tau)d\tau \qquad 17$$

and from Equation 16 the accumulated input is

$$q(t) = at - F \int_0^t C(\tau)d\tau \qquad 18$$

After integration of Equation 17 and substitution of

$$\int_0^t h(\tau)d\tau = H(t)$$

we have

$$q(t) = at - a \int_0^t H(\tau)d\tau \qquad 19$$

Integration by parts gives

$$q(t) = at[1 - H(t)] + a \int_0^t \tau h(\tau)d\tau \qquad 20$$

For large t, $H(t) \rightarrow 1$ and $\int_0^\infty \tau h(\tau)d\tau = \tilde{t} =$ mean transit time; thus, for large t Equation 20 becomes

$$q_\infty = a\bar{t}$$

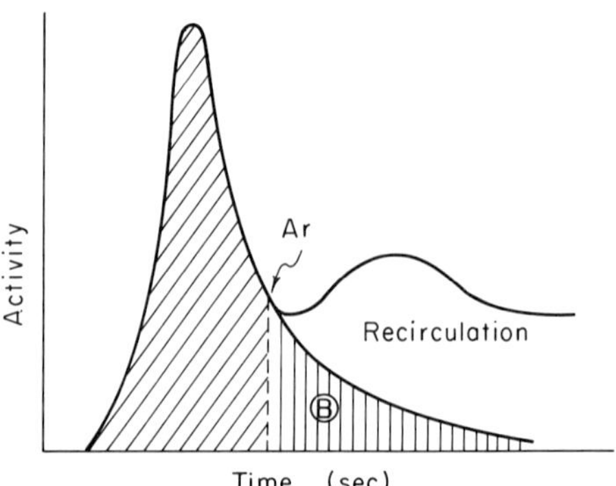

Figure 15. Uptake-washout curve typical of detected activity from brain, heart, kidney, etc., after intravenous injection of a tracer. Extension of washout curve from point Ar is done graphically or analytically in order to obtain the area under the curve $\int q(t)dt$.

36

This leads to the powerful relationship

$$\frac{F}{V} = \frac{a}{q_\infty} = \frac{\text{constant input rate}}{\text{plateau radioactivity}} \qquad 21$$

Note this is another statement of conservation of mass or amount because $F/V = $ (mean transit time)$^{-1}$; thus, accumulated amount $=$ input rate $\times$ mean transit time. The mean transit time can be determined from an impulse injection independent of Equation 21, since mean transit time for an impulse injection is the same as mean transit time for the continuous infusion for the same inert substance. These expressions for constant infusion are valid only if there is no recirculation.

Flow Determined From the Conservation of Mass

In this section we start with a somewhat rephrased principle of conservation of material and derive Equations 10, 21, and other important relations used in nuclear cardiology. A statement of the principle of conservation of material can be rephrased as:

change of amount/time $=$ flow (concentration in) $-$ flow (concentration out)

This is usually written as

$$\frac{dq(t)}{dt} = F[A(t) - B(t)] \qquad 22$$

where the change of quantity in the organ or tissue with time dq/dt is related to the flow times the difference in concentration of the input A and output B. Equation 22 is known as the *Fick principle*.

Another way of writing this equation for inert substances is to state that the change in quantity with time is proportional to the amount of a tracer which enters the system minus the specific volume flow times the quantity in the system at a particular time.

$$\frac{dq(t)}{dt} = FA(t) \quad \frac{F}{V}q(t) \qquad 23$$

This equation is particularly useful. Notice that $FA(t)$ is flow times concentration which gives amount per time, and flow/volume times amount also gives amount per time. If we use a tracer with a very short half-life, some of the detectable activity will disappear while we are measuring the tracer concentration. Thus, for short half-life tracers an additional term is required in Equation 23.

$$\frac{dq(t)}{dt} = FA(t) - \frac{F}{V}q(t) - \lambda q(t) \qquad 24$$

where the last term is the decay rate constant λ times the amount in the system at time t; that is, fraction per time multiplied by amount $=$ amount per time. It is particularly helpful to note the specific volume flow F/V is similar to the decay constant, as both are measures of fraction of something disappearing per time. The mean time is merely the reciprocal of the rate constant. (Note: The tracer half-life is related to the mean time as $T_{1/2} = \bar{t}\ln 2$.)

Below are five measurement conditions for which various modifications of Equations 21,

22, or 24 give useful relationships between experimental or clinical measurements and organ flow or perfusion.

Fick Principle

Recall the Fick principle as stated in Equation 22:

$$\frac{dq(t)}{dt} = F[A(t) - B(t)]$$

If we sum all the incremental changes $dq(t)/dt$, that is, integrate both sides of Equation 22, we have

$$\int_0^\infty \frac{dq}{dt}\, dt = F \int_0^\infty [A(t) - B(t)]dt$$

or

$$F = \frac{\text{total amount injected}}{\int_0^\infty [A(t) - B(t)]dt} \qquad 25$$

We saw a similar equation in Equation 7. An application of Equation 25 is the determination of myocardial perfusion by coronary artery injection of a tracer. Flow is equal to the sum of the differences between the coronary artery concentration and coronary sinus concentrations at various times divided into the amount of tracer injected into the coronary perfusion space.

It is not practical to monitor arterial and venous concentrations or to sample tissue concentrations being drained by venous blood. Thus, to simplify Equation 25, the following manipulations are made. Assume that the ratio of the concentration of a substance in blood (B) to concentration in the tissue just perfused by that blood is

$$p = \frac{q/V}{B} \qquad 26$$

where V is a unit volume, or $\rho V =$ density times volume is a unit mass. Here p is the partition coefficient which is frequently given as λ, but λ will be used for the radionuclide decay constant.

When the measurements of both the accumulation of a substance in an organ and the arteriovenous difference can be made independently, there is no concern about the partition coefficient; however, when venous blood is not available, we resort to the partition coefficient to relate tissue concentration to venous blood concentration so as to be able to remove the venous component from Equation 22. The basic problem is that this coefficient varies with tissue type[36] as well as with hematocrit.[37] The tissue heterogeneity problem is difficult to compensate for;[36,38,39] however, the hematocrit problem has been dealt with efficiently by Carlin and Chien.[37] The partition coefficient for xenon in myocardial tissue is 0.7.[40] Thus, for practical use, those equations used for xenon clearance should contain a coefficient p and F/pV should be substituted wherever F/V appears.*

---

*A second factor deserves mention: the extraction fraction E, which is the fraction of the injected substance actually involved in tissue perfusion; for now, this is equal to 1, that is, 100 percent extraction.

To show how the partition coefficient enters into the conservation of mass equation in a logical way, we rewrite Equation 22 and divide both sides by the unit volume

$$\frac{1}{V}\frac{dq(t)}{dt} = \frac{F}{V}[A(t) - B(t)]$$

then, substituting Equation 26 for $B(t)$ we get

$$\frac{1}{V}\frac{dq(t)}{dt} = \frac{F}{V}\left[A(t) - \frac{q(t)}{p\,V}\right] = \frac{F}{Vp}\left[pA(t) - \frac{q(t)}{V}\right] \qquad 27$$

Note that the venous concentration $B(t)$ was converted to an equivalent expression, $q(t)/V$, which is the tissue concentration amount per volume. If we assume the arterial concentration is negligible immediately after injection, as is the case with xenon washout after coronary artery injection, then Equation 27 reduces to

$$\frac{1}{V}\frac{dq(t)}{dt} = -\frac{F}{pV}\cdot\frac{q(t)}{V} \qquad 28$$

This equation is a statement that the change in the amount of a substance per unit volume present at any time is directly proportional to the amount of the substance per unit volume remaining and the proportionality constant is equal to the specific volume flow divided by the partition coefficient.

For simplicity we set $q(t)/V = C(t)$ and $F/pV = k$; thus

$$\frac{dC(t)}{dt} - k\cdot C(t) \qquad 29$$

This equation is a differential equation whose solution is given by

$$C(t) = C_0 e^{-kt} \qquad 30$$

where C_0 is the tissue concentration immediately after injection. The value of k is merely the slope of the clearance curve when $C(t)$ is plotted against time. Note that this k is precisely the result we reached in Equation 10 for a monoexponential clearance:

$$\frac{\text{flow}}{\text{volume}} = k = \frac{1}{\text{mean transit time}}$$

Thus, the slope of a xenon washout curve plotted on semilogarithmic paper gives the flow per tissue volume. If we divide this constant by the density and multiply the result by 100, we get flow per 100 gm of tissue.

The method we have just derived is attributed to Kety and Schmidt.[31] This method is used to calculate myocardial perfusion by external monitoring of the detected activity in the myocardium for a few minutes after intracoronary injection of radioactive diffusible tracers such as ^{133}Xe, ^{79}Kr, and ^{131}I-antipyrine. The method is also used to measure skin or muscle blood flow by injection of ^{133}Xe intradermally[41] or intramuscularly.[42] This method has been applied to the human myocardium in particular in studies by Cannon and coworkers,[43] Holman and coworkers,[36] and Maseri and coworkers.[38] The technique compares well with other techniques used to measure global flow and microsphere embolization techniques; however, there are some fundamental flaws in the method as it is usually applied which can lead to serious errors, namely, that the partition coefficient is not constant throughout the myocardium, and for a heterogeneous distribution of flow the specific volume flow is biased toward the high flow tissues.

Stewart-Hamilton Equation

Recall the basic Fick principle (Equation 22):

$$\frac{dq(t)}{dt} = F[A(t) - B(t)]$$

Assume now that there is no outflow, or that the amount of some substance which flows into the system stays in the field of view of an external counter or accumulates in an enclosure such as the bladder. Then $B(t) = 0$ and

$$\frac{dq(t)}{dt} = FA(t) \qquad\qquad 31$$

We can then sum all of the contributions by integration of each side of Equation 31:

$$\int_0^\infty \frac{dq(t)}{dt} \cdot dt = F \int_0^\infty A(t)dt,$$

$$q_0 = \text{total amount injected} = F \int_0^\infty A(t)dt,$$

$$F = \frac{q_0}{\displaystyle\int_0^\infty A(t)dt} \qquad\qquad 32$$

This expression is known at the Stewart-Hamilton equation and is similar to Equation 7, which we derived above by a different route. Here the denominator is the sum of the concentrations rather than the sum of the amounts as in Equation 7. Note the same expression can be derived from

$$C(t) = C_0 e^{-kt}$$

by merely integrating both sides to get

$$\int_0^\infty C(t)dt = C_0 \int_0^\infty e^{-kt}\, dt = C_0 \left[-\frac{e^{-k(\infty)}}{k} + \frac{e^{-k(0)}}{k} \right] = \frac{C_0}{k} \qquad\qquad 33$$

But k is equal to F/V if p equal 1; thus

$$\frac{F}{V} = \frac{C_0}{\displaystyle\int_0^\infty C(t)dt} \qquad\qquad 34$$

Here and in Equation 7 we have F/V on the left; Equation 32 had only F. The difference is that in the latter case the expression on the right is amount over concentration and concentration is expressed as amount per volume.

The tasks of the Stewart-Hamilton approach are to measure the concentration changes with time for a tracer entering or leaving the system. The approaches of both Kety-Schmidt and Stewart-Hamilton make the assumption that there is no recirculation and the input is instantaneously and completely diffused. Recirculation is not a problem for xenon and krypton because these gases are exhaled as soon as they reach the lungs and do not reenter the left heart. When recirculation cannot be avoided, a correction of the washout curve can be made by subtracting this recirculation contribution; this can be determined by the time-activity curve from a bolus injection at the organ outlet such as the right atrium in the case of the heart.[38]

Constant Infusion Rate

Let us return again to our fundamental equation and see what happens for a constant infusion of an inert tracer which does not recirculate. In this case the arterial concentration $= A(t) = S$, a constant. We can now rewrite Equation 27 and integrate to get

$$C(t) = pS(1 - e^{-kt})$$

where, as before, p is the partition coefficient and k is F/pV. This equation allows us to measure flow if the tissue input concentration is constant. Note for very long times $e^{-kt} \to 0$; thus, $C(t) = pS$, which is similar to Equation 21 for $p = 1$ if we rewrite that equation as

$$\frac{F}{V} = \frac{S \cdot F}{C(t) \cdot V}$$

where the constant input rate is amount/time = concentration $\times$ flow.

Variable Arterial Input

Recall once again our fundamental relation (Eq. 27):

$$\frac{1}{V} \frac{dq(t)}{dt} - \frac{F}{pV}\left[pA(t) - \frac{q(t)}{V}\right]$$

Solution of this equation by integration from time of injection to some particular time t gives

$$\frac{q(t)}{V} = \frac{F}{V} \cdot e^{-F/pVt} \int_0^t A(\tau) e^{\frac{F}{pV}\tau} d\tau$$

$$= pk \int_0^t A(\tau) e^{-k(t-\tau)} d\tau \qquad\qquad 35$$

where $k = F/pV$.

This equation gives us the flow if the concentration on the arterial input side is variable but measurable. An example of the application of Equation 35 would be the injection of a diffusible substance into the left atrium (Fig. 16). The aortic concentration of tracer is the input function $A(t)$. By performing numerical integration of measured values of $A(t)$ one can find which flow value best matches the tissue uptake $q(t)/V$ at

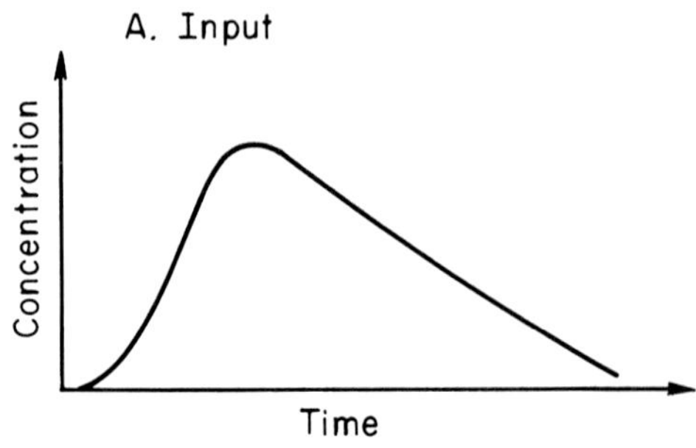

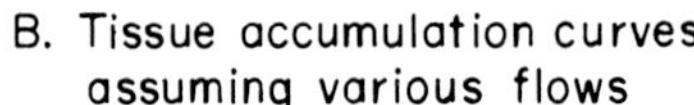

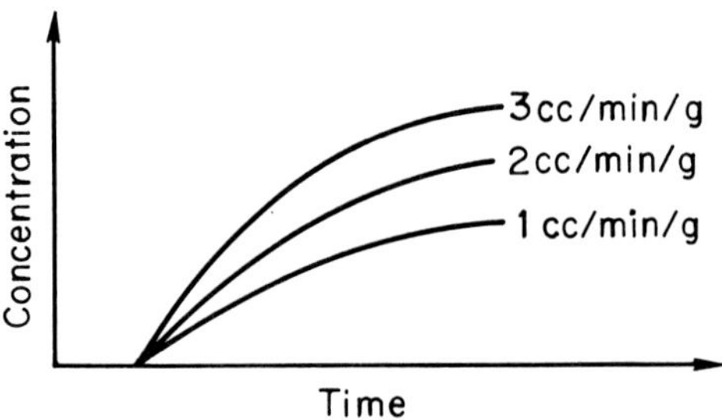

Figure 16. Deducing flow from the tissue accumulation curve after an infusion of tracer with a known input function (Eq. 35).

a particular time. For an arterial concentration curve such as the one in Figure 16A we can calculate the expected tissue concentration at any time from the onset of infusion as shown in Figure 16B.

Summary of Flow Principles

The underlying concept of flow physiology is the law of conservation of material which is expressed in various equations or methods such as the Kety-Schmidt method, the Stewart-Hamilton equation, the constant input equation, and the variable input equation. The Kety-Schmidt method is used extensively in nuclear cardiology for myocardial, skeletal muscle, and skin-blood flow using coronary or tissue injection of xenon or krypton as diffusible but inert tracers. In the following section we explore methods of flow evaluation using nondiffusible and diffusible tracers; we also review the concepts of flow-limited and diffusion-limited tracers. These principles can be applied to nuclear cardiology with new instrumentation for noninvasive, three-dimensional measurement of myocardial flow, amino acid metabolism, oxygen utilization, fatty acid metabolism, and glucose utilization. The basic principle which underlies most of these techniques is the statement that the change in amount of some substance with time is equal to the input rate minus the amount leaving the tissue by flow and through radioactive decay (Fig. 17). Recall Equation 24:

$$\frac{dq(t)}{dt} = FA(t) - \frac{F}{V}q(t) - \lambda q(t)$$

This equation will be used under three relatively new approaches in the next section. Note, in summary, that this equation is precisely the Fick principle and the starting point for derivations of the Stewart-Hamilton or Kety-Schmidt equations if radioactive decay, $\lambda q(t)$, is neglected and we relate amount/volume, $q(t)/V$, to output concentration, $B(t)$.

MYOCARDIAL PERFUSION AND METABOLISM

General Principles

This section presents an application of the methods for measuring flow discussed in the previous section. It is possible to measure coronary artery blood flow by implanting Doppler or electromagnetic flow meters on coronary arteries in animal preparations; however, total flow is seldom of prime importance. The crucial information being sought is how well the muscle cell is able to perform metabolically and functionally. We measure metabolic potentials or integrity by assessing how well the myocyte is being bathed by blood. The assumption is that if the muscle cells are well perfused, they are healthy. Muscle cell health can be inferred independently by examination of global or regional myocardial function, that is, how well the heart is pumping. Methods for evaluating function are discussed in the next section.

There are three major flow compartments: intravascular, interstitial, and intracellular. Agents which do not diffuse out of the intravascular spaces—such as red blood cells, human serum albumin, and microspheres—can be labeled with a host of radionuclides including ^{123}I, ^{99m}Tc, ^{111}In, ^{68}Ga, and ^{15}O-carboxyhemoglobin. Tracers which diffuse through the interstitial space and more or less through the intracellular space include inert gases such as the radioisotopes of xenon and krypton, ^{15}O-labeled water, labeled antipyrine, and ^{11}C-labeled alcohol. Some of the agents which become actively involved in the metabolic processes of the myocytes include potassium (^{43}K, ^{38}K) and the potassium analogues: rubidium (^{79}Rb, ^{81}Rb, ^{82}Rb, ^{84}Rb, ^{86}Rb), cesium (^{129}Cs, ^{130}Cs, ^{131}Cs), and thallium (^{201}Tl). Metabolic substrates of importance to nuclear cardiology are ^{13}N-ammonia, ^{11}C-glucose, ^{15}O$_2$, amino acids labeled with ^{11}C or ^{13}N, and fatty acids such as ^{11}C-palmitic acid, ^{11}C-oleic acid, and ^{123}I-hexadecanoic acid.

Nondiffusible Tracers

Radionuclide-labeled microspheres appropriate for use in humans are comprised of macroaggregated albumin labeled with iodine, usually ^{131}I, but ^{123}I is a better substitute for single photon imaging and ^{68}Ga for positron tomography. These microspheres of 20 to 30 μm diameter are injected in the coronary arteries during patient catheterization. Their size is such that they will be trapped in smaller capillaries of the myocardium; thus, the distribution of the radiolabeled microspheres embolized in the myocardium will be directly dependent on the flow distribution. The trapped particles are imaged by external detectors discussed earlier under Physics.

Four problems arise from this approach. First, since a coronary catheterization is involved, the procedure is invasive and necessarily embolizes some capillaries in an already compromised heart. The second problem is that the image seen is a projection of a three-dimensional activity onto a single plane. The third problem is the need for high doses in order to obtain sufficient statistics if tomography is used to overcome the second problem. Finally, since the basis of the method is that the number of microspheres trapped in a given region is equated to flow, the particles must be large enough to be trapped. The trapped particles must be larger than the smallest branches; thus, very small microspheres are needed to demonstrate flow in small capillaries.

The usual method for microsphere scintigraphy results in an image that is dependent not only upon blood flow per unit mass, but also upon the total mass in the region of interest. Thus, one would expect that for myocardial perfusion on a per mass basis,

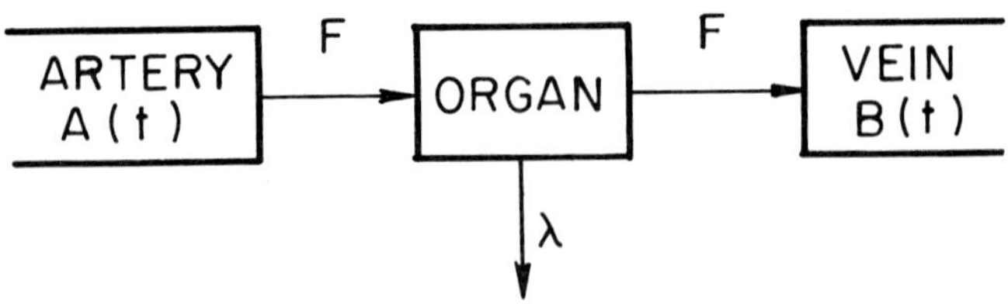

$$\text{Change in amount} = \text{Flow} \times \text{concentration}$$
$$- \text{Flow/volume} \times \text{amount}$$
$$- \text{radioactive decay}$$

$$\frac{dq(t)}{dt} = F\, A(t) - \frac{F}{V}\, q(t) - \lambda q(t)$$

Figure 17. Basic flow principle from which the equations of this chapter are derived. The change in amount of a substance with time is equal to the amount input minus the amount leaving through flow and the amount which has decayed.

the left ventricular myocardium would show up with far more activity. In fact, the right ventricle usually does not show up well at all because there is much less mass. When the myocardium is infarcted and replaced with fibroid tissue, less flow is available to that region, so that a decrease in regional activity occurs when this technique is employed.

Materials used consist of ^{99m}Tc-labeled human albumin microspheres prepared from a commercial kit. During the intracoronary injection, approximately 20,000 particles, 30 μm in size, are injected with an activity of approximately 3 mCi (total). Another type of injection is the I-131 macroaggregated albumin which is available in sizes up to 80 μm. The absorbed radiation dose to the heart for 3 mCi of technetium-labeled human albumin microspheres is 1.7 rads.

This method is currently in clinical use at Loma Linda University Medical Center[44] and the University of Washington[45] where information about viable myocardium is inferred from microsphere distribution images. The major use of the microsphere technique is in acute animal experiments where the regional distribution of minute amounts of radioactive microspheres are used to provide an experimental reference standard for flow.[35,46,130]

Labeled Red Blood Cells and Albumin

A well-known nondiffusible tracer is ^{131}I-human serum albumin (^{131}I-HSA), which is used for blood volume measurements. The method involves injection of a known amount of ^{131}I-HSA and, after a period for mixing, samples are drawn for counting. The volume is merely the quotient of the amount injected divided by the blood concentration after mixing. Because the ^{131}I-HSA leaks from the intravascular system, a few samples are usually drawn over 15 to 30 minutes; the volume for each sample is then extrapolated to the injection time for a better estimate of the plasma volume. More recently, a method of labeling human red blood cells has been developed[47] following the observation that the presence of tin adsorbed on red blood cells leads to a strong fixation of ^{99m}Tc-pertechnetate. An in vitro method was developed[48] which is very practical and convenient. Less than 1 ml of patient blood is needed for the blood volume work; however, 6 ml are drawn for blood pool imaging procedures discussed in a later section.

A second method for labeling RBCs involves inhalation of $C^{15}O$ to form ^{15}O-carboxy-hemoglobin. The ^{15}O ($T_{1/2} = 2$ min) is produced by a cyclotron, reduced to CO, and administered to the patient.[131]

The clinical application of quantitative whole-body blood pool imaging to nuclear cardiology using whole-body scanning techniques is not yet underway; however, important new approaches to old problems are afforded by new instrumentation and the stable RBC labeling techniques. One can now investigate blood volume redistribution during disease and in therapy situations.

Flow-Limited and Diffusion-Limited Tracers

Tracers which are extracted from the input blood network in proportion to the flow are known as flow-limited tracers; that is, the amount of tracer which diffuses through the tissue is limited only by the amount delivered. Water, alcohol, antipyrine, and the inert gases are flow-limited, although there may be a slight permeability barrier to water[49] and antipyrine. Tracers which are not flow-limited are diffusion-limited, in which case the capillary endothelium or the cellular membranes are diffusion barriers. Substances such as potassium, rubidium, palmitic acid, sugars, and amino acids are, strictly speaking, not flow-limited; that is, the accumulation of these substances in tissue is dependent on the state of the transport mechanism of the cell membrane which reflects the metabolic state of the cell. *In practice, however, the fact that potassium, rubidium, cesium, thallium, and ammonia are not flow-limited does not obviate the utility of these ions for evaluation of relative myocardial perfusion in normal myocardium and in regions surrounding experimentally induced infarction.* These notions are reflected in Figure 18, wherein the concept of extraction ratio as explained below is presented as a measure of how much a substance deviates from flow-limited behavior.[33]

Extraction Fraction

The fraction of activity which is removed by the tissue between input and output is known as the extraction fraction. For example, in Figure 18A we see that the amount of a nondiffusible tracer which appears in the myocardium outflow (coronary sinus) is greater than the amount of rubidium or potassium at a given time. Since the extraction fraction or ratio for rubidium is less than that for potassium, the concentration of Rb in the coronary sinus will be higher than that for K if equal concentrations are injected through the coronary arteries. For substances which are not flow-limited the extraction ratios will decrease as flow increases. For inert substances diffusion equilibrium is reached very rapidly and the diffusion rate can be considered infinite.[39,50] Thus, it would seem that potassium and its analogues are poor indicators of flow. Nevertheless, in the range of flow found in the myocardium, these cations do a good job of reflecting flow as documented from microsphere studies. These statements are true for healthy myocardium and appear practically sound for the ischemic myocardium.

Sapirstein Principle

For a substance which accumulates in tissue, such as potassium, rubidium, thallium, ammonia, palmitic acid, or glucose, we can estimate flow from the relation:

amount present in an organ $-$ flow $\times$ extraction $\times$ concentration

$$q(t) = F \cdot E \int_0^t C(t)\, dt \qquad\qquad 36$$

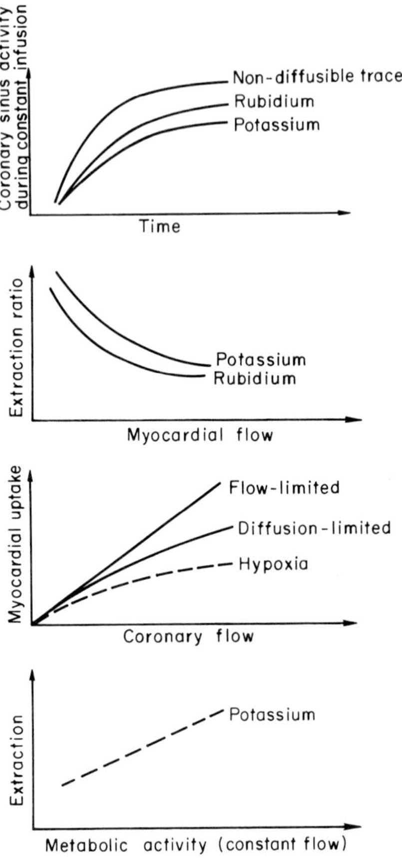

Figure 18. The proportion of injected potassium and its analogues extracted varies with the rate of delivery because of finite membrane permeability and, possibly, nutrient-versus-nonnutrient flow channel shunting.

This relation assumes that the amount which accumulates in the tissue on the first pass is proportional to flow, and that the fraction of injected activity found in an organ is proportional to the fraction of cardiac output to that organ.[51] If the extraction ratio is more or less constant throughout the body, and there is no redistribution, this method might work well. But as shown by the whole-body scans of the distribution of potassium in man (Fig. 19), the potassium distribution changes with time. Furthermore, potassium is a nondiffusible tracer for the brain since it does not pass the blood-brain barrier.

As our methods of quantitation improve we might find the simple diagnoses of good, poor, and no uptake for the potassium analogues (K^+, Rb^+, Cs^+, Tl^+, NH_4^+) can be improved upon because the uptake of these cations reflects capillary surface area and membrane permeability as well as flow. For this reason some of this physiology is reviewed here.

Diffusion of ions such as potassium, rubidium, cesium, ammonium, and thallium can be accounted for by considering that the travels of each ion involve two barriers and three tissue spaces. The ion moving through the intravascular or capillary space diffuses through or between clefts in endothelial cells of approximately four angstroms in diameter. Upon reaching the interstitial space there is a rapid diffusion to the myocardial muscle cell membrane known as the sarcolemma membrane. The passage through that membrane is an active process, probably involving ATPase transport. Indeed, it has

46

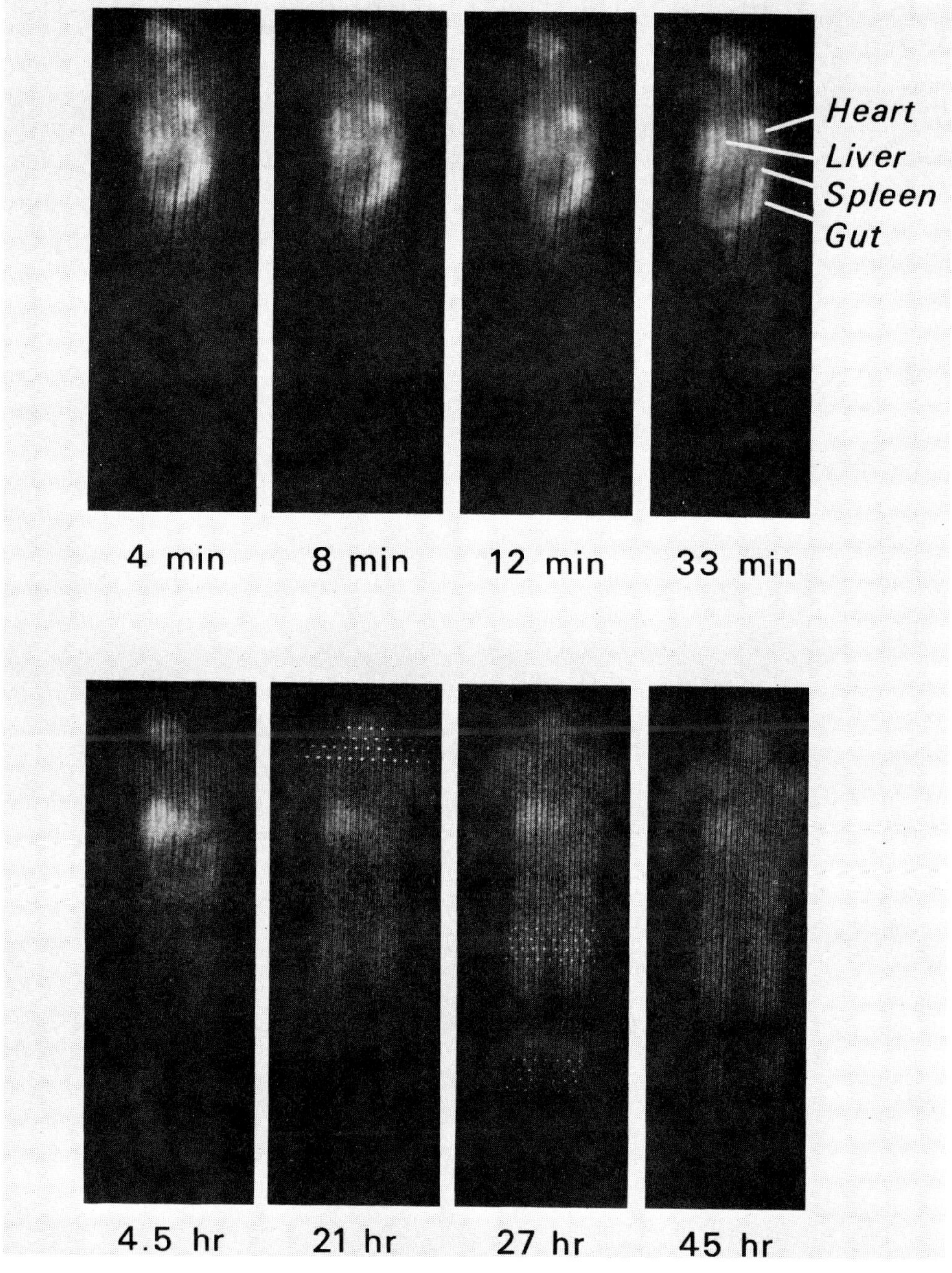

Figure 19. Distribution of potassium or its analogues reflects the fractional distribution of cardiac output for about 30 min after intravenous injection. Thereafter there is a gradual redistribution, as shown in the lower panels.

been shown that transport of thallium is also ATPase dependent.[52] Studies of Sheehan and Renkin[53] suggest that the capillary offers about 70 percent of the total resistance to potassium and the cell membrane about 30 percent. The resistance to rubidium at the capillary is about 50 percent, and about 50 percent at the cell membrane.

The transcapillary exchange, which is probably passive, is usually quantified by a term called "permeability-surface product." This product increases with flow; however, the size of the interstitial fluid compartment appears to be independent of flow. Thus, during increased flow a greater area of capillary surface opens, giving rise to an increase in potassium diffusion until a point at which the cell accumulation can no longer handle the amount of cation in the interstitial fluid and the effective extraction from the capillary concentration of cation appears to decrease. This diffusion dependence is illustrated in Figure 20.

At normal perfusion rates the heart actively removes more than 50 percent of the labeled rubidium or potassium entering the coronary circulation. In order to learn about the rates of transfer of ions across the capillary wall in sarcolemma membrane multiple tracers are used. For example, since potassium, rubidium, or thallium must flow through the interstitial space before entering the myocyte, an agent such as sucrose or sodium can be used to determine the interstitial flow per unit volume, since these substances, though diffusible, do not readily move through the myocyte barrier.

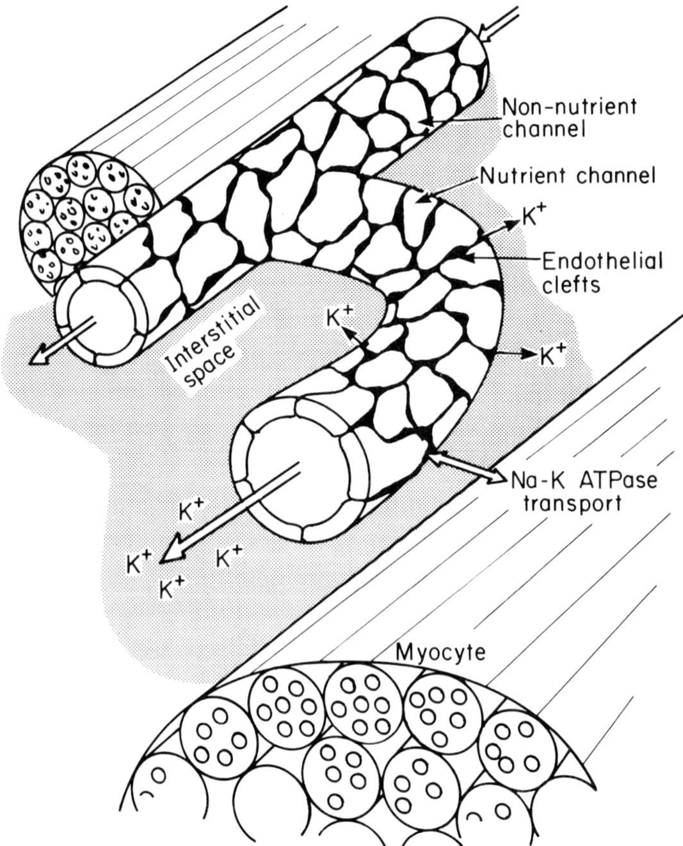

Figure 20. Potassium, rubidium, cesium, and thallium leak through capillary clefts, diffuse through interstitial spaces, and are transported into the myocyte interior by the Na-K-ATPase system.

The biological effect of rubidium is similar to that of potassium, and the intracellular rubidium pool is approximately the same size as the potassium pool.[54] There is little doubt that both the capillary wall and the cardiac muscle membrane act as barriers to the transport of rubidium. Although the rate constant at the capillary wall appears to vary with perfusion, it appears that the rate constant for the uptake of rubidium by cells is independent of flow but not independent of the cells' metabolic state.

The Effects of Hypoxia

The important factor in plasma potassium or rubidium and perhaps thallium clearance by the myocardium may not be the rate of coronary blood flow when the cellular flux of the ion is modified by the state of myocardial oxygenation. It is well known that potassium loss occurs during myocardial hypoxia induced by atherosclerosis as well as by pacing in normal human hearts. This is probably the result of interference with the ATP-dependent cation exchange mechanism caused by inadequate oxygen supply to the cell.[55] The usual escape of potassium from the heart muscle is about 1.6 percent per minute. Under normal range of flow the fractional extraction of potassium is about 0.6 ml/min/gm. As emphasized above, the amount of potassium or its analogues which accumulate in the myocardium shortly after injection is closely related to flow; however, the uptake by the sarcolemma might be reduced by hypoxia independent of flow.[56] It is not yet clear to what extent the thallium-rubidium-potassium accumulation data can be used to interpret ATPase activity as well as flow because it appears that even the truly compromised myocyte still accumulates these cations. Critical experiments in this area are presently underway in a number of laboratories.

The state of our knowledge about potassium movements in the myocardium is illustrated in Figure 20. The reason this basic physiological problem emerges in this chapter is the fact that with new instrumentation for emission tomography we have now at hand methods for measuring flow independent of extraction of potassium or its analogues as discussed below.

Organ Blood Flow Measured by ^{81}Rb/^{81m}Kr Activity Ratios

Rubidium-81 has a 4.7 hour half-life and decays to krypton-81m, which has a 13-second half-life. As mentioned above, krypton behaves similarly to xenon in that it is a diffusible tracer which can be used for flow measurements. After injecting ^{81}Rb into tissue it accumulates in the cell with a relatively slow turnover. The amount of ^{81m}Kr which is transported out of cell by flow can be deduced from the ratio of ^{81}Rb/^{81m}Kr activity; this relationship is depicted in Figure 21. Since the number of ^{81}Rb atoms which decay per unit of time is equal to the number of ^{81m}Kr atoms formed, the ratio of activity between the photons from each of these atoms will be equal to 1 if there is no flow. Since ^{81}Rb gives off photons of energies in the range of 450 to 520, and ^{81m}Kr gives off photons at 190 keV, one can examine simultaneously with two windows of a spectrum analyzer the ratio of the number of photons from each of these radionuclides. Under conditions of flow the ^{81m}Kr that is generated within the tissue where the ^{81}Rb is concentrated is washed out at a rate proportional to the flow; thus, the remaining ^{81m}Kr activity that can be detected within the organ is smaller than it would be in the no-flow situation. Consequently, the ratio of ^{81}Rb to ^{81m}Kr is higher for the flow situation than for the no-flow situation (Fig. 21). Flow is calculated as

$$\frac{F}{V} = 2.88 \text{ ml/min/gm} \cdot \text{activity ratio} - 1 \qquad\qquad 37$$

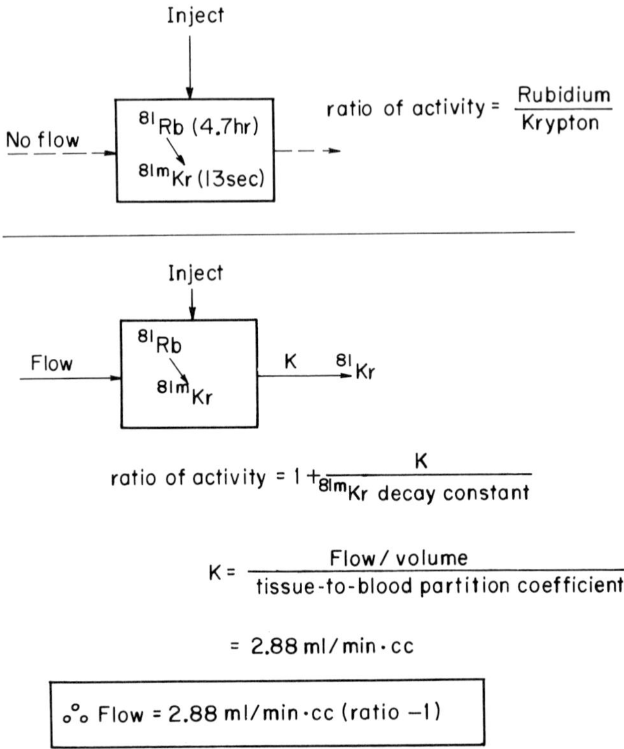

Figure 21. The ratio of detected activity for ^{81}Rb and its daughter ^{81m}Kr can be used to measure flow because the diffusible krypton is removed from the tissue by organ blood flow, but the rubidium has a greatly reduced physiologic clearance.

The flow term F/V is the specific volume flow; that is, the flow per unit volume perfused. The partition coefficient of 0.9 is contained in the constant 2.88 ml/min/gm. It can be shown that this particular value of 2.88 represents a blood flow causing a ^{81m}Kr disappearance rate equal to physical decay. This technique has been used by Van Herk to examine the flow in animal models.[57] Limited human heart studies done by our group and by others[58] have not been successful because of the inability to separate the superposition of ^{81m}Kr activity being discharged from the myocardium into the lung tissues and air surrounding the patient. A second difficulty has been with the lack of purity of ^{81}Rb; as much as 30 percent of the isotope injected into animals and patients is contaminated with ^{82m}Rb. The latter has abundant high energy photons which penetrate the collimator systems and lead to poor contrast images. The ^{81}Rb/^{81m}Kr ratio method is extremely viable if one perfects a rapid-scanning, single-photon or single-photon and positron tomographic system. The concept for a possible future method of measuring myocardial flow using this technique is illustrated in Figure 22.

Flow Determined From Equilibrium Imaging

As discussed above, the constant infusion of a tracer for sufficient time to reach equilibrium leads to a simple relation between flow and the quantity in the tissue:

$$\frac{F}{V} = \frac{\text{input rate}}{\text{plateau activity}}$$

This equation requires knowledge of the input rate and assumes that the half-life of the tracer is long and that there is no recirculation. An interesting relationship between perfusion and detected activity can be derived for the case of constant administration of short-lived radionuclides until steady state or equilibrium is reached. The importance of this approach is that the rapid decay of some tracers effectively removes the activity from recirculation. Also, because the constant infusion can be maintained for a long time, emission-computed tomography can be performed by instruments which require data acquisition times longer than the dynamic biological change times. The method is also applicable to projection imaging devices such as Anger or multicrystal cameras as long as the input can be separated from the amount present. Derived below are the basic relationships for the relatively new and extremely powerful technique previously used by Fazio and Jones[59] for lung ventilation studies using inhaled ^{81m}Kr (13 sec) and cerebral perfusion by intracarotid injection of ^{81m}Kr.[60] Recall our fundamental relation (Eq. 24) between the change in amount to input, output, and decay (Fig. 17):

$$\frac{dq(t)}{dt} = FA(t) - \frac{F}{V}q(t) - \lambda q(t)$$

At equilibrium $dq(t)/dt = 0$, because by steady state we mean there is no change in amount $q(t)$. Thus

$$0 = FA(t) - \frac{F}{V}q(t) - \lambda q(t) \qquad 38$$

and

$$q(t) = \frac{FA(t)}{\frac{F}{V} + \lambda} \qquad 39$$

Note also that

$$\lambda = \frac{ln2}{T_{1/2}} \qquad 40$$

If the half-life is very short compared to the mean transit time, e.g.,

$$\lambda = \frac{ln2}{T_{1/2}} \gg \frac{F}{V} = \frac{1}{\bar{t}} \qquad 41$$

then Equation 39 can be rewritten by neglecting the F/V term, since this is much smaller than λ:

$$F = \frac{\lambda q(t)}{A(t)} \qquad 42$$

This relationship has been used to measure brain blood flow by intracarotid infusion of ^{81m}Kr ($T_{1/2} = 13$ sec). We turn now to two situations where the specific volume flow is about equal to the half-life.

Equilibrium Flow Imaging Using Highly Diffusible Tracers

If we infuse a tracer such as $H_2^{15}O$, wherein the tissue permeability is great, we have—on rearranging Equation 38 and noting that $q = C \cdot V$ where C is the tissue concentration—

$$F = \lambda \cdot \frac{C}{A} \cdot V + \frac{F}{V} \cdot \frac{C}{A} \cdot V \qquad 43$$

$$\frac{F}{V}\left(1 - \frac{C}{A}\right) = \lambda \frac{C}{A} \qquad 44$$

$$\frac{F}{V} = \frac{\lambda}{\left(\dfrac{A}{C} - 1\right)} \qquad 45$$

The physical decay for oxygen is

$$\frac{ln2}{2.03} = 0.34 \ \text{min}^{-1}$$

Thus, flow can be determined if we have a means of measuring A and C, the input concentration and the tissue concentration, respectively. A method of implementing flow using Equation 45 and emission computed tomography is illustrated in Figure 23. This method has not been implemented as yet in clinical investigations, though the instrumentation is available.

For a tracer which is extracted from the perfusate the arguments of Equation 38 become slightly more complicated. We assume the tissue of interest has two components; a rapidly perfusing component and a more slowly exchanging component represented by the fractional loss of the tracer from the intracellular pool. For example, the amount of rubidium injected into the coronary capillaries which is available for rapid perfusion is $(1 - E)$, where E is the extraction ratio. Though this ratio varies with flow, we assume

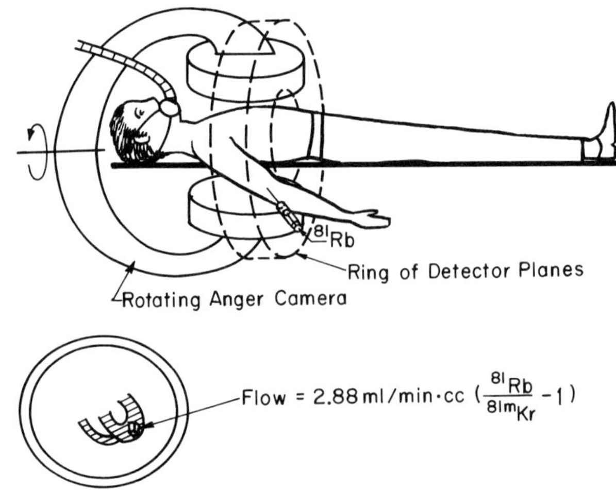

Figure 22. A method of measuring myocardial flow using single photon emission transverse section tomography and intravenously injected rubidium-81.

it is sufficiently constant for the following derivation (Eq. 46). The conservation of material equation 24 is modified to become:

$$\frac{dq(t)}{dt} = FA(t) - FA(t)(1 - E) - \gamma q(t) - \lambda q(t) \qquad 46$$

where E is the extraction ratio and γ is the slow turnover of the extracted rubidium from the system per unit time. At equilibrium $dq(t)/dt = 0$; thus

$$F = \frac{(\gamma + \lambda)q}{EA} \qquad 47$$

where A is $A(t)$ at equilibrium and q is $q(t)$ at equilibrium. But the fractional loss of potassium and its analogues is very small relative to the rate of decay of rubidium of 0.55 min^{-1} (i.e., 0.016 min^{-1}); thus Equation 47 becomes

$$F = \frac{\lambda q}{EA} \qquad 48$$

Note: This is similar to Equation 42 except for the presence of the extraction ratio E.

An example of the application of this technique is the use of short half-lived [82]Rb which can be administered by constant infusion from a tabletop [82]Rb generator.[61] As we discussed above, the extraction ratio for rubidium varies with flow; nevertheless, a practical constant value over the flow ranges we encounter is 0.5 to 0.7. Thus,

$$F(\text{cc/min}) = \frac{(0.55 \text{ min}^{-1})\, q(\text{equilibrium})}{(\sim 0.6)\, A(\text{equilibrium})} \simeq \frac{q}{A} \qquad 49$$

In order to use Equation 49 quantitative information from volumes of interest must be obtained using emission-computed tomography. The input concentration A at equilibrium is obtained by a region of interest in the left ventricular blood volume, and the amount present at equilibrium is determined by the number of events in a volume of myocardium (Fig. 23). This method has not been implemented to date with patients, although preliminary inquiries have been conducted by Harper[58] and by our team at Donner Laboratory. On cessation of constant input the amount detected in the myocardium should decrease as a monoexponential, as can be seen in Equation 46 when zeroes are substituted for each term having an $A(t)$; note that the solution of the resulting equation is the same as Equation 30. Recent experiments show this not to be the case for the detected activity from the heart of dogs, probably because the circulating concentration of rubidium $A(t)$ does not go to zero immediately after cessation of infusion.

NUCLEAR CARDIOLOGY WITH OXYGEN, FATTY ACIDS, GLUCOSE, AND AMINO ACIDS

This section explores some specific techniques on the frontiers of nuclear cardiology relating to assessment of tissue state or metabolism. The following are specific areas of recent or shortly anticipated applications for imaging the myocardium:

1. Myocardial imaging with potassium and its analogues.
2. Oxygen-15 diffusible and nondiffusible tracer imaging.

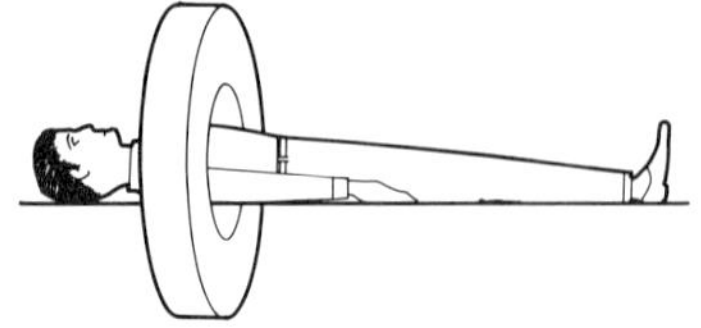

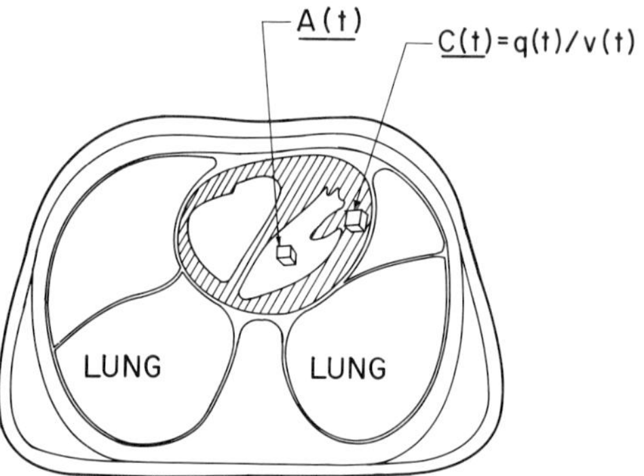

Figure 23. Inhaled ^{15}O-labeled CO_2 converts to $H_2{}^{15}O$; at equilibrium myocardial flow can be calculated using Equation 45 where the left ventricle activity $A(t)$ and myocardial activity $C(t)$ are measured by emission computed tomography.

3. Positive infarct imaging with ^{99m}Tc- and ^{68}Ga-labeled phosphates.
4. Glucose uptake imaging.
5. Fatty acid accumulation imaging.
6. Amino acid uptake imaging.

Myocardial Imaging with Potassium and Its Analogues

The uptake of potassium analogues such as rubidium, cesium, thallium, and ammonia is related to both regional myocardial blood flow and the active transport of these agents across the cell membrane. These agents also accumulate in skeletal muscle, liver, and stomach. The ions of potassium, rubidium, cesium and thallium not only have similar physical, electrical, and chemical properties, but are known to be transported through the sarcolemma by sodium-potassium ATPase.[52] Thallium is not strictly a potassium analogue, but myocardial extraction is similar to that of potassium;[62] similarly, ^{13}NH$_4{}^+$, though not an analogue, has an extraction similar to potassium. The amount of these ions which enters the myocyte depends on the transport of these ions through the myocyte as well as the permeability of the capillary endothelium. These ions are believed to pass through clefts in the capillary endothelium (Fig. 20). The size of the hydrated ions can lead to a difference in passive diffusion across the capillary wall.[63]

A graphic synopsis of comparative studies of the myocardial accumulation and washout of ^{43}K, ^{129}Cs, ^{81}Rb, and ^{201}Tl is given in Figure 24. ^{42}K is no longer used in nuclear cardiology because radiation doses are high and the 1.5 MeV photons require special collimation; however, the method of ^{42}K constant arterial input by controlled intravenous infusion is worthy of note, because flow can be determined by the slope of the regional uptake curves as discussed above under "constant infusion rate." The maximum myocardium/blood ratio occurs earlier for potassium than rubidium, thallium or cesium. Cesium accumulation in the myocardium is delayed relative to the other cations; however, cesium persists in the myocardium for much longer times than potassium, rubidium, or thallium. Thus, the optimum imaging period for potassium is 5 to 20 min after injection; for ^{201}Tl and ^{81}Rb it is 10 to 45 min; and for ^{129}Cs it is 30 to 120 min.[64-68]

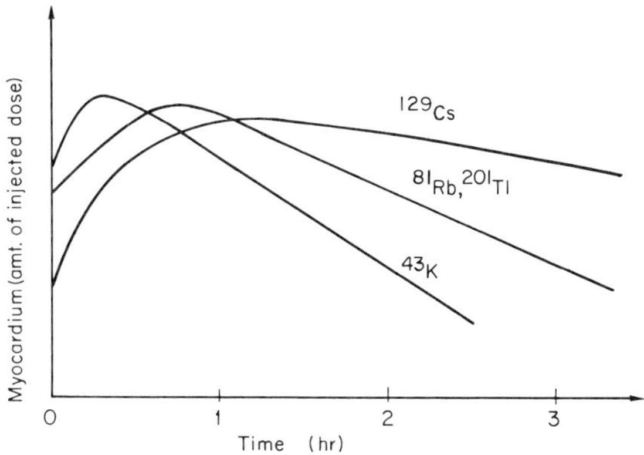

Figure 24. Time-varying concentrations for potassium, rubidium, cesium, and thallium in the human myocardium.

The myocardial image after injection of potassium or its analogues is an ovoid or U-shaped annulus of activity representing the projection of the left ventricle, with the right ventricle appearing only with right ventricular hypertrophy. Frequently the apex shows diminished uptake due to a diminished muscle mass. A similar image appears after injection of $^{13}NH_4^+$ ion. In the latter case the ^{13}N ($T_{1/2} = 10$ min) is produced by a cyclotron. Use of $^{13}NH_4^+$ has been supplanted by the more convenient ^{201}Tl for myocardial imaging using single photon imaging devices.[69] However, $^{13}NH_4^+$ appears to have great potential for sizing infarction and detecting ischemia when used with positron emission tomographs.[133] Infarction is detected as a decrease in the uptake; ischemia is detected by an absence of uptake after exercise.[70] Regional or generalized coronary artery disease might not be detected in the myocardial images unless the heart is stressed by exercise. In addition, the pattern of uptake can appear abnormal in hypertrophy and valvular disease.[71] A review of the history of potassium analogue imaging is given by Strauss.[72]

Using nonimaging, external, quantitative detection of the positron emitter ^{84}Rb with detection instruments, Knoebel and coworkers[73] have been able to provide valuable clinical assessments about the normalcy of the myocardium. Rubidium-84 has a long half-life—33 days—and interfering high energy gamma photons. It is conceivable that a similar coincidence detection system with ^{82}Rb ($T_{1/2} = 75$ sec) from a bedside generator[61,74] could provide a clinical tool for bedside moment-to-moment global Rb-uptake evaluation. Further discussion of cations for myocardial perfusion is given in this volume in the chapter by Beller.

Positive Imaging of Myocardial Infarcts Using Labeled Phosphate Compounds

Many years ago it was discovered that fluorescein and tetracycline accumulate in necrotic muscle tissue, and, that with the appropriate radionuclide label they could be used to delineate myocardial infarctions.[75] In 1974 a new method of radionuclide imaging of myocardial infarcts using ^{99m}Tc-labeled, phosphorus-based pharmaceuticals was developed by Bonte and coworkers.[76,77] The method of evaluating the presence of myocardial infarction by injection of technetium pyrophosphates 24 hours after the acute episode is now used widely.

Positive imaging of myocardial infarcts using positron emitters is accomplished by use of ^{68}Ga chelated to polymethylene phosphates. Both ^{68}Ga-EDTMP and ^{68}Ga-DTMPM can be used in the same fashion as ^{99m}Tc-PYP for infarction imaging.[78] The ^{68}Ga compounds are readily obtained at any facility using the germanium-gallium generator. Germanium-68 has a half-life of 275 days; its daughter, gallium-68, decays by positron emission 88 percent of the time and has a half-life of 68 min. The gallium is eluted from an alumina column on which [68]Ge is adsorbed. The elution is a simple procedure and gives ample quantities of a convenient positron emitter for use with positron emission tomographic devices discussed in the instrumentation section.

The mechanism responsible for the ^{99m}Tc-pyrophosphate uptake in acute myocardial infarcts and other tissues appears to be associated with the ability of technetium pyrophosphate to adsorb to calcium hydroxyapatite and amorphous calcium phosphate. Calcium in various forms accumulates in necrotic tissue with selective formation of calcific deposits in the mitochondria and elsewhere in the cell; recent studies suggest that this is the physiologic mechanism involved.[79] The number of such adsorption sites is a function of the degree and duration of preceding ischemic cellular injury. The appearance of a positive label in an infarcted heart region is dependent on the number of adsorption sites as well as delivery of ^{99m}Tc-PYP to the necrotic region. Thus, the physiology of uptake is a complex relationship between flow, concentration, cell permeability, and sites of adsorption. If there are numerous sites of adsorption but little flow, then a poor uptake can be the result. If there is moderate flow, a moderate number of sites of adsorption, but a rapid disappearance of the injected [99m]T-pyrophosphate into bone and kidneys, then one can expect a relatively low uptake. On the other hand, if a region of necrotic cells is well perfused, then we expect a vigorous accumulation of ^{99m}Tc-PYP. Recall the basic Sapirstein relationship (Eq. 36) which is applicable to this situation:

$$q(t) = F \cdot E \int_0^t A(\tau)d\tau$$

where $q(t)$ is the amount of ^{99m}Tc-PYP which accumulates in the tissue, F is the flow, E is the extraction ratio, and $A(\tau)$ is the arterial concentration which varies with time.

If $A(\tau)$ decreases at an extraordinarily rapid rate due to extraction from the blood by bone and kidney, then $q(t)$ will be lower than it would be under normal blood clearance, given constant flow and extraction ratio.

If flow is very low, then $q(t)$ will be low for a given history of arterial concentration and extraction ratios.

The critical parameter in the model is the extraction ratio E. This parameter is a measure of how much ^{99m}Tc-PYP can be adsorbed in the myocardium. If we assume the amount of ^{99m}Tc-PYP which can be extracted is proportional to the number of calcium sites for adsorption and the cell permeability, and if we assume also that these two factors are dependent on the duration of anoxia, we can formulate a relationship between E and the previously mentioned severity of ischemia as

$$E = 1 - e^{-\sum_i^T \Psi i} \qquad\qquad 50$$

where

$$\Psi_i = \Theta \left\{ \int_a^b [F_{th} - F(t)]dt \right\}_i \qquad\qquad 51$$

In Equation 51, F_{th} is the threshold flow below which the myocardial metabolic demands exceed nutrient supply and $F(t)$ is the flow at any time during the ischemic period from onset of subnormal flow t_a to the time t_b when $F(t)$ returns to normal. Here θ is a proportionality constant which relates the degree of ischemia to the extraction function. Each period of subnormal flow is designated by i. The other condition on the model is that the product of the low flow period between b and a and the degree of low flow

$$\int_a^b [F_{th} - F(t)]dt$$

must be greater than some ischemia value K. For example, if $F(t) \rightarrow 0$ for a short time, the transient ischemia will not be significant, so that

$$\int_a^b [F_{th} - F(t)]dt = \begin{cases} \Psi, & \text{if} > K \\ 0, & \text{if} < K \end{cases}$$

This model attempts to present a mathematical model for equating the accumulation of ^{99m}Tc-PYP or ^{68}Ga phosphates in the injured myocardium—based on the severity and duration of flow below the metabolic demands—and the flow conditions at the time of positive infarct imaging. Thus, if there are a few severe ischemic periods over a 24-hour period before a ^{99m}Tc or ^{68}Ga-phosphate study, accumulated damage can be ascertained by a value of E between 1 and 0. A similar value of E can occur if there is a longer period of less ischemia because the duration of diminished flow results in the same insult as two brief ($\sim$30 min) periods of severe ischemia. The basic model, Equation 50, implies that the lower the flow, the greater E; or the larger the time duration for a flow below F, the greater E.

The model emphasizes the fact that what we see after injecting a labeled phosphate for positive imaging is related to the previous history of compromised flow as well as the flow and blood concentration of the label reaching the ischemic and necrotic tissue. This model can be improved by including a provision for recovery.

^{15}O-Labeled Water and Hemoglobin

Oxygen-15 is a positron emitter which can be used for studies of the blood pool, body water, organ perfusion, and oxygen utilization. The half-life of ^{15}O is only 2 min; thus, a cyclotron is required for its use in nuclear cardiology. Nevertheless, the potentials of ^{15}O for noninvasive blood pool and myocardial oxygen utilization are so great that the $1.5 million cost might become a tractable investment for the cyclotron and imaging system which can handle oxygen as well as isotopes of nitrogen and carbon. Oxygen-15 is produced by bombarding stable nitrogen with deuterons. The deuteron enters the nitrogen nucleus and reacts with the 7 protons and 7 neutrons to produce a nucleus with 8 protons and 7 neutrons, which is oxygen of atomic weight 15. This unstable nuclide decays by emission of a positron (electron with a plus charge) and the positron combines with an electron in the tissue to yield two annihilation photons as discussed in the first section, "Physics."

Before it is injected into the body ^{15}O can be put into hemoglobin by converting the ^{15}O$_2$ produced from the cyclotron to carbon monoxide using charcoal at 900°C. The C^{15}O is mixed with the patient's blood to form ^{15}O-carboxyhemoglobin, creating an effective label for the blood pool.[131]

Alternatively, C^{15}O$_2$ can be formed by merely passing ^{15}O$_2$ over charcoal at 400°C. The C^{15}O$_2$ becomes H$_2$^{15}O immediately on inhalation in accordance with the reaction

$$C^{15}OO + H_2O \rightleftharpoons H_2CO_2^{15}O \rightleftharpoons H_2^{15}O + CO_2$$

which is mediated by carbonic anhydrase. Every time this fast reaction sequence occurs there is a 33 percent probability that labeled oxygen will be transferred to water. In aqueous solution the amount of dissolved CO_2 is small and at equilibrium; almost 100 percent of the labeled oxygen is in the form of water.[78]

A third form of ^{15}O is ^{15}O-oxyhemoglobin which can be used to infer oxygen utilization.

Flow Imaging with $H_2{}^{15}O$

After prolonged inhalation of the radioactive carbon dioxide, a steady state is reached such that the regional counting rates remain constant in the myocardium. Since the flow per unit volume is approximately 0.5 min^{-1} and the physical decay constant of ^{15}O is 0.34 min^{-1}, the situation is depicted by Equation 39 in the previous section. That is, at equilibrium the amount appearing in the myocardium at any time is proportional to the arterial concentration times flow divided by the myocardial flow plus the decay rate. Since the clearance rate is about the same magnitude as the decay rate, we cannot make the simplification of Equation 42, but need to use Equation 45:

$$\frac{F}{V} = \frac{\lambda}{\left(\dfrac{A}{C} - 1\right)}$$

In Figure 23 a method of extracting the arterial and myocardial concentrations (A and C, respectively) is shown using transverse section emission tomography techniques discussed in the first section. Using emission tomography, then, it is theoretically possible to measure specific volume flow of diffusible water in the myocardium. However, this exciting development has not reached clinical trials or experimental verification.

Once flow has been measured, it is possible to infer O_2 utilization by investigating the ratio between the flow data and the accumulation of activity from ^{15}O-oxyhemoglobin; this can be done with an approach similar to recent studies of brain ^{15}O-utilization which employed the MGH positron camera.[80]

Glucose Metabolism

An assessment of myocardial glucose metabolism can be made by injecting glucose labeled with ^{11}C, or better, 2-deoxyglucose labeled with ^{18}F or ^{11}C.

Glucose transport is a passive carrier-mediated process involving an equilibration of intracellular and extracellular glucose concentrations[81] with subsequent redistribution of labeled metabolites throughout the body. There is no preferential accumulation of glucose in the myocardium since, as increasing quantities are metabolized, there is a corresponding loss of activity from the heart.

The only difference between 2-deoxyglucose and glucose is the replacement of the —OH group on the second carbon by $=O$. Deoxyglucose is transported between blood and the myocyte by the same saturable carrier that transports glucose. In the tissues it competes with glucose for hexokinase which phosphorylates sugars to their respective hexose-6-phosphates. Deoxyglucose-6-phosphate is not further metabolized because it is not a suitable substrate for other enzymes in the glucose metabolic pathways; thus, labeled deoxyglucose accumulates in the myocardium without any loss for as long as 90 min.[82] In addition, there is considerable accumulation in brain, liver, and lungs. About 3.5 percent of the injected activity of ^{18}F-deoxyglucose is localized in the dog heart, which is comparable to the percent of injected ^{201}Tl which localizes in the myocardium. The persistent accumulation of ^{18}F-deoxyglucose in the heart is greater than any other viable imaging agent. The primary energy source for the normal myocardium under physiologic conditions is β-oxidation; however, under stress the anoxic or ischemic myocardium receives energy from glycolysis. Thus, if we combine studies of fatty acid metabolism with those of glucose transport in the cell, we have the potential for understanding the metabolic state of the myocardium in vivo.

Flow is usually the major controlling factor, so that in vivo biochemistry using nuclear medicine techniques suggested in this and succeeding paragraphs necessitates an independent evaluation of flow.

Imaging of Fatty Acid Accumulation in the Myocardium

Fatty acids provide the major energy source for the normal nonischemic heart and are efficiently extracted from the blood by myocardial cells. Nuclear cardiology studies using labeled fatty acids commenced in 1964 with the use of ^{131}I oleic acid by Evans[83] and subsequently, ^{123}I labeling of the 16 carbon fatty acid 16-iodo-9-hexadecanoic acid using radioiodine exchange with bromine.[84] The latter compound has a good myocardial extraction efficiency similar to ^{11}C fatty acids such as steric, oleic, and palmitic acids. However, in comparative studies it was found that the iodine-labeled fatty acid has a more rapid clearance half-life than the oleic or steric acid by a factor of three, probably a result of beta oxidation of the fatty acid with hydrolysis of the final radioiodinated metabolic product iodoacetate. Physical characteristics of ^{123}I make it suitable—in lieu of [201]Tl—for imaging viable tissue and even for single photon transverse section tomography because the dose to the whole body is 0.03 rad per mCi;[85] thus, 5 to 10 mCi can be injected.

Palmitic acid labeled with [11]C ($T_{1/2} = 20$ min) has been used in conjunction with positron emission tomography to quantitate myocardial infarction.[86] The palmitic acid is synthesized by the carboxylation of a Grignard reagent (pentadecyl magnesium bromide) using cyclotron-produced [11]CO_2. The water insoluble palmitic acid is complexed with serum albumin before injection. ^{11}C-palmitic acid accumulation in the myocardium appears to be stable over hours as evidenced by a half-life disappearance of 5.5 hours, whereas palmitic acid disappears from the blood with a half-life of 2 to 5 min. Transverse sections showing the distribution of palmitic acid in the dog myocardium show a quality similar to that obtained for [13]NH_4^+, with a similar target-to-nontarget ratio of 5 to 1.

The rationale for investigating ^{11}C-palmitic acid accumulation is that its distribution reflects the aerobic conditions of the myocardium as palmitic acid extraction is decreased under conditions of hypoxia even if normal perfusion prevails.[87] Thus, it might appear that this substance would reflect the metabolic state of the myocardium independent of flow. Unfortunately, the delivery of ^{11}C-palmitic acid to the healthy tissue is dependent on flow, and under poor flow conditions the lack of accumulation of a fatty acid is not strictly an indicator of the lack of β-oxidation.

In addition, while this long-chain, free fatty acid is known to be a major fuel of the mammalian myocardium, the intracellular sequestration of palmitate is inversely proportional to the arterial lactate concentration.[88] Nevertheless, ^{11}C-palmitic acid myocardial uptake is an exciting approach to the study of the integrity of the myocardium using emission tomographic techniques.

Amino Acid Uptake by the Myocardium

With the availability of positron emission tomographic devices discussed in the first section and methods of labeling amino acids with ^{11}C and ^{13}N, there has been an interest in ascertaining whether amino acid uptake by the myocardium might provide a method for assessment of protein synthesis and degradation. The prospects for this approach are quite exciting; however, there has not been sufficient clinical application to provide a methodology. The potential for amino acid distribution studies is great, however, and warrants some discussion of the physiologic principles involved.

The amino acid with the highest demonstrated uptake in the myocardium is asparagine; its uptake of 13.7 percent of the injected dose is about four times higher than that of potassium and its analogues, glucose, and fatty acids.[89] However, the application of ^{13}N-asparagine to nuclear cardiology is limited by the fact that the enzyme asparagine synthetase is not readily available.

Skeletal and myocardial muscle are the major sites for degrading the branched-chain amino acids: leucine, isoleucine, and valine. These amino acids are the only essential amino acids which are oxidized mainly by muscles.[90] It is achieved that these amino acids provide a carbon chain for intermediates in the tricarboxylic acid cycle, and that amino groups released upon degradation of the branched amino acids are transferred to pyruvate derived from glycolysis to form alanine and perhaps glutamine.[91]

The oxidation of the branched-chain amino acids is stimulated by the presence of free fatty acids which apparently inhibit the pyruvate dehydrogenase system in muscles. It has been found that by infusing the rat myocardial tissues with octanoate, oleate, or palmitate, the uptake of the branched-chain amino acid leucine in the myocardium is increased by a factor of two and that of alanine and pyruvate is decreased. The oxidation of histidine, ornithine, glutamate, and threonine is not affected by adding fatty acids to the perfusate. Of interest is the fact that the pool of the fatty acids is not increased; what increases is the rate of oxidation after transport into the cell. Thus, it is not clear that the branched-chain amino acid incorporation would be an indirect measure of pyruvate dehydrogenase activity from in vivo nuclear cardiology studies.

The alanine and perhaps the glutamine which are produced in the myocardium are transported to the liver for hepatic gluconeogenesis. In addition, alanine[20] and glutamic acid[92] have been shown to incorporate in the myocardium of man. Of great relevance to nuclear cardiology is the observation that the myocardium of patients with coronary artery disease extracts more glutamate than the normal myocardium.[134] The potentials of this observation have not yet been demonstrated. In the past few years there has been considerable interest in the fact that glutamine and glutamic acid are not incorporated to any significant amount in the myocardium of dogs but demonstrate a significant uptake in man and monkeys.[93] Whether a similar situation holds true for valine has not yet been demonstrated; however, recent studies at our laboratory suggest that neither dl-carbon-11 valine nor nitrogen-13 valine have a significant uptake in the myocardium of dogs, whether they are in a fed or a fasting state. There is no significant increase in uptake after fatty acid infusion.

Finally, it is significant to observe that the $^{13}NH_4^+$ imaging studies performed in animals and man, in addition to reflecting flow, demonstrate the incorporation of the ammonia into glutamine synthesis by the myocardium. Thus, the appearance of ammonia in the myocardium might have as its basic biochemical basis the synthesis of glutamine by the myocardium, even though its overriding physiologic basis is the delivery of ammonia by flow through intact capillary bed.[133]

METHODS OF RADIOANGIOCARDIOGRAPHY

There are distinct approaches to imaging of the blood pools:

1. Rapid sequential image recording and playback
2. Initial pass end-systole, end-diastole imaging
3. Gated blood pool imaging
4. Multigated radioisotope angiography
5. Functional imaging of ejection fraction and stroke volume

Sequential Image Recording

The qualitative assessment of isotope distribution images of the blood pool in infants and adults is an effective and useful adjunct to diagnosis of congenital and acquired hemodynamic defects that reflect heart pathology. The method generally involves collection of images from a flow study and examination of the time sequence of radionuclide movement through the heart chambers and outflow tracts using video tape techniques or digitized images from the gamma camera or multicrystal camera (System 77). An example of results from Kriss and coworkers[94] is shown in Figure 25. The information contained in these images provides measures of cardiac performance that in fact yield all the data now derived from right and left heart catheterization, with the exception of blood gases and pressures. Wesselhoeft and coworkers used the pinhole collimator to magnify the images from very small infants.[95] Stop-action capability of the video tape system and permanent information storage allow selected portions to be viewed at chosen intervals. Summation images of sequential frames can be recorded. The isotopes and radiopharmaceuticals most commonly used are free ^{99m}Tc-pertechnentate, ^{99m}Tc-albumin, ^{111}In, and ^{99m}Tc-labeled red blood cells. The latter is probably the best, as the label remains in the blood pool. Labeling can be done in vivo by injection of tin-pyrophosphate followed by ^{99m}Tc-pertechnetate,[47] or in vitro by labeling the patient's red blood cells by combining the RBCs with stannous chloride and ^{99m}Tc-pertechnetate.[48]

Initial Pass Angiography

The gamma camera or multicrystal camera is placed to view the left ventricle in the left anterior oblique (LAO) or modified LAO by tilting the camera 30 degrees cephalad (Fig. 26). A bolus injection of 10 to 20 mCi of ^{99m}Tc-pertechnetate is made in an antecubital vein, and the X-Y position of each event is recorded by digital computer on a fast data acquisition device such as video tape or magnetic disc. This method is called "list mode accumulation." A time mark is recorded every ten milliseconds. After approximately 15 seconds the initial pass is complete and the data from 4 to 15 seconds

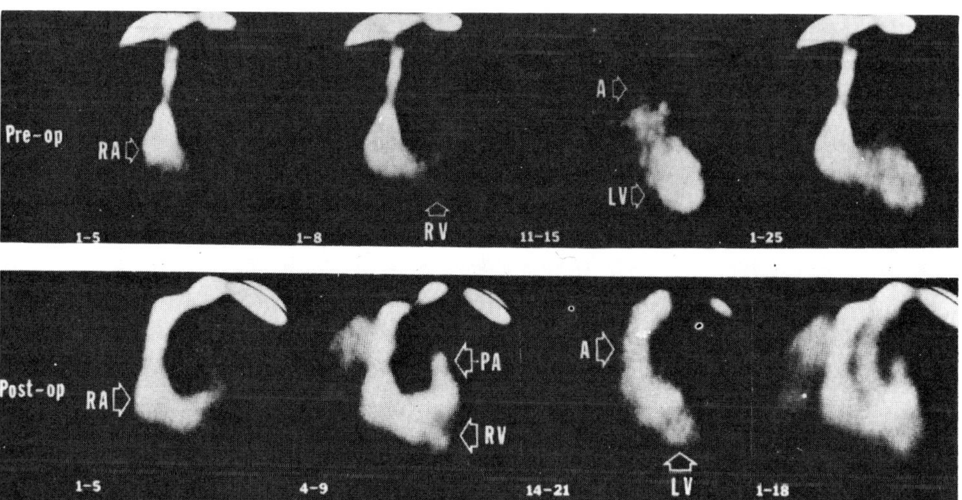

Figure 25. Qualitative radioangiographic procedures of Kriss and coworkers[94] involve rapid imaging of the sequential flow through cardiac chambers after bolus injection of an intravascular tracer. An example of tetralogy of Fallot before and after correction is shown. (Reproduced with permission.)

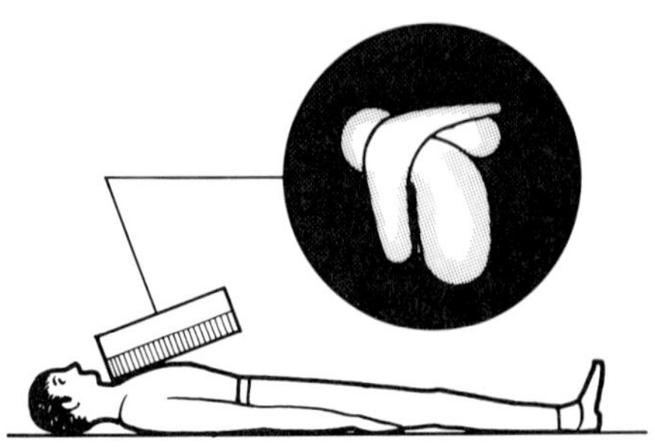

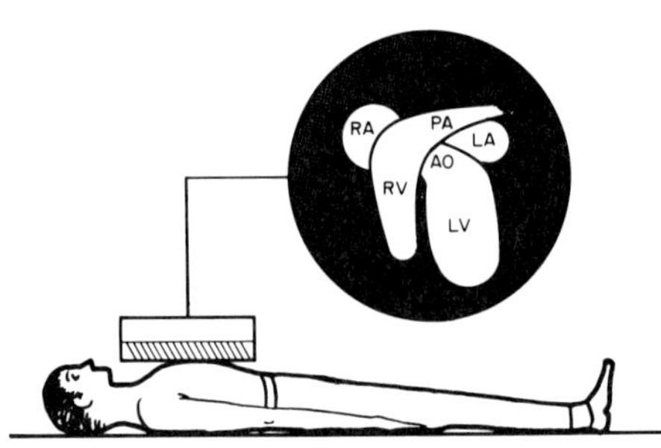

Figure 26. The modified left anterior oblique view involves a 30 degree left and caudally-directed tilt of the Anger camera or other imaging device. By machining the cephalad-caudal tilt of the collimator, the image resolution is improved because the distance between the heart and camera is decreased.[12]

after injection are organized into 50-msec frames (images). An image is framed for the period from 8 to 15 sec after injection. On that image the left ventricle can be delineated. All that is necessary is knowledge of where the left ventricle appears on the frames. The image picture elements (pixels) are flagged with a light pen, and using this as a region of interest a curve of changing activity is produced from 100 to 200 of the 50-msec frames (Fig. 27). The frames of maximum counts represent end-diastole and the frames of minimum activity, end-systole. The frames corresponding to the times of end-diastole are summed to give an image of end-diastole; the same is done with the frames corresponding to the times of end-systole. The trick here is to automatically sort all of these 100 to 200 images. This is done by differentiating, with respect to time, the smoothed time-activity curve, then automatically grouping frames belonging to the zeroes of the first derivative curve to high or low count categories. Usually, 150 msec (or three frames) are used for each high or low count category. This process is done with computer software. The result is two images, one of end-diastole and the other of end-systole.

Gated Blood Pool Imaging

The data acquisition can be physiologically controlled by using the ECG to electronically gate (turn on and off) the imaging device or data recording system. Each period comprises 50 to 70 msec. End-diastole occurs at the R-wave and end-systole occurs at the latter portion of the T-wave. Images of end-systole and end-diastole can be obtained by controlling the moment when the camera is on (shutter control) or by controlling the periods during which data are transferred to some recording device. A method of obtaining the physiologic signal involves use of the ECG to give the R-wave and a predetermined electronic delay after the R-wave for end-systole. A second more accurate method of determining end-systole is by triggering on the second heart sound phonocardiographically.[96] The images are produced by summing the end-systole (50 msec) and

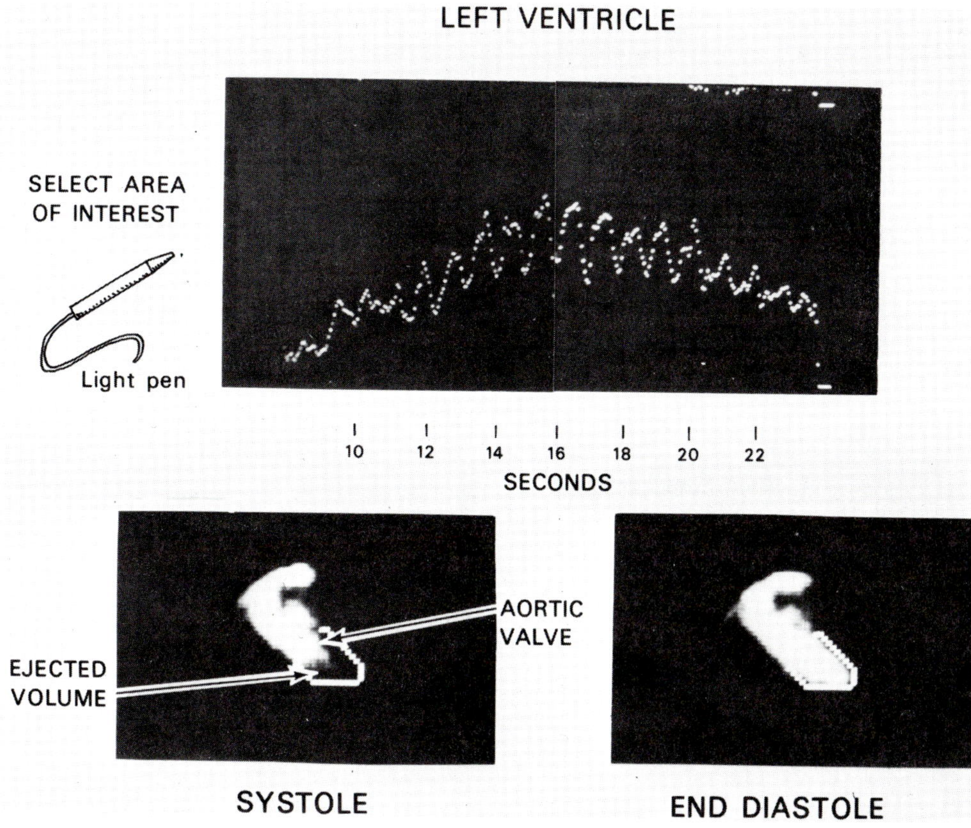

Figure 27. Example of retrospective gating wherein the time-activity curve was generated at 50-msec intervals by using an area over the left ventricle for count integration. The sum of the frames at maximum activity gives an image of end-diastole; the sum of the frames at minimum activity gives an image of end-systole.[100]

end-diastole images over a period of 5 to 15 minutes after injection of a blood pool agent such as ^{99m}Tc-albumin or ^{99m}Tc-labeled red blood cells.

Multigated Radioisotope Angiography

Rather than limit the cardiac imaging to only two images representing end-systole and end-diastole, more elaborate computer-based techniques have been developed whereby a sequence of consecutive images representing the entire cardiac cycle are prepared. This procedure can be called radiocineventriculography, wherein 12 or more images per cycle are formed.[97,98] Previously, researchers were limited to mass storage devices such as magnetic tape memory. Now minicomputers with only 24K of 16-byte words can perform these tasks by use of data sorting in the computer "on-the-fly" using a large active memory wherein 12 or more images can be created from the list mode data. The present approach allows one to view the images as an endless movie during and after the study.[97]

Functional Stroke Volume and Ejection Fraction Imaging

An image representing kinetic parameters such as time rate of change is known as a functional, parametric, or rate image and has been used in nuclear medicine,[99,100] but has been used only rarely in nuclear cardiology until recently. J. A. Parker and co-workers,[12] and D. E. Maddox[102] have presented a method of displaying the relative magnitude of the stroke volume or the ejection fraction as one image by subtraction of the end-systole image from the end-diastole image or division of that difference by the end-diastole image, respectively. The result is a gray scale distribution which immediately gives information on regional wall motion as well as ventricular performance.

Applications of Blood Pool Imaging and Ejection Fraction Imaging

Analysis of left ventricular wall motion by viewing the sequentially summed frames in movie format or the functional image described above leads to four major applications of blood pool imaging:

1. Dyskinesia evaluation
2. Congenital and acquired blood pool abnormalities (e.g., ventricular aneurysm)
3. Valvular performance
4. Ejection fraction estimation

A major defect of radioangiography is lack of sufficient resolution to provide the anatomic definition enjoyed by two-dimensional contrast angiography. However, some quantitative measurements can be made once the data are "controlled" and sequenced into frames representing phases of the cardiac cycle using either video tape systems or computer storage.

The techniques involve measurement or calculation using single plane techniques of contrast cineangiography[103] or biplane methods of Dodge and Sandler[104,105] and modifications which allow delineation of akinetic or dyskinetic left ventricular segments.

Quantitative Radiocardiography

The combination of image control discussed above and quantitative measures of the change in activity of various chambers, discussed in this section, leads to powerful methods of quantitative evaluation of cardiovascular function dynamics. Selective quantitative radiocardiography pioneered by Donado[106] offers the potentials of noninvasive, nonmorbid, and convenient evaluation of the following:

Cardiac output (CO)
Stroke volume (CO/heart rate)
Ejection fraction of right and left heart
Ejection volume
Right-to-left shunts
Left-to-right shunts
End-diastolic volume of right and left heart
Mitral regurgitation
Aortic regurgitation
Combined mitral and aortic regurgitation
Rate of volume changes
Intracavitary transit times

These evaluations take approximately five minutes of patient time and one-half to four hours of analysis, depending on the analysis algorithms and associated computing hardware. The evaluations can be done by a single camera view and one or two isotope injections. The imaging view accepted by many investigators and clinicians is the modified left anterior oblique (Fig. 26) because the longitudinal rotation axis in this view is parallel to the ventricular septum. This view allows resolution of the right and left sides of the heart and good separation of the atria from the ventricles.

Cardiac Output and Stroke Volume

The cardiac output is calculated by observing the time-activity curves of the isotope movement through the left ventricle in the anterior or modified left anterior oblique view. The data are acquired by flagging the region of the left ventricle using a computer system, or by selecting the field of interest on video playback using a photodiode or PMT to transfer the activity data to a strip chart recorder.[107,108] The cardiac output is given by

$$CO = \frac{C_{eq} \times \text{blood volume}}{A + B} = \frac{\text{total amount}}{\displaystyle\int_0^t C(t)\, dt}$$

where C_{eq} is the count rate at equilibrium, $A + B$ is the area under the uptake-washout curve, and B is an extrapolated portion of the curve (Fig. 15). When using a strip chart recorder, the practical implementation of this equation is given as

$$CO\ (\text{l/min}) = \frac{\text{curve height at equilibrium} \times \text{chart speed} \times \text{blood volume}}{\text{area}}$$

In practice it has not been easy to obtain the area under the recirculation curve of Figure 15; however, as discussed in the section on left-to-right and right-to-left shunts, this problem can be overcome and automated using the gamma variate fit to the time-activity curve. Cardiac output can be determined by this technique using either a gamma camera or collimated scintillation probe.[108,109] The scintillation probe provides a convenient portable means of evaluating cardiac output at the patient's bedside; however, the position of probe placement relative to the left ventricle is critical in most situations, and the selected area advantage of the camera would seem to be preferred if a small portable camera with area integration capabilities were available.[108,110,111] Technically and economically a portable camera, 14 cm in diameter, is adequate for most cardiac function analyses.

Regardless of how the cardiac output is determined the stroke volume is calculated as cardiac output/heart rate; end-diastolic volume (EDV) can then be calculated from stroke volume and a measure of ejection fraction (EF): EDV = stroke volume/EF.

Ejection Fraction

The fraction of left ventricle blood ejected through the aortic valve is the forward ejection fraction. It is the difference between end-diastolic volume (EDV) and end-systolic volume (ESV) divided by the end-diastolic volume:

$$EF = \frac{EDV - ESV}{EDV}$$

Presently, the single most valuable functional measure of ventricular performance in myocardial disease is the ejection fraction. The major cause of a diminished ejection fraction is probably a reduction in the ischemic muscle fiber shortening during systole, resulting in an increased end-systolic volume. End-diastolic volume increases have also been noted during infarction.[112] The physiologic importance of the ejection fraction is the fact that ejection fraction can be measured noninvasively and can therefore be used to evaluate the progress of patients undergoing intervention therapy for myocardial infarction.

Ejection fraction can be determined by external monitoring using either of two methods. The first is to calculate the ratio of volume at end-diastole to volume at end-systole.[113,114] The other method was developed by Zaret, Strauss, and coworkers. ^{99m}Tc-albumin is injected intravenously with the patient in the RAO view, and images of the heart at end-systole and end-diastole are obtained using a scintillation camera and an electronic gate triggered by the patient's electrocardiogram. Each image is composed of 300,000 counts, representing end-systole and end-diastole of 200 to 400 heartbeats. An outline of the left ventricle free wall is hand-drawn from the end-diastole and end-systole images. The position of the aortic and mitral valve planes is determined using a radionuclide angiogram obtained at the time of tracer injection. Left ventricle ejection fraction is calculated from the area and length of the long axis of the ventricular outline at end-systole and end-diastole.

A second technique involves the measurement of the peak-to-valley count difference of the undulating activity curve as the isotope moves into and out of the left ventricle.[108,110,115-117] This technique can be employed with a high speed camera-computer system or a single-crystal probe and strip chart recorder. The latter method requires the area of interest to be as large as the ventricle in end-diastole; however, unwanted activity from the right ventricle, left atrium, aorta, and tissue blood pool contributes significantly to the time-activity curve of the left ventricle. Thus, the ejection fraction must be calculated from

$$EF = \frac{\text{diastole counts} - \text{systole counts}}{\text{diastole counts} - \text{background}} \qquad 52$$

The magnitude of this background or cross-talk contribution must be determined to describe the effective baseline for the peak to background measurement (Fig. 28). Two

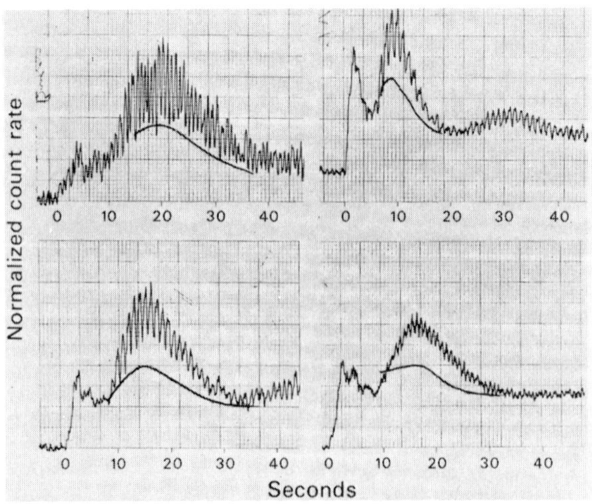

Figure 28. Strip chart recordings of the change in activity over the left ventricle with each heartbeat for four separate studies. The solid line indicates the background correction curve. (Reproduced with permission.[108])

methods have been used. The baseline can be established by placing a ring or angular region of interest around the left ventricle (LV) area of interest. This new time-activity curve is used to compensate for background. Usually the area of the annulus is chosen to be the same as that for the left ventricle.[108,115,116] The background curve is adjusted such that its activity is the same as that at the nadir of the LV curve, but before the beginning of recirculation, as shown by the solid lines in Figure 28. The technique can be applied in the anterior, MLAO, LAO, or RAO views. In the latter view, Kirch and coworkers[115] use an ellipse rather than an annulus to avoid atria and great vessels.

The second method[117] of background subtraction has a good anatomic basis in that the background time-activity curve is derived by the contributions from the area where the ventricle is during diastole, but not during systole. Thus, the computer forms an image of diastole minus systole, and within the region that was flagged as the left ventricle, the computer is programmed to select the picture element with the maximum activity in the subtraction image. Then any element within the previously defined LV area of interest which has half this activity or more is selected for a new area of interest. The background area of interest is thus defined. Both procedures give good correlation with ejection fraction determined from LV contrast cineangiography.

Right-Heart Mean Transit Time and Ejection Fraction

The uptake-washout curves from the right atrium and right ventricle can be determined by sequential right anterior oblique scintillation camera images after rapid injection of ^{99m}Tc-pertechnetate, ^{99m}Tc-labeled RBCs or ^{99m}Tc-labeled human serum albumin. The technique described below is limited to right heart evaluation because left heart projection images include activity outside the left ventricle and atrium and a compact bolus injection to the left atrium is difficult to achieve unless a Swan-Ganz catheter is wedged in a peripheral pulmonary artery.

Recall our basic discussion on conservation of material. Rephrasing this principle for the right heart leads to

$$q_r(t) = F \int_0^t C_a(\tau)d\tau - F \int_0^t C_r(\tau)d\tau \qquad 53$$

That is, the amount in the right ventricle, q_r, at any time t is equal to the amount input from the atrium (i.e., flow $\times$ concentration $= FC_a$) minus the amount output from the ventricle (i.e., flow $\times$ concentration $= FC_r$). In other words, the amount present is the difference between the input and the output. Note that the amount in the right ventricle q_r is merely the concentration in the right ventricle C_r times the volume:

$$q_r = C_r V_r = F \int_0^t C_a(\tau)d\tau - F \int_0^t C_r(\tau)d\tau \qquad 54$$

$$\frac{F}{V_r} = \frac{C_r(t)}{\int_0^t C_a(t)dt - \int_0^t C_r(t)dt} = \frac{1}{\tilde{t}} \qquad 55$$

where $\tilde{t}$ is the mean transit time. Equation 55 states that the ratio between the amount present at any time t after injection divided by the difference between the input and output areas of concentration vs. time curves is a constant, and equal to the reciprocal of mean transit time. To use this concept for cascaded compartments such as the right atrium and right ventricle, we must make the assumptions that the flow in the right atrium equals that in the right ventricle, which must be the same as the systemic flow. Thus,

$$F = F_{\text{systemic}} = F_a = F_r$$

We assume that the flow is nonpulsatile, that chamber volumes are constant, and that there is perfect mixing. Also, for Equation 53 to hold there must be no shunts. The presence of shunts can be detected by examination of the constancy of Equation 55:

$$\frac{F}{V} = \frac{C_r}{\Delta C_r} = \text{constant} \qquad\qquad 56$$

where ΔC_r is the denominator in Equation 55. This is a powerful concept. Note that the difference between input and output is merely the shaded area of Figure 29. To implement this procedure we must accumulate data at 50-msec intervals after a bolus injection in the RAO position. Regions of interest representing the right atrium and right ventricle are flagged with a light pen or other electronic flagging device. The uptake-washout curves are generated from these flagged regions.

The major trick to this method is normalizing the curves of right atrium and ventricle activity by arriving at some common basis. This is done by noting that since the input must equal the output, the point of maximum activity from the right ventricle must intersect the activity curve of the right atrium. Thus, either the right atrium or right ventricle curves are multiplied by whatever factor is needed to cause the point of maximum concentration from the right ventricle to intersect the washout limb of the right atrium curve.

This method was proposed by Ishii and MacIntyre[118] and has been perfected by Freedman and coworkers.[119]

The ejection volume (EV) of the right ventricle can be determined from the exponential washout slope of the time-activity curve of the right ventricle. The EV is merely the cardiac output divided by the decay constant (EV_r = cardiac output/λ = cardiac output $\times$ mean time). This useful formula cannot be applied to the left ventricle because the washout does not follow a single exponential; however, the most likely rate constant could be derived from $\Sigma A(t)/\Sigma A(t)t$ after the washout curve is analytically extended either as in Figure 15 or using the gamma variate discussed below. This detailed discussion of a method for determining right-heart mean transit time is given to show how these methods relate to our previous analysis of the physiology of flow.

Ventricular Volumes

As mentioned above, assessment of left ventricular volumes can be made from single or biplane radioangiocardiography using the gamma camera and techniques similar to those for contrast cineangiography. The isotope techniques do not have any discomfort or morbidity and do not interfere with heart function as contrast dye studies do. In contrast angiography the right anterior oblique view is preferred because this projection is perpendicular to the long axis of the left ventricle during diastole, and there is less foreshortening in this plane than in either the anteroposterior or lateral projections. From measurements of Sandler and Dodge,[105] the heart movement during systole results in a foreshortened left ventricular axis for the RAO position, and thus the calculation of end-systolic volume is likely to be too small if done from the RAO position; nevertheless, calculation of volumes from a single view gives reasonably accurate results

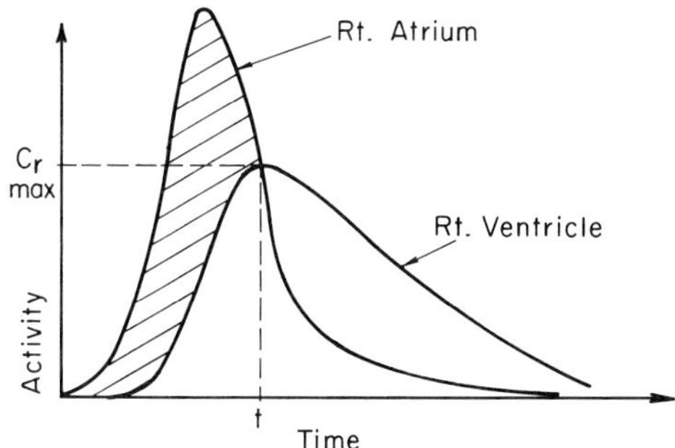

Figure 29. Curves of detected intraventricular activity after rapid intravenous injection of an intravascular tracer such as ^{99m}Tc. After adjusting curve heights, the areas are used to calculate specific volume flow of the right ventricle.

if the values are corrected by a calibration factor based on comparison between single and biplane studies or model phantom studies.

The biplane technique pioneered by Dodge and coworkers[104] for contrast studies was first applied to radioangiography by Mullins and colleagues[120] who obtained two orthogonal views (anterior and lateral) of the flow of ^{99m}Tc-pertechnetate through the canine heart after direct intracavitary injections. Volume estimates for end-diastole were calculated by performing three measurements on the outline of the left ventricle for the two views. The assumption made is that the left ventricle at end-diastole can be approximated by a solid ellipsoid whose volume is

$$V = \frac{4}{3}\,\pi(l/2)bc \qquad\qquad 57$$

where l is the major axis and b and c are the minor semiaxes of the ellipsoid. These minor axes are determined from the planimetered area X of the outline from each view. Thus

$$b = 2X_{ant}/\pi l, \qquad c = 2X_{lat}/\pi l_s \qquad\qquad 58$$

where l_s is the shorter of the two long axes. Substitution of these equations into Equation 57 gives a concise expression for ventricular volume:

$$V = \frac{0.849 X_{ant} \cdot X_{lat}}{l_s} \qquad\qquad 59$$

Volumes at end-diastole and end-systole can be calculated from gated studies. This biplane technique has not been actively pursued, perhaps because of the need for two views. It is possible to combine two views in one imaging system and use labeled red blood cells as well as multigated blood pool imaging to give both MLAO and RAO.

The left ventricular volume (LVV) can be calculated from one view by two methods. Sullivan and coworkers[121] demonstrated that a single RAO view gave sufficient information for calculation of the volume using the longest axis for l and the widest axis perpendicular to l for $2b$ and $2c$. In this case Equation 57 becomes

$$V = \frac{4}{3}\,\pi(l/2)bc - \frac{4}{3}\,\pi(l/2)(w/2)^2 = (\pi/6)lw^2 \qquad\qquad 60$$

where w is the widest measurement perpendicular to the long axis l. The correlation of results from this technique with those of left ventricle contrast cineangiography is impressive.[121] The data in this case are from average isotope distributions in the left ventricle during systole and diastole.

Volumes at end-diastole and end-systole can be calculated from gated studies wherein summation images of the heart at end-diastole and end-systole are obtained by collecting data in specified time intervals relative to the ECG as discussed above.

An indirect method of determining volume arises from the relation

$$V = \text{flow} \times \text{mean transit time} = \text{cardiac output} \times T_i \qquad 61$$

where T_i is the time constant determined by parameter fit to the Laplace transform function, a technique proposed by Ishii and MacIntyre[118] which gives good results as long as there is no regurgitation.

The four methods of LV volume measurement outlined above are

1. EDV = stroke volume/EF
2. EDV by Equations 59 or 60
3. LVV by Equation 60
4. Volume by Equation 61

Left-To-Right and Right-To-Left Shunts

Using the pulmonary time-activity curve, Maltz and Treves[122] have presented a clever method for determining the pulmonary-to-systemic ratios (Qp/Qs) which reflect quantitatively the degree of left-to-right shunts. Their method involves fitting the pulmonary curve to a gamma variate of the form

$$A(t) = t^{\alpha}e^{-t/\beta} \qquad 62$$

using a simple least squares technique. Here α and β are arbitrary constants. The fit is made to data from the top of the pulmonary time-versus-activity curve to avoid recirculation. The complete curve given by Equation 62 using the derived parameters α and β is then subtracted from the observed uptake-washout curve and the resulting curve again fitted by the gamma variate. The areas under these two curves are used to calculate the Qp/Qs ratio, as illustrated in Figure 30. This method can separate those shunts with Qp/Qs greater than 1.2 and can quantitate the amount of shunting when Qp/Qs is between 1.2 and 3.0.

Another method analyzing the presence of left-to-right shunts using the radionuclide pulmonary time-activity curve[123,124] involves a simple ratio of the maximum activity to that on the downslope curve; this ratio corresponds to a time increment between the maximum activity and the first appearance of isotope in the uptake-washout curve for a region of interest over the lung. This technique relies heavily on the correct determination of the appearance time and time of maximum concentration.

Another technique that allows quantitative evaluation of both left-to-right and right-to-left shunts involves comparison of the time-activity curves from two isotope injections, ^{133}Xe or ^{127}Xe and ^{99m}Tc, and is based on the fact that xenon is almost completely washed out in the lungs before returning to the left heart unless there exists an intercavitary shunt whereas ^{99m}Tc passes sequentially from right to left heart.[125] The patient is positioned in the MLAO position (Fig. 26) and ^{133}Xe in saline solution is injected intravenously; about 1 min later ^{99m}Tc-sulfur colloid or some other blood pool agent is injected without moving the patient. The data are collected and stored for each injection. The flow pattern of ^{99m}Tc-sulfur colloid gives visualization of the heart chambers, and is used for selecting regions of interest over various heart chambers during the playback of the data

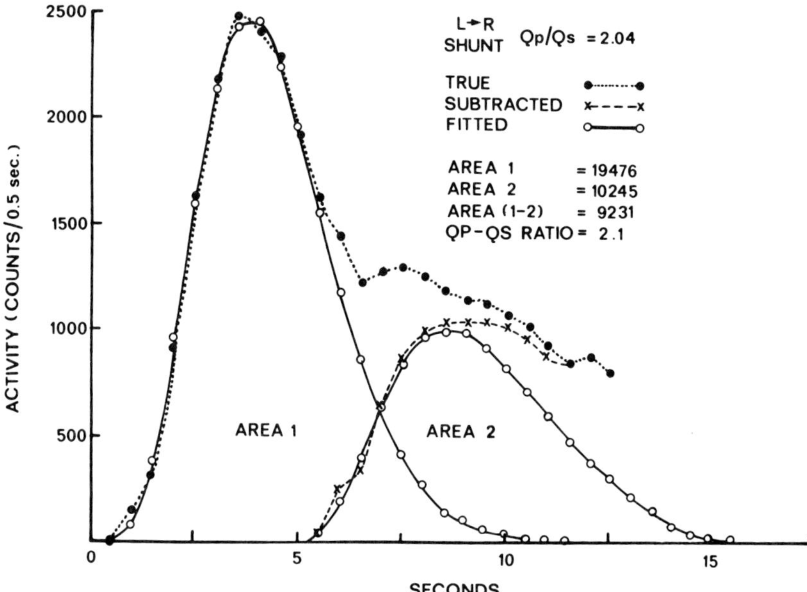

Figure 30. An effective method for the noninvasive detection of left-to-right shunts involves fitting a gamma variate function to the uptake-washout curves of isotope flowing through the lungs, and subtracting the fitted curve from the observed curve; a second gamma variate fit is then calculated. The ratio of the areas of these curves is used to detect the presence of a shunt and its severity.

for generation of time-activity curves for both ^{99m}Tc and ^{133}Xe flow series since the patient remains stationary for the two injections. The normal ^{99m}Tc time-activity curve will show a single peaked curve for a region over either the right heart or the left heart (Fig. 31). The normal ^{133}Xe curve will be a single-peaked curve for the region over the right heart only since all the xenon ejected from the right ventricle will be cleared by the lungs before reaching the left ventricle, unless, of course, there is a right-to-left shunt (Fig. 31). In left-to-right intracardiac shunts the ^{133}Xe in saline curve will be normal, but the ^{99m}Tc-sulfur colloid curve will show a characteristic double-peaked curve over the right heart (Fig. 31). Previous quantitative work using this technique was limited by camera speed and the change in the spatial impulse response with isotope energy.[126]

Another technique can be used to measure shunting by evaluating the discrepancy between diastolic volume calculated by the area-length technique (Eq. 60), and the ventricular volume from the stroke volume (Eq. 61). Weber and coworkers[116] give the quantitative estimate of fraction regurgitated as

$$\frac{\text{volume by area} \cdot \text{EF} - \text{stroke volume}}{\text{volume by area} \cdot \text{EF}}$$

Aortic and Mitral Regurgitation

The problem of regurgitation from the aortic or mitral valves has been dealt with successfully by Kirch and coworkers[115] who have analyzed the pulsatile movement of isotope in the left atrium and left ventricle using a single view in the RAO (40°) position with the scintillation camera. Previous attempts to solve this problem using transform approaches were not successful because the Laplace transform as it was used does not adequately describe the pulsatile flow of the heart. Their technique consists

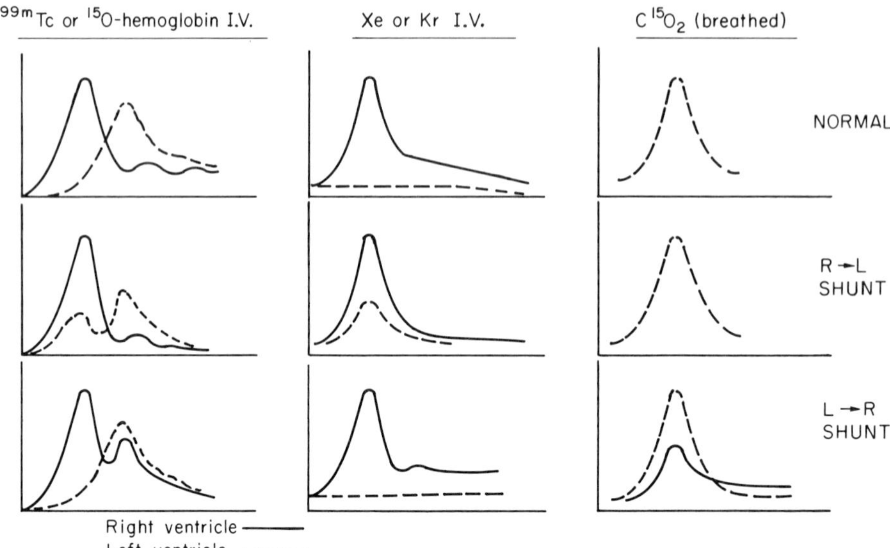

Figure 31. Injected intravascular tracers appear first in the right, then the left cardiac chambers, whereas inhaled tracers appear first in the left, then the right chambers. Cardiac shunts can be detected by noting the sequence of tracer arrival in each chamber or chambers and lung after injection or inhalation of tracers.

of solving a set of difference questions, which involve four parameters that describe flow into and out of the two chambers—left atrium and left ventricle. The solution is facilitated by the Z transform (digital Laplace transform), and is easily implemented using the left atrial and left ventricular time-activity curves obtained from injection of approximately 12 mCi of technetium from a Swan-Ganz catheter in the pulmonary wedge position.

The forward ejection fraction for the two chambers and regurgitant fractions for two valves on the left side of the heart have been determined with good correlation to findings from conventional left ventricular cineangiography studies. The regurgitant flow on the right side of the heart can be determined in a similar fashion by placing the catheter in the superior vena cava.

The background is removed by using a semiannular (ellipse) region of interest surrounding the heart, but excluding the atria and great vessels. The statistics are adequate for these studies; the only problem encountered is that in the early phases of the study there is incomplete mixing which can lead to some difficulty in detecting combined mitral and aortic regurgitation.[115] A more serious problem is that this technique involves placement of a pulmonary catheter.

Dynamic Emission Tomographic Cardiology

The techniques of this last section suffer from the fact that measurements are limited to projections so that the true volume of interest is not measured. The problem is to image the heart in motion in three dimensions for both biochemical studies and pump function observations using a noninvasive method. Using multiple layers of crystals surrounding the body and the positron emitters discussed in the first two sections and by Harper,[127] it should be possible to evaluate not only the perfusion and biochemical state of the myocardium, but the regional or segmental ejection fraction as well as wall motion.

SUMMARY

This chapter is a primer on the physics of radionuclide detection, flow physiology, and methods of in vivo evaluation of myocardial metabolism and intercavitary flow by noninvasive methods of intravenous isotope injection. This summary presents key concepts for the application of currently available instrumentation as well as future directions of nuclear cardiology.

1. Quantitative information is obtained in nuclear cardiology at the cost of high resolution imaging for two reasons: (a) the intrinsic resolution of the detecting systems is limited by available technology, and (b) the statistics required to achieve a high resolution image necessitate doses and imaging times far in excess of those which can be tolerated. Image resolution for both projection images as well as transverse sections are limited to the range of 5 to 20 mm, depending upon the configuration and instrument involved.

2. The second important concept is the fact that nuclear cardiology gives quantitative information regarding the amount of radiopharmaceutical which has accumulated in or is flowing through the cardiovascular system. This information allows one to deduce the dynamics of flow as well as actual metabolic rates.

3. The major emphasis for future work might well lie in the multiple transverse section imaging of the myocardium using both rotating and static devices. The key feature of this approach is the fact that the volume of interest can be localized and actual concentrations of radiopharmaceuticals can be measured by external detection using reconstruction tomography images.

4. Quantitative data on the distribution of a metabolite which accumulates in the myocardium is of little value if regional blood flow is not also known.

5. Finally, it is shown in this chapter that specific volume flow can be evaluated using short half-life isotopes and equations derived from the principle of conservation of mass. In principle it is now possible to obtain quantitative values delineating endo- and epicardial flow for the heart of man without invasive catheterization or high radiation doses. These procedures involve constant inhalation of carbon dioxide labeled with ^{15}O which converts to labeled water and can be used for evaluating myocardial perfusion; bolus injection of ^{82}Rb, a short half-life analogue of potassium, for repeated (every 5 min) imaging of the evolution of myocardial infarction size; evaluation of the accumulation of labeled fatty acids, amino acids, and sugars in the myocardium; presentation of images which reflect the magnitude of ejection fraction; and noninvasive evaluation of cardiac shunts.

It is now possible to perform on the same patient during a few hours the following studies of myocardium: cation perfusion evaluation; water perfusion; uptake of fatty acids, amino acids, and glucose; oxygen utilization of the myocardium; and even measurement of the quantity of lung water. We now have the tools and methods to evaluate the in vivo biochemistry of the ischemic,[128] infarcting, repairing, and hypertrophic myocardium.

ACKNOWLEDGMENTS

The manuscript was prepared with the assistance of Mary Graham, Robert Stevens, William Greenberg, Brian Moyer, and Grant Gullberg.

REFERENCES

1. ANGER, H. O.: *Scintillation camera.* Rev. Sci. Instrum. 29:27, 1958.

2. BENDER, M. A., AND BLAU, M.: *The autofluoroscope.* Nucleonics 21:52, 1963.

3. JONES, R. H., AND SCHOLZ, P. M.: *Data enhancement techniques for radionuclide cardiac studies.* Symposium on Medical Radionuclide Imaging, IAEA-SM-210/97, Los Angeles, Calif., October 25–29, 1976.

4. Brownell, G. L., Correia, J. A., and Zamenhof, R. G.: *Positron instrumentation.* In Lawrence, J. H., and Budinger, T. F. (eds.): *Recent Advances in Nuclear Medicine*, vol. 5, Grune & Stratton, New York, 1978.

5. Cho, Z. H., Chan, J. K., and Eriksson, L.: *Circular ring transverse axial positron camera for 3-dimensional reconstruction of radionuclides distribution.* IEEE Trans. Nucl. Sci. NS-23:613, 1976.

6. Phelps, M. E.: *Emission computed tomography.* Semin. Nucl. Med. VII (4):337, 1977.

7. Ter-Pogossian, M. M., Mullani, N. A., Hood, J., et al.: *A multislice positron emission computed tomograph (PETT IV) yielding transverse and longitudinal images.* Radiology 128:477, 1978.

8. Derenzo, S. E., Budinger, T. F., Cahoon, J. L., et al.: *High resolution computed tomography of positron emitters.* IEEE Trans. Nucl. Sci. NS-24:544, 1977.

9. Dwyer, E. M., Jr.: *Radioisotope studies in coronary artery disease.* Cardiovasc. Clin. 7(2):101, 1975.

10. Evans, R. D.: *The Atomic Nucleus.* McGraw-Hill Book Co., New York, 1955.

11. Budinger, T. F., and Rollo, F. D.: *Physics and instrumentation.* Prog. Cardiovasc. Dis. 20(7):19, 1977.

12. Parker, J. A., Uren, R. F., Jones, A. G. et al.: *Radionuclide left ventriculography with the slant hole collimator.* J. Nucl. Med. 18:848, 1977.

13. Holman, B. L., Idoine, J. D., Sos, T. A., et al.: *Tomographic scintigraphy of regional myocardial perfusion.* J. Nucl. Med. 18:764, 1977.

14. Perez-Mendez, V., Kaufman, L., Lim, C. B., et al.: *Multiwire proportional chambers in nuclear medicine: Present status and perspectives.* Int. J. Nucl. Med. Biol. 3:29, 1976.

15. Mulder, H., and Pauwels, E. K. J.: *A new nuclear medicine scintillation camera based on image-intensifier tubes.* J. Nucl. Med. 17:1008, 1976.

16. Kirch, D. L., Steel, P. P., Lefree, M. T., et al.: *Application of a computerized image-intensifier radionuclide imaging system to the study of regional left ventricle dysfunction.* IEEE Trans. Nucl. Sci. NS-23:507, 1976.

17. Kaufman, L., Lorenz, V., Hosier, K., et al.: *Two-detector, 512-element high purity germanium camera prototype.* IEEE Trans. Nucl. Sci. NS-25:189, 1978.

18. Muehllehner, G., Buchin, M. P., and Dudek, J. H.: *Performance parameters of a positron imaging camera.* IEEE Trans. Nucl. Sci. NS-23:258, 1976.

19. Ter-Pogossian, M. M.: *Basic principles of computed axial tomography.* Semin. Nucl. Med. 7:109, 1977.

20. Cohen, S.: Personal communication, 1978.

21. Budinger, T. F., and Gullberg, G. T.: *Transverse section reconstruction of gamma-ray emitting radionuclides in patients.* In Ter-Pogossian, M. M., et al. (eds.): *Reconstruction Tomography in Diagnostic Radiology and Nuclear Medicine.* University Park Press, Baltimore, 1977, pp. 315–342.

22. Budinger, T. F., Derenzo, S. E., Greenberg, W. L., et al.: *Quantitative potentials of dynamic emission computed tomography.* J. Nucl. Med. 19:309, 1978.

23. Budinger, T. F., Derenzo, S. E., Gullberg, G. T., et al.: *Emission computer assisted tomography with single photon and positron annihilation photon emitters.* J. Comp. Assist. Tomog. 1:131, 1977.

24. Phelps, M. E., Hoffman, E. J., Mullani, N. A., et al.: *Application of annihilation coincidence detection to transaxial reconstruction tomography.* J. Nucl. Med. 16:89, 1975.

25. Budinger, T. F.: *Three-dimensional imaging of the myocardium with isotopes.* In Harrison, D. C., et al. (eds.): Volume 72—*Cardiovascular Imaging and Image Processing, Theory and Practice 1975.* Society of Photo-Optical Instrumentation Engineers, Palos Verdes Estates, Calif., 1975, pp. 263–271.

26. Budinger, T. F., Cahoon, J. L., Derenzo, S. E., et al.: *Three-dimensional imaging of the myocardium with radionuclides.* Radiology 125:433, 1977.

27. Keyes, Jr., J. W., Leonard, P. F., Svetkoff, D. J., et al.: *Myocardial imaging using emission computed tomography.* Radiology 127:809, 1978.

28. Lewis, M., Buja, L. M., Safler, S., et al.: *Experimental infarct sizing using computer processing and a three-dimensional model.* Science 197:167, 1977.

29. Stewart G. N.: *Researches on the circulation time and on the influences which affect it. IV: The output of the heart.* J. Physiol. 22:159, 1897.

30. Meier, P., and Zierler, K. L.: *On the theory of the indicator-dilution method for measurement of blood flow and volume.* J. Appl. Physiol. 6:731, 1954.

31. Kety, S. S., and Schmidt, C. F.: *The nitrous oxide method for the quantitative determination of cerebral blood flow in man: Theory, procedure and normal values.* J. Clin. Invest. 27:476, 1948.

32. Holman, B. L., McNeil, B. J., and Adelstein, S. J.: *Regional blood flow studies with radioisotopes.* In *Dynamic Studies with Radioisotopes in Medicine*, Vol. II. IAEA, Vienna, 1975, pp. 3–18.

33. Bassingthwaighte, J. B.: *Flow estimation from tracer washout.* Prog. Cardiovasc. Dis. 20:83, 1977.

34. KLOCKE, F. J., AND WHITTENBERG, S. M.: *Heterogeneity of coronary flow in human coronary artery disease and experimental myocardial infarction.* Am. J. Cardiol. 24:782, 1969.

35. ADELSTEIN, S. J., AND MASERI, A.: *Radioindicators for the study of the heart: Principles and applications.* Prog. Cardiovasc. Dis. 20:3, 1977.

36. HOLMAN, B. L., ADAMS, D. F., JEWITT, D., ET AL.: *Measuring regional myocardial blood flow with ^{133}Xe and the Anger camera.* Radiology 112:99, 1974.

37. CARLIN, R., AND CHIEN, S.: *Effect of hematocrit on the washout of xenon and iodoantipyrine from dog myocardium.* Circ. Res. 40:505, 1977.

38. MASERI, A., PESOLA, A., L'ABBATE, A., ET AL.: *Contribution of recirculation and fat diffusion to myocardial washout curves obtained by external counting in man: Stochastic vs. monoexponential analyses.* Circ. Res. 35:826, 1974.

39. BASSINTHWAIGHTE, J. B., STRANDELL, T., AND DONALD, D. E.: *Estimation of coronary blood flow by washout of diffusible indicators.* Circ. Res. 23:259, 1968.

40. CONN, H. L., JR.: *Equilibrium distribution of radio-xenon in tissue: Xenon-hemoglobin association curve.* J. Appl. Physiol. 16:1065, 1961.

41. CHIMOSKEY, J. E.: *Skin blood flow by ^{133}Xe disappearance validated by venous occlusion plethysmography.* J. Appl. Physiol. 32:432, 1972.

42. LINDBJERG, I. F.: *Leg muscle blood flow measured with $^{133}Xenon$ after ischaemia periods and after muscular exercise performed during ischaemia.* Clin. Sci. 30:399, 1966.

43. CANNON, P. J., WEISS, M. B., AND CASARELLA, W. J.: *Studies of regional myocardial blood flow: Results in patients with left anterior descending coronary artery disease.* Semin. Nucl. Med. 6:279, 1976.

44. KIRK, G. A., ADAMS, R., JANSEN, C., ET AL.: *Particulate myocardial perfusion scintigraphy: Its clinical usefulness in evaluation of coronary artery disease.* Semin. Nucl. Med. 7:67, 1977.

45. RITCHIE, J. L., HAMILTON, G. W., WILLIAMS, D. L., ET AL.: *Myocardial imaging with radionuclide-labeled particles.* Radiology 121:131, 1976.

46. UTLEY, J., CARLSON, E. L., HOFFMAN, J. I. E., ET AL.: *Total and regional myocardial blood flow measurements with 25μ, 15μ, 9μ, and filtered $1-10\mu$ diameter microspheres and antipyrine in dogs and sheep.* Circ. Res. 34:391, 1974.

47. STOKELY, E. M., PARKEY, R. W., BONTE, F. J., ET AL.: *Gated blood pool imaging following ^{99m}Tc-stannous pyrophosphate imaging.* Radiology 120:433, 1976.

48. SMITH, T. D., AND RICHARDS, P.: *A simple kit for the preparation of ^{99m}Tc-labeled red blood cells.* J. Nucl. Med. 17:126, 1976.

49. ROSE, C. P., GORESKY, C. A., AND BACH, G. G.: *The capillary and sarcolemmal barriers in the heart.* Circ. Res. 41:515, 1977.

50. PAPPENHEIMER, J. R.: *Passage of molecules through capillary walls.* Physiol. Rev. 33:387, 1953.

51. SAPIRSTEIN, L. A.: *Regional blood flow by fractional distribution of indicators.* Am. J. Physiol. 193:161, 1958.

52. GEHRING, P. J., AND HAMMOND, P. B.: *The interrelationship between thallium and potassium in animals.* J. Pharmacol. Exp. Ther. 155:187, 1967.

53. SHEEHAN, R. M., AND RENKIN, E. M.: *Capillary, interstitial and cell membrane barriers to blood-tissue transport of potassium and rubidium in mammalian skeletal muscle.* Circ. Res. 30:588, 1972.

54. LOVE, W. D., ROMNEY, R. B., AND BURCH, G. E.: *A comparison of the distribution of potassium and exchangeable rubidium in the organs of the dog, using rubidium.* Cir. Res. 2:112, 1954.

55. PARKER, J. O., CHIONG, M. A., WEST, R. O., ET AL.: *The effect of ischemia and alternations of heart rate on myocardial potassium balance in man.* Circulation 42:205, 1970.

56. LEVENSON, N. I., ADOLPH, R. J., ROMHILT, D. W., ET AL.: *Effect of myocardial hypoxia and ischemia on myocardial scintigraphy.* Am. J. Cardiol. 35:251, 1975.

57. VAN HERK, G.: *Dynamic blood flow measurement with the $^{81}Rb/^{81m}Kr$ ratio.* Ph.D. thesis, University of Groningen, Netherlands, 1976.

58. HARPER, P. V.: *Personal communication,* 1978.

59. FAZIO, F., AND JONES, T.: *Assessment of regional ventilation by continuous inhalation of radioactive krypton-81m.* Br. Med. J. 3:673, 1975.

60. FAZIO, F., MARDINI, M., FIESCHI, C., ET AL.: *Assessment of regional cerebral blood flow by continuous carotid infusion of krypton-81m.* J. Nucl. Med. 18:962, 1977.

61. YANO, Y., CHU, P., BUDINGER, T. F., ET AL.: *Rubidium-82 generators for imaging studies.* J. Nucl. Med. 18:46, 1977.

62. STRAUSS, H. W., HARRISON, K., LANGAN, J. K., ET AL.: *Thallium-201 for myocardial imaging. Relation of thallium-201 to regional myocardial perfusion.* Circulation 51:641, 1975.

63. KERNAN, R. P.: *Cell K.* Butterworths, Washington, D.C., 1965.

64. NISHIYAMA, H., SODD, V. J., ADOLPH, R. J., ET AL.: *Intercomparison of myocardial imaging agents:* ^{201}Tl, ^{129}Cs, ^{43}K, *and* ^{81}Rb. J. Nucl. Med. 17:880, 1976.

65. WACKERS, F. J., SHOOT, J. B., SOKOLE, E. B., ET AL.: *Noninvasive visualization of acute myocardial infarction in man with thallium-201.* Br. Heart J. 37:741, 1975.

66. ZARET, B. L., STRAUSS, H. W., MARTIN, N. D., ET AL.: *Non-invasive regional myocardial perfusion with radioactive potassium.* N. Eng. J. Med. 288:809, 1973.

67. POE, N. D.: *Comparative myocardial uptake and clearance characteristics of potassium and cesium.* J. Nucl. Med. 13:557, 1972.

68. GORTEN, R. J.: *Evaluation of myocardial function with potassium and cesium nuclides.* Semin. Nucl. Med. 7:37, 1977.

69. WALSH, W. F., FILL, H. R., AND HARPER, P. V.: *Nitrogen-13-labeled ammonia for myocardial imaging.* Semin. Nucl. Med. 7:59, 1977.

70. ZARET, B. L., STENSON, R. W., MARTIN, N. D., ET AL.: *Potassium-43 myocardial perfusion scanning for the non-invasive evaluation of patients with false-positive exercise tests.* Circulation 48:1234, 1973.

71. POE, N. D., EBER, L. M., NORMAN, A. S., ET AL.: *Myocardial images in nonacute coronary and non-coronary heart diseases.* J. Nucl. Med. 18:18, 1977.

72. STRAUSS, H. W.: *Cardiovascular nuclear medicine: A new look at an old problem.* Radiology 121:257, 1976.

73. KNOEBEL, S. B., McHENRY, P. L., STEIN, L., ET AL.: *Myocardial blood flow in man as measured by a coincidence counting system and a single bolus of* $^{84}RbCl$. Circulation 36:187, 1967.

74. BUDINGER, T. F., YANO, Y., AND HOOP, B.: *A comparison of* $^{82}Rb^{+}$ *and* $^{13}NH_3$ *for myocardial positron scintigraphy.* J. Nucl. Med. 16:429, 1975.

75. MALEK, P., RATUSKY, J., VAVREJN, B., ET AL.: *Ischaemia detecting radioactive substances for scanning cardiac and skeletal muscle.* Nature 214:1130, 1967.

76. BONTE, F. J., PARKEY, R. W., GRAHAM, K. D., ET AL.: *A new method for radionuclide imaging of myocardial infarcts.* Radiology 110:473, 1974.

77. PARKEY, R. W., BONTE, F. J., MEYER, S. L., ET AL.: *A new method for radionuclide imaging of acute myocardial infarction in humans.* Circulation 50:540, 1974.

78. WELCH, M. J., AND WAGNER, S. J.: *Preparation of positron-emitting radio-pharmaceuticals.* In Lawrence, J. H., and Budinger, T. F. (eds.): *Recent Advances in Nuclear Medicine,* vol. 5. Grune & Stratton, New York, 1978.

79. BUJA, L. M., TOFE, A. J., KULKARNI, P. V., ET AL.: *Sites and mechanisms of localization of technetium-99m phosphorus radiopharmaceuticals in acute myocardial infarcts and other tissues.* J. Clin. Invest. 60:724, 1977.

80. SUBRAMANYAM, R., ALPERT, N. M., HOOP, B., JR., ET AL.: *A model for regional cerebral oxygen distribution during continuous inhalation of* $^{15}O_2$, $C^{15}O$, *and* $C^{15}O_2$. J. Nucl. Med. 19:48, 1978.

81. KONES, R. J.: *Insulin, adenyl cyclate, ions, and the heart.* Trans. N.Y. Acad. Sci. 36:738, 1974.

82. GALLAGHER, B. M., ANSARA, A., ATKINS, H., ET AL.: *Radiopharmaceuticals XXVII.* ^{18}F-labeled 2-deoxy-2fluoro-d-glucose as a radiopharmaecutical for measuring regional myocardial glucose metabolism in vivo: Tissue distribution and imaging studies in animals.* J. Nucl. Med. 18:990, 1977.

83. EVANS, J. R., GUNTON, R. W., BAKER, R. G., ET AL.: *Use of radioiodinated fatty acid for photoscans of the heart.* Circ. Res. 16:1, 1965.

84. POE, N. D., ROBINSON, G. D., AND MACDONALD, N. S.: *Myocardial extraction of labeled long-chain fatty acid analogs.* Proc. Soc. Exp. Biol. Med. 148:215, 1975.

85. MYERS, W. G.: *Radioiodine-123 for medical research and diagnosis.* In Lawrence, J. H. (ed.): *Recent Advances in Nuclear Medicine,* vol. 4. Grune & Stratton, New York, 1974, pp. 131–160.

86. HOFFMAN, E. J., PHELPS, M. E., WEISS, E. S., ET AL.: *Transaxial tomographic imaging of canine myocardium with* ^{11}C-palmitic acid.* J. Nucl. Med. 18:57, 1977.

87. WEISS, E. S., HOFFMAN, E. J., PHELPS, M. E., ET AL.: *External detection and visualization of myocardial ischemia with* ^{11}C-substrates in vivo and in vitro.* Circ. Res. 39:24, 1976.

88. ROSE, C. P., AND GORESKY, C. A.: *Constraints on the uptake of labeled palmitate by the heart.* Circ. Res. 41:534, 1977.

89. GELBARD, A. S., CLARKE, L. P., AND LAUGHLIN, J. S.: *Enzymatic synthesis and use of* ^{13}N-labeled L-asparagine for myocardial imaging.* J. Nucl. Med. 15:1223, 1974.

90. ODESSEY, R., KHAIRALLAH, E. A., AND GOLDBERG, A. L.: *Origin and possible significance of alanine production by skeletal muscle.* J. Biol. Chem. 249:7623, 1974.

91. BUSE, M. G., BIGGERS, J. F., FRIDERICI, K. H., ET AL.: *Oxidation of branched-chain amino acids by isolated hearts and diaphragms of the rat.* J. Biol. Chem. 247:8085, 1972.

92. GELBARD, A. S.: Personal communication, 1978.

93. GELBARD, A. S., McDONALD, J. M., REIMAN, R. E., ET AL.: *Species differences in myocardial localization of N-13-labeled amino acids.* J. Nucl. Med. 16:529, 1975.

94. KRISS, J. P., ENRIGHT, L. P., HAYDEN, W. G., ET AL.: *Radioisotopic angiocardiography: Findings in congenital heart disease.* J. Nucl. Med. 13:31, 1972.

95. WESSELHOEFT, H., HURLEY, P. J., WAGNER, H. N., JR., ET AL.: *Nuclear angiocardiography in the diagnosis of congenital heart disease in infants.* Circulation 45:77, 1972.

96. BERMAN, D. S., SALEL, A. F., DeNARDO, G. L., ET AL.: *Clinical assessment of left ventricular regional contraction patterns and ejection fraction by high-resolution gated scintigraphy.* J. Nucl. Med. 16:865, 1975.

97. BACHARACH, S. L., GREEN, M. V., BORER, J. S., ET AL.: *A real-time system for multi-image gated cardiac studies.* J. Nucl. Med. 18:79, 1977.

98. STRAUSS, H. W., SINGLETON, R., BURLOW, R., ET AL.: *Multiple gated acquisition (MUGA). An improved noninvasive technique for evaluation of regional wall motion (RWM) and left ventricular function.* Am. J. Cardiol. 39:284, 1977.

99. KAIHARA, S., NATARAJAN, T. K., MAYNARD, C. D., ET AL.: *Construction of a functional image from spatially localized rate constants obtained from serial camera and rectilinear scanner data.* Radiology 93:1345, 1969.

100. BUDINGER, T. F., AND HARPOOTLIAN, J.: *Developments in digital computer implementation in nuclear medicine imaging.* Comput. Biomed. Res. 8:26, 1975.

101. PARKER, J. A., SECKER-WALKER, R. H., HILL, R. L., ET AL.: *A new technique for the calculation of left ventricular ejection fraction.* J. Nucl. Med. 13:649, 1972.

102. MADDOX, D. E.: *Radionuclide ventriculography and its applications to myocardial infarction.* Adv. Cardiol. 23:119, 1978.

103. GREENE, D. G., CARLISLE, R., GRANT, C., ET AL.: *Estimation of left ventricular volume by one-plane cineangiography.* Circulation 35:61, 1967.

104. DODGE, H. T., SANDLER, H., BALLEW, D. W., ET AL.: *The use of biplane angiocardiography for the measurement of left ventricular volume in man.* Am. Heart J. 60:762, 1960.

105. SANDLER, H., AND DODGE, H. T.: *The use of single plane angiocardiograms for the calculation of left ventricular volume in man.* Am. Heart J. 75:325, 1968.

106. DONADO, I.: *Radiocardiographic determinations in man of diastolic and residual blood volumes.* Minerva Nucl. 2:12, 1958.

107. BITTER, F., BESCH, W., SCHAFER, N., ET AL.: *Integrierte Herz-Kreislaufanalyse mit Hilfe der quantitativen Funktionsszintigraphie.* In Horst, W. (ed.): *Frontiers of Nuclear Medicine.* Springer-Verlag, Berlin, 1971, pp. 250–261.

108. VAN DYKE, D., ANGER, H. O., AND SULLIVAN, R. W.: *Cardiac evaluation from radioisotope dynamics.* J. Nucl. Med. 13:585, 1972.

109. MacINTYRE, W. J., STORAASLI, J. P., KRIEGER, H., ET AL.: *^{131}I-labeled serum albumin: Its use in the study of cardiac output and peripheral vascular flow.* Radiology 59:849, 1952.

110. WAGNER, H. N., JR., WAKE, R., NICKOLOFF, E., ET AL.: *The nuclear stethoscope. A simple device for generation of left ventricular volume curves.* Am. J. Cardiol. 38:747, 1976.

111. BACHARACH, S. L., GREEN, M. V., BORER, J. S., ET AL.: *ECG-gated scintillation probe measurement of left ventricular function.* J. Nucl. Med. 18:1176, 1977.

112. RIGO, P., MURRAY, M., TAYLOR, D., ET AL.: *Hemodynamic and prognostic findings in patients with transmural and nontransmural infarction.* Circulation 51:1064, 1972.

113. ZARET, B. L., STRAUSS, H. W., HURLEY, P. J., ET AL.: *Left ventricular ejection fraction and regional myocardial performance in man without cardiac catheterization.* Circulation 42(Suppl. III):120, 1970.

114. STRAUSS, H. W., ZARET, B. L., HURLEY, P. J., ET AL.: *A scintiphotographic method for measuring left ventricular ejection fraction in man without cardiac catheterization.* Am. J. Cardiol. 28:575, 1971.

115. KIRCH, D. L., STEELE, P. P., AND METZ, C. E.: *Quantitative radioisotope angiocardiography.* In *Proceedings of the Third Symposium on Sharing of Computer Programs and Technology in Nuclear Medicine.* USAEC Conference 703627, Miami, Florida, 1973, pp. 16–30.

116. WEBER, P. M., DOS REMEDIOS, L. V., AND JASKO, I. A.: *Quantitative radioisotopic angiocardiography.* J. Nucl. Med. 13:815, 1972.

117. PARKER, J. A., SECKER-WALKER, R., HILL, R., ET AL.: *A new technique for the calculation of left ventricular ejection fraction.* J. Nucl. Med. 13:649, 1972.

118. ISHII, Y., AND MacINTYRE, W. J.: *Measurement of heart chamber volumes by analysis of dilution curves simultaneously recorded by scintillation camera.* Circulation 44:37, 1971.

119. FREEDMAN, G. S., DWYER, A., AND WOLBERG, J.: *Radionuclide determination of cardiac chamber flow/volume characteristics.* J. Nucl. Med. 17:84, 1976.

120. MULLINS, C. B., MASON, D. T., ASHBURN, W. L., ET AL.: *Determination of ventricular volume by radio-isotope-angiography.* Am. J. Cardiol. 24:72, 1969.

121. SULLIVAN, R. W., BERGERON, D. A., VETTER, W. R., ET AL.: *Peripheral venous scintillation angiocardiography in determination of left ventricular volume in man.* Am. J. Cardiol. 28:563, 1971.

122. MALTZ, D. L., AND TREVES, S.: *Quantitative radionuclide angiocardiography. Determination of Qp:Qs in children.* Circulation 47:1049, 1973.

123. ALAZRAKI, N. P., ASHBURN, W. L., HAGAN, A., ET AL.: *Detection of left-to-right cardiac shunts with scintillation camera pulmonary dilution curve.* J. Nucl. Med. 13:142, 1972.

124. FOLSE, R., AND BRAUNWALD, E.: *Pulmonary vascular dilution curves recorded by external detection in the diagnosis of left-to-right shunts.* Br. Heart J. 24:166, 1962.

125. BOSNJAKOVIC, V. B., BENNETT, L. B., GREENFIELD, L. D., ET AL.: *Dual-isotope method for diagnosis of intracardiac shunts.* J. Nucl. Med. 14:514, 1973.

126. ZIMMERMAN, R. E., AND HOLMAN, B. L.: *Modulation transfer function for the Pho/Gamma III and Pho/Gamma HP scintillation cameras using ^{99m}Tc and ^{133}Xe.* J. Nucl. Med. 13:481, 1972.

127. HARPER, P. V.: *Potentials and problems of short-lived radionuclides in medical imaging applications.* Int. J. Appl. Radiat. Isot. 28:5, 1977.

128. SOBEL, B. E.: *Salient biochemical features in ischemic myocardium.* Circ. Res. 34 & 35(Suppl. III):173, 1974.

129. VOGEL, R. A., KIRCH, D., LEFREE, M., ET AL.: *A new method of multiplanar emission tomography using a seven pinhole collimator and an Anger scintillation camera.* J. Nucl. Med. 19:648, 1978.

130. HEYMANN, M. A., PAYNE, B. D., HOFFMAN, J. I. E., ET AL.: *Blood flow measurements with radionuclide-labeled particles.* Prog. Cardiovasc. Dis. 20:135, 1977.

131. BROWNELL, G. L., AND COCHAVI, S.: *Transverse section imaging with carbon-11 labeled carbon monoxide.* J. Compt. Assist. Tomog. 2:533, 1978.

132. SMITH, R. O., LOVE, W. D., LEHAN, P. H., ET AL.: *Delayed coronary blood flow detected by computer analysis of serial scans.* Am. Heart J. 84:670, 1972.

133. GOULD, L., SCHELBERT, H., PHELPS, M., AND HOFFMAN, E.: *Identification of 47% diameter coronary stenosis by non-invasive computed tomography of $^{13}NH_4{}^+$ during dipyridamole coronary vasodilation.* Circulation 58:II-64, 1978.

134. MUDGE, G. H., JR., MILLS, R. M., JR., TAEGTMEYER, H., ET AL.: *Alterations of myocardial amino acid metabolism in chronic ischemic heart disease.* J. Clin. Invest. 58:1185, 1976.

Scintigraphic Evaluation
of Left Ventricular Function*

Donald Twieg, Ph.D., Robert W. Parkey, M.D., and James T. Willerson, M.D.

One of the most important problems in the evaluation of cardiac patients is assessment of ventricular function, particularly left ventricular function. The most satisfactory methods for direct assessment of left ventricular function involve some form of imaging. The traditional method for imaging the ventricles has been contrast angiography and cardiac catheterization. More recently, left ventricular imaging has been performed by echocardiography and scintigraphy.

While it cannot provide all the information available through cardiac catheterization, scintigraphic angiography has become increasingly popular because it is more convenient, less expensive, and involves less hazard to the patient.

Ventricular images, whether scintigraphic or contrast angiographic, are evaluated in two ways. The functional status of various regions of the myocardium is generally evaluated qualitatively by observing the contractions of the margins of the ventricle. Total ventricular function may be assessed quantitatively through contrast and radioisotope ventricular angiograms by measuring the fraction of ventricular volume ejected at each heartbeat (ejection fraction).

Depressed regional myocardial function is of interest because it is often a manifestation of regional ischemia or regional damage from an infarction. The location and extent of diminished wall motion due to infarction damage or ischemia may often be judged from contrast injection or radioisotope images of the ventricle.

Initially, radionuclide blood pool scintigrams were treated in much the same way as contrast angiograms, in spite of some major differences between the two. In contrast angiograms contrast material is injected directly into the ventricle, which is visualized at various stages of contraction and relaxation by filming serial fluoroscopic images at a rapid rate throughout the cardiac cycle. In radioisotope angiograms, the entire blood pool is "labeled" with a gamma-ray–emitting tracer in place of contrast medium, and the cardiac blood pool is imaged by means of a gamma camera.

In contrast angiography ventricular volume at a given cardiac cycle may be estimated by assuming that the ventricle is well approximated by a geometrical form, usually a rotationally symmetrical ellipsoid. The outline of the ventricle is then assumed to correspond to the projection of an ellipsoid in the plane of its major axis. Standard methods use the length and width, or the length and area of the elliptical outline in one or two views to determine the volume of the ellipsoid.[1,2]

*Supported in part by NIH Ischemic SCOR HL-17669. Dr. Willerson is an Established Investigator of the American Heart Association.

The first methods used for measuring ventricular volumes in scintigraphic angiography involved outlining the ventricle and computing the volume by the same geometric methods developed for contrast angiography.[3,4] However, radioisotopic studies offer alternative methods for measuring ejection fraction. These are termed "count-volume" methods because their basic premise is that the number of counts detected from within the left ventricle is proportional to its volume throughout systolic contraction.

There is another significant difference between contrast and radioisotope angiography. Rather than rapid rate filming during a few cardiac cycles, scintigraphic angiography generally requires imaging over many cardiac cycles in order to produce satisfactory images. By "gating" data acquisition in synchrony with features of the patient's electrocardiogram, it is possible to obtain composite images corresponding to distinct portions of the cardiac cycle. End-diastolic and end-systolic images (Fig. 1) are of particular interest; they are sufficient for visual assessment of net ventricular wall motion during systole and quantitative assessment of total ventricular performance by ejection fraction determination.

Scintigraphic methods allow quick, nonhazardous, and inexpensive assessment of ventricular function in a clinical setting. As these methods become more widely used, many cardiologists and general practitioners will base decisions upon scintigraphic assessments of ventricular function. Our purpose here is to provide some insight into the basic technical aspects and clinical applications of the scintigraphic methods.

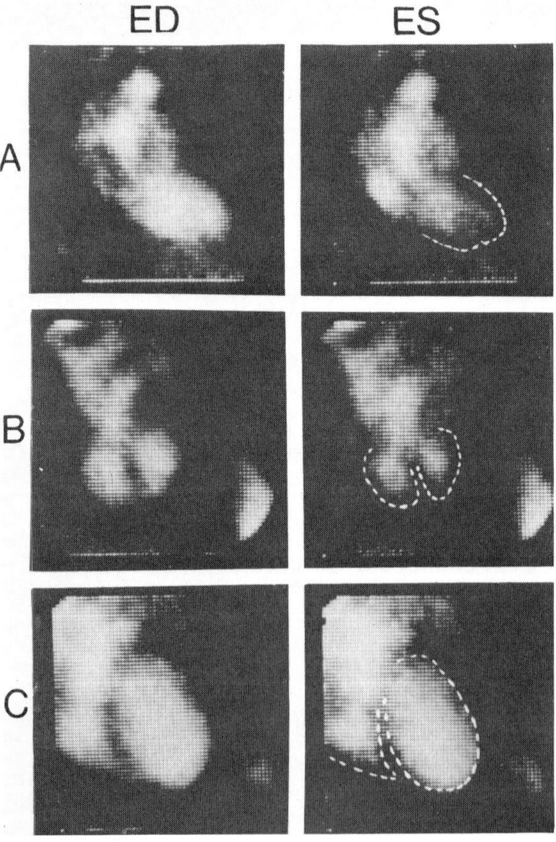

Figure 1. Two-frame gated blood pool studies. A, Patient with normal cardiac function, RAO view. Frame on left is end-diastolic, frame at right is end-systolic. End-systolic boundary is indicated by dashed line. B, Same patient, MLAD view. C, Patient with apical akinesis, MLAO view.

FIRST-PASS METHODS

Scintigraphic first-pass studies involve imaging of the cardiac blood pool immediately after introduction of tracer into the venous blood. Separate images are acquired sequentially as the radioactive tracer first passes through the heart (Figs. 2 and 3). These images are typically recorded for 20 to 40 seconds following peripheral intravenous injection (usually via the basilic median antecubital vein). Although they are usually performed with a scintillation camera, first-pass studies have also been performed using a nonimaging, single-crystal scintillation probe.[5] This latter method requires that accurate estimation of the position of the left ventricle be made before the study to obtain accurate ejection fraction estimates.

As can be appreciated from the serial images shown in Figures 2 and 3, the morphology of the cardiac chambers and great vessels may be readily visualized by means of a first-pass study. Furthermore, if the images are acquired and digitized with a small computer, left ventricular ejection fraction may be computed by monitoring the activity within the ventricle as the tracer passes through it. It is difficult, however, to judge ventricular wall motion from a first-pass study unless a highly efficient scintillation camera is used (i.e., a multicrystal camera rather than an Anger camera[6]).

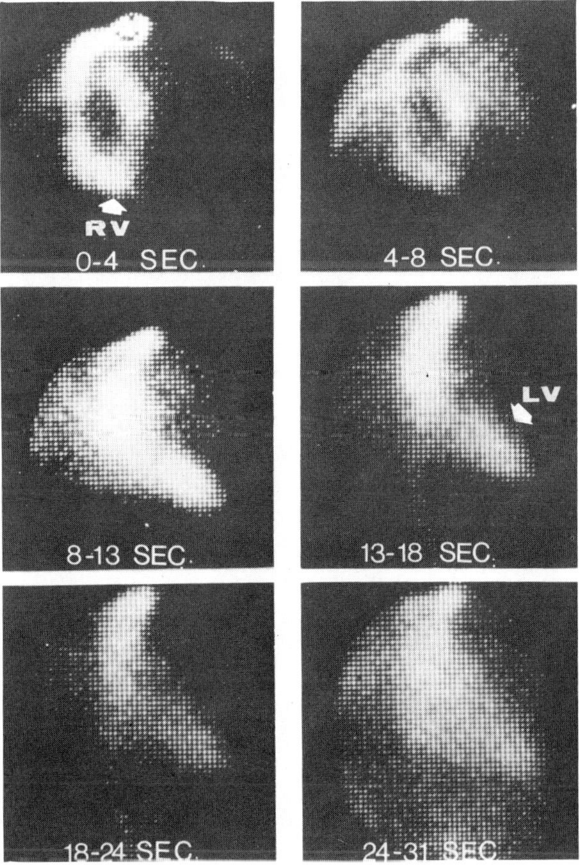

Figure 2. First-pass study (RAO view). Images acquired sequentially during the passage of a tracer bolus through the heart. Each frame represents activity detected during the times indicated. Note that the right heart and left heart may be visualized separately in the first and fourth frames.

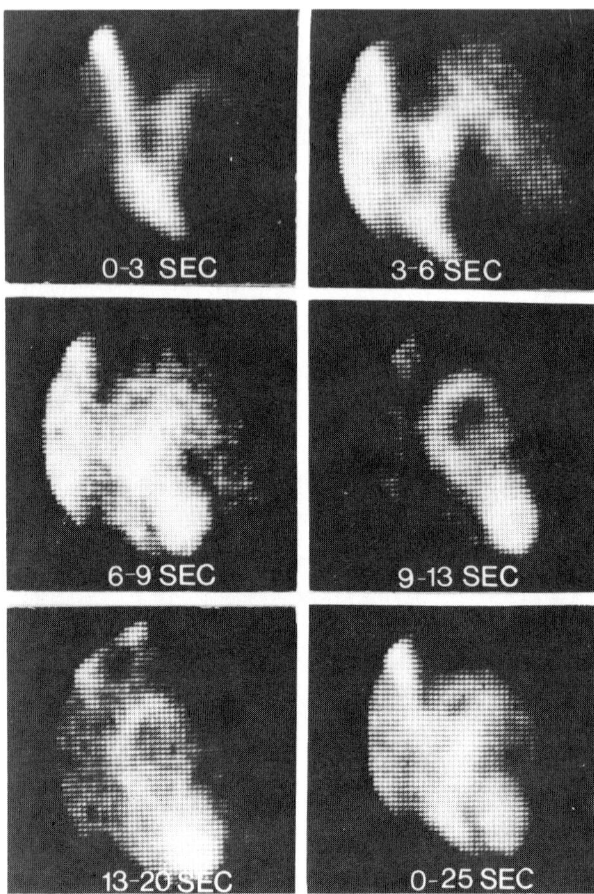

Figure 3. First-pass study (MLAO view). The last frame represents the sum of activity throughout the study. As in the RAO first-pass, labeled blood in the lungs and extracardiac tissue is responsible for the cloudy appearance of the summed image.

As with gated studies, a small computer is a virtual necessity for convenient quantitative interpretation of the data. Commercially available nuclear medicine computers suitable for cardiac studies have a light pen or some other device by which the user can specify a region of interest within the scintigraphic image. The region of interest (ROI) is an area of the image corresponding to some structure in which one wishes to know the tracer content. In the case of ventricular function study, the ROI is specified over the ventricle and temporal variations in activity within the ROI are plotted as a ventricular time-activity curve (TAC) (Fig. 4).

The left ventricular TAC consists of a slowly varying curve upon which more rapid oscillations are superimposed. The cyclic component represents changes in ventricular volume. Assuming a constant ventricular tracer concentration during systole allows ejection fraction to be computed as the fractional decrease in activity during systole (Fig. 4). Measuring ejection fraction by this method would be an extremely simple matter if it were not for the fact that there are other contributions to the left ventricular TAC. These contributions represent activity from blood-containing structures within the left ventricular region of interest—the lungs, left atrium, aorta, etc. To minimize this contribution, a combination of spatial and temporal separation of other chambers from the left ventricle is

82

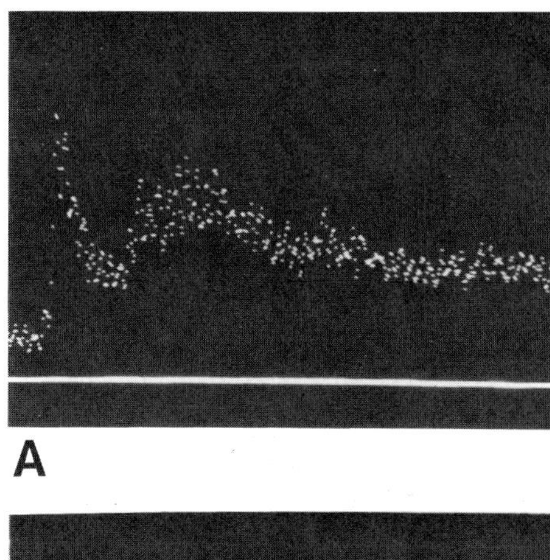

A

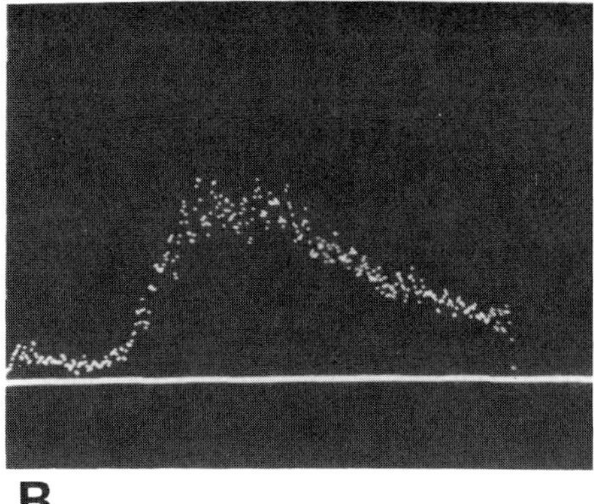

B

Figure 4. Time-activity curves representing activity over the left ventricle during passage of the tracer bolus through the heart. A, Curve from RAO view. Initial peak reflects right ventricular activity, second broader peak reflects left ventricular activity. B, Curve from MLAO view. Right ventricular peak is absent, since area over left ventricle does not include right ventricle.

used. Either a right anterior oblique (RAO) or modified left anterior oblique (MLAO) view is used. If the tracer is injected rapidly, the right ventricle, overlying the left ventricle in the RAO view, does not contribute substantially to the TAC at the time of peak left ventricular activity. In the MLAO view, the left ventricle is spatially separated from the left atrium and the right ventricle (Fig. 3).

In this way the major nonventricular contributions to the left ventricular TAC are removed. However, there will be some nonventricular activity from the overlying and underlying blood-containing structures; these must be estimated and subtracted in order to obtain an accurate ventricular TAC in an accurate ejection fraction estimate. For first-pass studies this background subtraction process consists of several steps. First, a "background" region of interest, separate from the left ventricle itself, is designated; this should have a count density representative of the regions behind and before the ventricle. Second, the computer is used to generate a TAC representing the history of activity within the background region during the first-pass study. Next, each point of this curve is normalized by the ratio of ventricular ROI to background ROI and the resulting normalized background curve is subtracted from the ventricular TAC; the resulting curve represents the variations in ventricular activity.

The choice of the background area is of crucial importance in determining the accuracy of ejection fraction estimates based on the first-pass method. There are several methods for choosing the background area; all of these exclude the major vessels and the other cardiac chambers from the background area. One method is to choose an area of the lung which is well removed from the heart; another method is to choose a semiannular region closely surrounding the left ventricle but not including the right ventricle. The choice of the left ventricular ROI is of equal importance. Schelbert and colleagues[7] have shown that if this area extends to include the root of the aorta, a significant error in ejection fraction estimate may result. While the left ventricular ROI should include only the left ventricular cavity along the free margin of the left ventricle, it is wise to extend the region slightly beyond where the boundary of the left ventricle is believed to be. This will help prevent inadvertent exclusion of tracer counts from the ventricular radionuclide blood pool.

GATED BLOOD POOL METHODS

Tracers

Tracers to be used in gated blood pool imaging should remain almost entirely within the vasculature during the study. Human serum albumin labeled with ^{131}I or ^{99m}Tc or red blood cells labeled in vitro or in vivo[8,9] with ^{99m}Tc are often used for this purpose.

A recently developed technique for in vivo labeling of red blood cells with ^{99m}Tc involves intravenous injection of stannous pyrophosphate (5 to 15 mg), and about 30 minutes later injection of ^{99m}Tc sodium pertechnetate (15 mCi). After about 20 minutes, 85 to 90 percent of the injected technetium has labeled the red blood cells. The blood pool remains labeled for 4 to 6 hours and is suitable for serial imaging throughout this period.

Gated Blood Pool Imaging

Gated blood pool imaging is usually performed in two views to facilitate the identification of wall motion abnormalities. At least one view is chosen so that the ventricular volume can be estimated either by the geometric or the count-volume method. The views most suitable for geometric volume and ejection fraction estimation are the 30° RAO and the MLAO.[9] In both views the long axis of the ventricle is easily visualized.

The MLAO view may also be used for count-volume ejection fraction measurement, since it separates the left ventricle from the other cardiac chambers. Gated blood pool images in both the RAO and MLAO views are often obtained (Figs. 1 and 2). The RAO view allows visualization of the apical and anterior walls of the left ventricle. This view usually permits visualization of the inferior margin of the left ventricle, but the atrioventricular valve plane and posterior portions of the inferior wall are difficult to see. Examining the left ventricular phase of the RAO first-pass study—in which the overlying right ventricle does not interfere as it does in the gated blood pool study—may help to delineate the border of the left ventricle (Fig. 1). In some cases an unusual view may reveal wall motion defects. For some patients with inferior myocardial infarcts, for example, the left anterior oblique view with a cephalad tilt of 10 to 20° affords the necessary view of the diaphragmatic aspect of the heart to allow inferior wall motion to be visualized.

Gated blood pool images are actually composite images which have been accumulated over hundreds of cardiac cycles. The end-diastolic image, for instance, is the sum of hundreds of brief "exposures" (typically of 0.05-sec duration), each of which was made in response to a gating signal produced just prior to the electrocardiographic R wave. For two-frame gated blood pool studies the gating device produces an additional gating signal at the time of end-systole to permit acquisition of the end-systolic image.

Multiple gated studies,[10,11] in which many (typically 14 to 28) composite frames are acquired corresponding to a composite cardiac cycle, may allow identification of more subtle wall motion defects than are apparent in two frames alone (Fig. 5). The cine-type playback of these studies on the computer screen usually gives a much more pleasing and easily interpreted display than does alternate viewing of the end-systolic and end-diastolic frames alone.

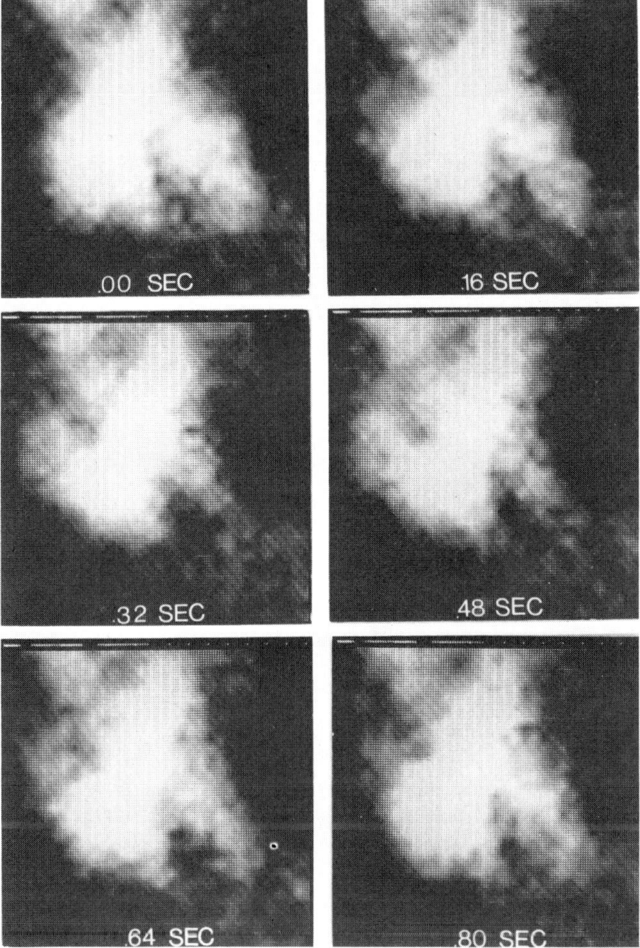

Figure 5. Multiple-frame gated blood pool study (LAO view). Six frames from the total 28 are shown. This patient has subnormal RV function, but normal LV function. When displayed sequentially at a rapid rate, a "movie" of cardiac contraction and relaxation is produced.

Quantitation of Wall Motion Defects

At present, techniques for quantitative measurement of ventricular wall motion are of most value in clinical research. When sufficient clinical correlates of quantitative wall motion measurement have been established for a standard quantitative measurement of segmental wall motion, these methods may be of even greater clinical use. Methods of quantitative wall motion evaluation rely on establishing a coordinate frame with reference to which the motion of the ventricular outlines are plotted. The ventricular wall is divided into segments or sectors, and the degree of shortening of each segment during systole is taken to represent the function of that segment of the myocardial wall. The ventricular outlines may be superimposed in any of several ways. For instance, outlines may be superimposed in their original orientation, or the end-systolic outline may be shifted so that the aortic valve planes of each outline coincide. Superimposing the ventricular outlines in their original orientation probably gives less meaningful results, in that both systolic contraction and the motion of the heart as a whole contribute to apparent wall motion during systole.

A fundamental limitation in assessing wall motion by superimposing outlines is that a poorly contracting region of myocardium might not lie directly on the periphery visible in a given view. This might lead to underestimation of its functional significance, or even failure to detect it, especially if the region is small.

Methods are being developed which use cyclic activity changes in small regions of the ventricular area of interest to assess regional contraction. The appeal of these methods is their ability to reliably detect and measure small regions of abnormal myocardium, regardless of whether or not they are on the periphery of the ventricle as it is viewed. Nonetheless, much work remains to be done before these methods can be considered clinically useful.

Geometric Volume and Ejection Fraction Computation

The most important determinants of accuracy in the geometric method of computing volume and ejection fraction are (1) the choice of a geometric model of left ventricular chamber, and (2) the process of outlining the ventricle.

As mentioned previously, the geometric methods of determining ventricular volume involve assuming a given form for the ventricle. The particular geometric model then specifies a relationship between the dimensions of the projected ventricular outline in a scintigraphic image and the volume of the ventricle. For instance, the length-area method treats the ventricle as an ellipsoid. The single plane formulation of this method[1] leads to the volume formula:

$$V = 8A^2/3\pi L,$$

where A is the area within the ventricular outline, and L is the major axis length of the ventricle. Volumes are determined at end-diastole (EDV) and end-systole (ESV), and then used to compute ejection fraction (EF):

$$EF = \frac{EDV - ESV}{EDV}$$

Depending on the resemblance of the actual ventricle to the form assumed, there is an inherent error produced by applying this geometric assumption to a given ventricular image.

Because of the limited resolution and noisiness of scintigraphic blood pool images it is difficult to precisely identify the position of the ventricular boundary even when the image quality is good. In poor quality images, such as when the blood pool label is inadequate, the edges of the ventricular blood pool are often obscured by background activity. Computer methods for automatically detecting boundaries in noisy images have been applied to gated blood pool images in an effort to remedy this problem, but these methods are generally less accurate and reliable than visual ventricular boundary determinations by an experienced observer.

Count-Volume Method

Geometric volume estimation requires accurate boundary detection and, because of the geometric assumption, is less likely to be accurate for patients with irregular ventricles and significant wall motion defects. Since this is a situation commonly encountered in studies of cardiac patients, an alternative method, the count-volume method, is preferable. The count-volume method makes no geometric assumptions, and is thus unaffected by irregularities in the shape of the ventricle. This method has been applied to two-frame and multiple-frame gated blood pools as well as to first-pass studies. While it is impossible to measure volumes directly from the number of counts detected from the ventricular area, it is possible to compute ejection fraction by comparing ventricular counts at end-diastole and at end-systole. When a multiple-gated study has been done, the counts observed in the ventricle at intermediate times during the cardiac cycle can be used to construct a ventricular volume curve (Fig. 6).

Left ventricular ejection fraction is readily calculable from the volume curve, as are other indices of ventricular function. For example, the maximum systolic ejection rate (dV/dt/EDV) may be a useful index of contractility.[12] It seems likely that the ventricular volume curve contains additional information regarding ventricular function which will be available with further clinical research.

Defining the ventricular edge is less crucial in relative ventricular volumes by count-volume methods than it is with geometric methods. However, the left ventricular ROI must not include other cardiac chambers or the aorta. The procedures for excluding background activity are essentially the same as those for first-pass studies. As in analysis of first-pass data, subtraction of this background is crucial to the accuracy of the method. In gated blood pool studies, this background ROI is often placed just beyond the end-systolic ventricular boundary, but within the end-diastolic boundary, avoiding the left atrium and right ventricle. The same constant background level is then subtracted from the measured ventricular activity at end-systole and end-diastole, and in the case of multiple-frame studies it is subtracted throughout the cardiac cycle.

One policy for choosing the left ventricular ROI designates has the ROI change during the phases of the cardiac cycle, following the contracting ventricular border; this prevents

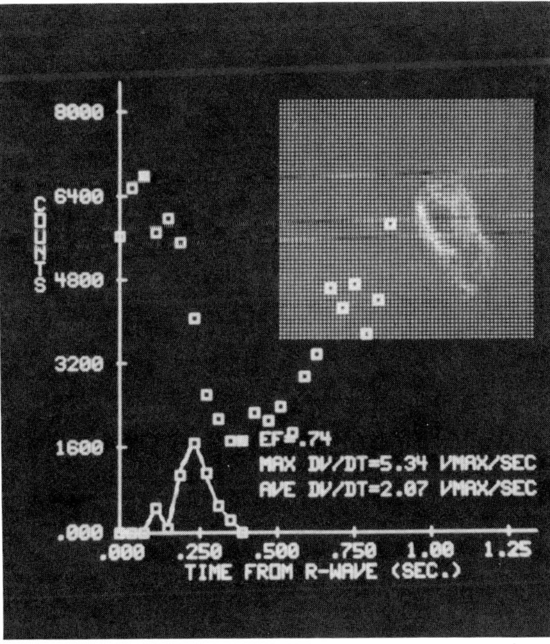

Figure 6. Left ventricular volume curve representing activity in each of the sequential multiple-gated frames. The lower, connected curve represents ejection rates during the systolic portion of the cycle. Outlines at the upper right represent computer-determined ventricular boundaries throughout the cardiac cycle.

87

the intrusion of the left atrium into the left ventricular ROI during systole. Recent findings[11] indicate that this method, together with choice of a remote lung region as a background area, leads to more reliable ejection fraction estimates. It is important that a user of these methods determine which policy for choosing areas of interest provides the most satisfactory results.

CLINICAL APPLICATIONS

Because they are not invasive and are relatively inexpensive, scintigraphic methods are well suited to studies of ventricular function in very ill patients, including those with recent infarction. Several investigators have scintigraphically evaluated ventricular functions serially in patients following myocardial infarction.[13-15] Scintigraphic measurements of ventricular function may prove to be of some value in establishing prognosis in patients following infarction. The location and extent of infarction may be appreciated by scintigraphic observation of wall motion abnormalities and, furthermore, low or serially decreasing ejection fractions following infarction may suggest continuing deterioration of ventricular function. Such a sequence of events would be expected to occur with infarction extension.

Scintigraphic techniques allow ventricular function to be monitored not only over periods of days or weeks, but also serially over minutes to hours following a major clinical event such as acute myocardial infarction. Thus, these techniques enable clinical investigators to examine short-term ventricular function changes resulting from pharmacological and physical interventions.

One of the most promising areas of intervention studies is the use of serial ventricular function tests during the exercise tolerance test.[16] Exercise stress taxes the ability of diseased coronaries to respond to increased myocardial metabolic demands. Stress-induced signs of coronary insufficiency are angina, electrocardiographic alterations, and decreased myocardial contractility segmentally and sometimes globally in the left and right ventricles. Heretofore, chest pain and ECG alterations have been the only available means of relatively easily identifying developing myocardial ischemia during stress in patients. Thus, the ability to scintigraphically visualize altered myocardial contraction during stress would be expected to increase the sensitivity and specificity of exercise stress testing in patients with significant coronary artery disease. Results of scintigraphic exercise stress studies at our institution[17] and elsewhere[16] suggest that this may indeed be the case. If these results are substantiated by further research, scintigraphic gated blood pool imaging may become a valuable and widely used adjunct to the exercise tolerance test.[18]

SUMMARY

Scintigraphic studies, performed with a scintillation camera and a small computer, can provide information concerning total and regional left ventricular function (Table 1). Of the two types of studies (first-pass and gated blood pool) the gated blood study is more appropriate for studies of regional function than is the first-pass method. Ejection fraction, an index of global ventricular function, is measurable from first-pass or gated blood pool studies. Gated blood pool studies allow visual assessment of net ventricular function as well as the local functional status of the myocardium. Multiple gated blood pool studies allow computation of ventricular volume curves, from which ejection fraction and ejection rate may be determined.

There are several areas of application for these relatively inexpensive and virtually noninvasive studies. In patients with recent myocardial infarction, visual assessment of regional ventricular wall motion and quantitative measurement of ejection fraction provide additional prognostic information, especially when measured serially. In patients with coronary artery disease, the development of regional ventricular wall motion abnormalities may in-

Table 1. Scintigraphic ventricular function assessment

Index of Ventricular Function	Analytical Method	Data
Total Ventricular Function		
EDV, ESV, ejection fraction	Geometric volume	Two-frame or multiple-frame gated blood pool
Ejection fraction	Count-volume determination, requiring LV boundary detection	First-pass or two-frame gated blood pool
LV volume curve, ejection fraction, ejection rate	Count-volume methods, requiring subtraction	Multiple-frame gated blood pool
Regional Ventricular Function		
Qualitative regional wall motion assessment	Visual comparison of ventricular borders, requiring boundary detection	Two-frame or multiple-frame gated blood pool
Quantitative regional wall motion assessment	Measurement motion of ventricular border during systole, requiring boundary detection and model for frame of reference	Two-frame or multiple-frame gated blood pool

dicate local myocardial ischemia. Consequently, scintigraphic imaging may add significantly to the usefulness of exercise stress testing.

REFERENCES

1. SANDLER, H., AND DODGE, H. T.: *The use of a single plane angiocardiogram for the calculation of left ventricular volume in man.* Am. Heart J. 75:325, 1968.

2. GREEN, D., CARLISLE, R., GRANT, C., ET AL.: *Estimation of left ventricular volume by one-plane cineangiography.* Circulation 35:61, 1967.

3. MULLINS, C. B., MASON, D. T., ASHBURN, W. L., ET AL.: *Determination of ventricular volume by radioisotope angiography.* Am. J. Cardiol. 24:72, 1972.

4. STRAUSS, H. W., ZARET, B. L., HURLEY, P. J., ET AL.: *A scintigraphic method for measuring left ventricular ejection fraction in man without cardiac catetherization.* Am. J. Cardiol. 28:575, 1970.

5. FOLSE, R., AND BRAUNWALD, E.: *Determination of fraction of left ventricular volume ejected per beat and of ventricular end diastolic and residual volumes.* Circulation 25:674, 1962.

6. MARSHALL, R. C., BERGER, H. J., COSTIN, J. C., ET AL.: *Assessment of cardiac performance with quantitative radionuclide angiocardiography. Sequential left ventricular ejection fraction, normalized left ventricular ejection rate, and regional wall motion.* Circulation 56:820, 1977.

7. SCHELBERT, H. R., VERBA, J. W., JOHNSON, A. D., ET AL.: *Nontraumatic determination of left ventricular ejection fraction by radionuclide angiocardiography.* Circulation 51:902, 1975.

8. STOKELY, E. M., PARKEY, R. W., BONTE, F. J., ET AL.: *Gated blood pool imaging following ^{99m}Tc stannous pyrophoshate imaging.* Radiology 120:433, 1976.

9. ALDERSON, P. O., BERNIER, D. R., LUDBROOK, P. A., ET AL.: *Serial radionuclide determinations of the ejection fraction with ^{99m}Tc-labeled red blood cells.* Radiology 119:729, 1976.

10. GREEN, M. V., OSTROW, H. G., DOUGLAS, M. A., ET AL.: *High temporal resolution EGC-gated scintigraphic angiocardiography.* J. Nucl. Med. 16:95, 1975.

11. BUROW, R. D., STRAUSS, H. W., SINGLETON, R., ET AL.: *Analysis of left ventricular function from multiple gated acquisition cardiac blood pool imaging. Comparison to contrast angiography.* Circulation 56:1024, 1977.

12. HAMMERMEISTER, K. E., BROOKS, R. C., AND WARBASSE, J. R.: *The rate of change of left ventricular volume in man. I. Validation and peak systolic ejection rate in health and disease.* Circulation 49:729, 1974.

13. KOSTUK, W. J., EHSAMI, A., KARLINER, J. S., ET AL.: *Left ventricular performance after myocardial infarction assessed by radioisotope angiography.* Circulation 47:242, 1973.

14. RIGO, P., MURRAY, M., STRAUSS, H. W., ET AL.: *Left ventricular function in acute myocardial infarction evaluated by gated scintiphotograpy.* Circulation 50:678, 1974.

15. SCHELBERT, H. R., HENNING, H., ASHBURN, W. L., ET AL.: *Serial noninvasive measurement of the left ventricular ejection fraction early and late after myocardial infarction.* Am. J. Cardiol. 38:407, 1976.

16. BORER, J. D., BACHARACH, S. L., GREEN, M. V., ET AL.: *Real-time radionuclide cineangiography in the noninvasive evaluation of global and regional left ventricular function at rest and during exercise in patients with coronary artery disease.* N. Engl. J. Med. 296:839, 1977.

17. PULIDO, J., DOSS, J., TWIEG, D., ET AL.: *Submaximal exercise testing in patients following acute myocardial infarction: myocardial scintigraphic and ECG observations. 50th Scientific Sessions of the American Heart Association,* Miami Beach, Fla., Nov. 28 to Dec. 1, 1977.

18. HOLMAN, B. L.: *Promising new radiotracer techniques for the diagnosis of coronary artery disease.* N. Engl. J. Med. 296:876, 1977.

Noninvasive Radionuclide Assessment
of Right Ventricular Performance in Man*

Harvey J. Berger, M.D., and Barry L. Zaret, M.D.

Left ventricular performance has been evaluated in detail with both invasive and non-invasive imaging techniques. In contrast, far fewer studies have focused upon right ventricular pump performance. There are two distinct reasons for the relative paucity of investigations involving this cardiac chamber. First, since the early studies of Starr and coworkers[1] it has been assumed that primary right ventricular damage does not significantly impair global cardiac function. Right ventricular infarction was thought to be a relatively uncommon problem. However, recent clinical studies involving hemodynamic monitoring of patients with either respiratory failure[2,3] or myocardial infarction[4] have suggested that functional impairment of the right heart occurs more frequently than was believed, and that occasionally it may impose severe hemodynamic complications. In addition, subsets of stable patients with right ventricular dysfunction have been identified among those with chronic obstructive pulmonary disease,[5,6] as well as those with valvular or congenital heart disease.[7-11] Precise measurement of right ventricular performance may provide clinically relevant diagnostic, prognostic, and therapeutic information. Second, the irregular shape and internal trabeculations of the right heart have made geometric modeling of the right ventricle difficult. While the left ventricle can be approximated closely by an ellipsoid, the right ventricle does not correspond to any conventional geometric shape, making contrast angiographic or conventional echocardiographic volume analysis dependent upon numerous geometric assumptions.

Right ventricular ejection fraction, or that percentage of maximal (end-diastolic) ventricular volume ejected with each heart beat, can be obtained reliably by radionuclide techniques without the geometric constraints imposed by conventional analysis. First-pass quantitative radionuclide angiocardiography allows calculation of this parameter of right heart performance without assumptions concerning the shape of the right ventricle and thus may be the optimal technique for directly assessing the function of this chamber. The high frequency components of the right ventricular time-activity curve are analyzed during the first passage of the radionuclide bolus through the central circulation.[5] In this way changes in radioactive count rates reflect proportional changes in chamber volume. The same first-pass study can be used to evaluate left ventricular performance by virtue of the temporal separation of the right and left heart phases of the initial radionuclide transit (Fig. 1).[12] This chapter focuses on the application of radionuclide angiocardiography to assessment of right ventricular ejection fraction and the relationship

* Supported in part by NHLBI Contract N01HV52977 and NHLBI Grant R01HL21690-1. Dr. Zaret is an Established Investigator of the American Heart Association.

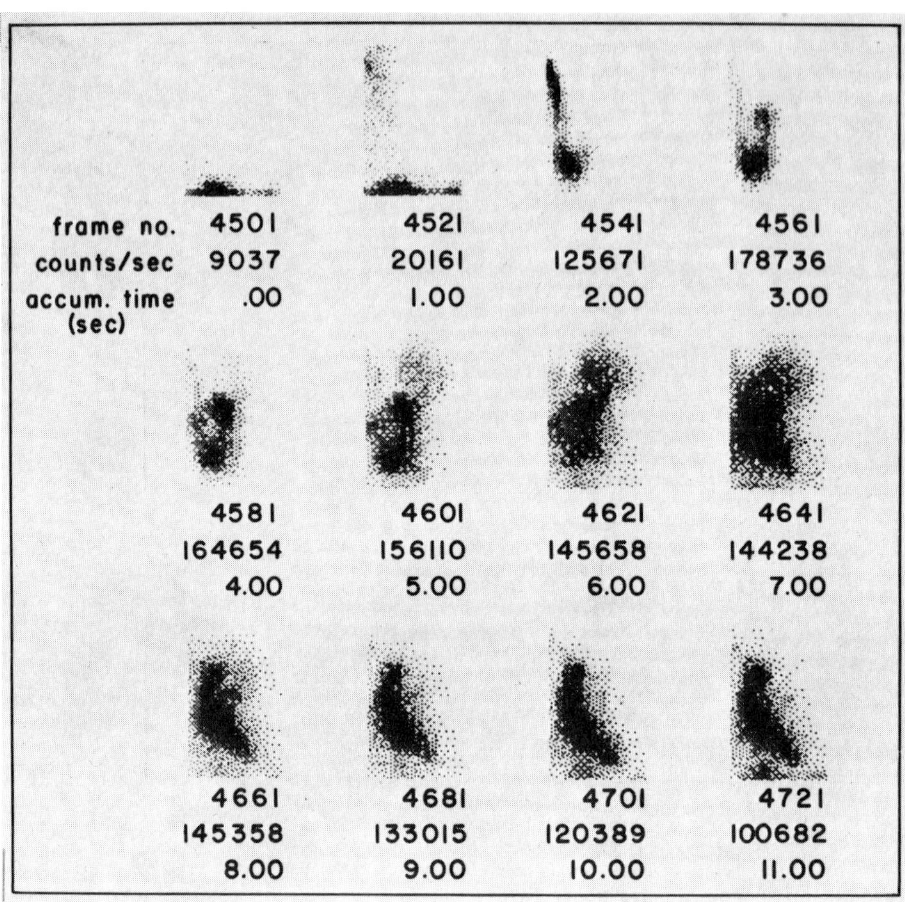

Figure 1. Serial analog display of the radionuclide bolus as it traverses the central circulation. Each image represents 20 summed 50-msec frames. Images 4561 and 4661 are used for defining the right and left ventricular regions of interest, respectively.

between right and left heart performance demonstrated by this technique. Prior to discussion of the clinical applications, the functional geometry of the two ventricles will be reviewed. In addition, the first-pass radiotracer techniques will be compared to other radionuclide imaging approaches of the right heart.

GEOMETRIC CONSIDERATIONS

The triangular, crescent-shaped right ventricular chamber is bounded by the convex interventricular septum and the concave free wall. A relatively narrow space is confined between these two broad surfaces, such that the surface area of the right ventricle is large in comparison to its volume. Three separate but simultaneous contractile movements are involved in right ventricular systolic performance. First, contraction of the trabeculae and papillary muscles moves the tricuspid valve plane downward towards the apex, shortening the longitudinal axis of the chamber, but accounting for only minimal effective ejection. Second, the right ventricular free wall approaches the convex surface of the septum. This mechanism accounts for the majority of right ventricular output. Third, the deep circular fibers of the left ventricle contract and increase the curvature of the septum. This traction contributes to the "bellows" action of the free wall. The relative importance of the thin free wall and the thick interventricular septum to overall right ventricular performance has not been well defined.[13]

The geometric configuration of the right ventricle makes this chamber ideally suited for ejection of large volumes of blood with minimal amounts of myocardial shortening. However, when pulmonary vascular resistance is increased, the right ventricle is not as readily adaptable to the development of high intracavitary pressure as is the left ventricle under comparable circumstances. The right ventricular myocardium would need to develop tension far greater than that of the left ventricle. Fortunately, under normal physiologic conditions, the pulmonary vasculature provides modest resistance to right ventricular outflow. However, rapid or sustained increases in pulmonary artery pressure inevitably lead to right heart failure and cor pulmonale.

In contrast, the left ventricle has a small surface area relative to its intracavitary volume. Left ventricular contraction involving the circumferential muscle bundles generally results in reduction of the shorter diameter of its cylindrical contour. Shortening of the longitudinal axis is far less effective in left ventricular ejection and plays a minimal role in left heart dynamics. Thus, the left ventricle is designed geometrically as a high pressure pump and is able to eject blood against the high resistance systemic circulation.[13]

Despite the large differences in shape, wall thickness, and outflow resistance of the two ventricles, each chamber ejects the same volume of blood in the absence of intracardiac shunts. The superficial spiral muscles of the right ventricular free wall and the deep constrictor muscles of the left ventricle and the interventricular septum are contracting in different directions and describe circles of different diameters. Obviously, equal myocardial shortening in the two chambers would produce a far larger right than left ventricular stroke volume. Since this cannot be the case, the degree of shortening and the ejection fractions of the two ventricles are different.

CONTRAST ANGIOGRAPHIC APPROACHES

Although volume analysis of the right ventricle has been attempted using contrast angiographic data, no technique has become generally accepted. The difficulties encountered result mostly from the geometric considerations described above. Various angiographic studies have approximated the three-dimensional shape of the right ventricle as a prism,[7] pyramid,[14] parallelopiped,[8] or ellipsc.[8] Biplanar area-length measurements, which are used widely for determination of left ventricular volume, also have been applied to the right ventricle.[8,10,11] An alternative approach which does not require assumption of a particular shape for the entire right ventricle, involves the use of Simpson's rule.[7,10,11] In this technique the right ventricle is divided into a series of small sections which are assumed to be of a regular shape. The summation of the volume of these sections along the length of the ventricle results in the total volume of the chamber. Although volume determinations based upon each of these approaches have correlated well with data derived from postmortem casts, there has been wide variation in the reported normal range. These validation studies have relied upon casts from hearts without cardiac disease. This issue becomes critical when studying a dilated abnormal heart, since in association with pathologic states, the trabeculae become relatively less important and the crescent-shaped right ventricle fills out to a more ovoid cross-section, often with little if any, change in the outline of the ventricular silhouette (Fig. 2). This introduces a variable error depending upon the extent of the dilatation. Jaffe and Ellis[15] have suggested that the irregular shape of the right ventricle may preclude accurate volume analysis from two projections alone using any geometric approach. Other technical problems involved in contrast angiographic studies include the propensity of right ventricular contrast injections to provoke premature ventricular contractions and the difficulty in reproducibly defining the ventricular

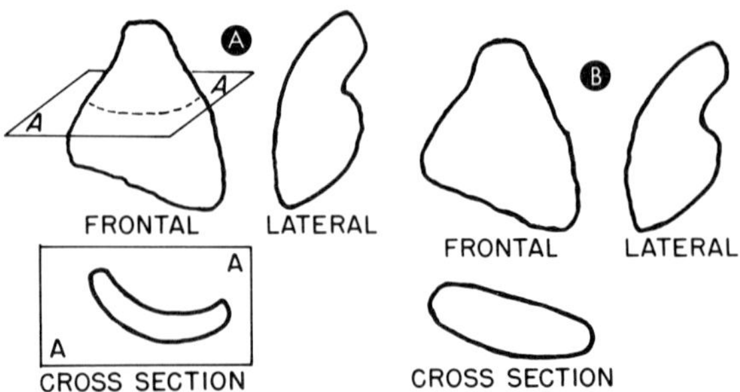

Figure 2. Frontal, lateral, and cross-sectional silhouettes of the normal (A) and dilated (B) right ventricle. As the right ventricle dilates in pathologic states, its crescent-shaped cross section fills out to become more ovoid. However, the frontal and lateral projections of the silhouette do not change comparably, giving rise to significant errors in volume measurements obtained with contrast angiography. (Reproduced from Jaffe and Ellis[15] with permission of Year Book Medical Publishers, Inc.)

silhouette with its extensive trabeculations. Even with these striking problems, contrast angiographic determination of right ventricular ejection fraction has been used in the study of patients with congenital, valvular, and coronary heart disease. The majority of the patients studied have been children with congenital abnormalities.[7-9,11] Only 17 normal adults have been included in all published reports to date.[11,14]

FIRST-PASS RADIONUCLIDE ANGIOCARDIOGRAPHY

Background

The use of radiocardiography in evaluation of circulation times and cardiac output dates back to the work of Blumgart and Weiss in 1927[16] and Prinzmetal and coworkers in 1948.[17] However, it was not until the recent studies of Van Dyke and coworkers[18] that quantitative measurements of ventricular performance were obtained from the high frequency components of the radionuclide angiocardiogram. With this approach, beat-to-beat variations in radioactive count-rates were evaluated during the first passage of the radionuclide bolus through the central circulation. In their studies, analysis was limited to the left ventricular phase of the angiocardiogram. Using a conventional scintillation camera and relatively simple computer processing, they obtained accurate measurements of left ventricular ejection fraction following peripheral venous injection of radiotracer. Their approach involved identification of a left ventricular region of interest, as well as a background zone, which was selected empirically as a ring around the entire left ventricle. Superimposed activity in the lungs, left atrium, and aorta comprised the background contribution. Time-activity curves were generated from each of the regions of interest, and the smoothed background time-activity curve was subtracted temporally from that of the left ventricle. Ejection fraction was calculated directly from the final background-corrected time-activity curve. Subsequent studies have used modifications of this original technique to measure left ventricular ejection fraction, mean circumferential fiber-shortening velocity, and systolic ejection rate.[12,19-22] Each of these radionuclide parameters of left ventricular performance has been found to correlate well with comparable indices obtained with contrast angiography.

Radionuclide approaches to assessment of right ventricular function have been more limited than those of left ventricular performance. Early studies of the right ventricle involved use of the scintillation probe for determination of pulmonary transit time and pulmonary blood volume.[23] Recently, first-pass techniques for evaluation of right ventricular ejection fraction have been described using a conventional gamma scintillation camera equipped with a simple dedicated computer system.[24,25] In these studies, time-activity curves were generated from a right ventricular region of interest, as well as from a background region surrounding the right ventricle. Right ventricular ejection fraction was calculated directly from the background-corrected time-activity curve in a way similar to that used for the left ventricle.

Combined Assessment of Right and Left Ventricular Performance

During the past four years first-pass techniques have been developed in this laboratory for combined assessment of right and left ventricular function from a single study.[5,12,26] All studies have been performed using a commercially available, computerized, multicrystal scintillation camera which allows accumulation of higher count rates than conventional cameras. This characteristic is particularly important for first-pass angiocardiography. The detector of this instrument consists of a mosaic of 294 crystals integrated through computer methods. Indices of global and regional ventricular performance are derived from time-activity curve analysis. Radionuclide angiocardiograms are obtained following peripheral venous injection of ^{99m}Tc radiotracers (pertechnetate or sulfur colloid). Data accumulation at 50-msec intervals requires less than 25 seconds at resting heart rates. Images representing the right and left ventricles are identified readily by study of the anatomic configuration and temporal appearance of the radioactive bolus as it passes through the central circulation (Fig. 1).

The anterior position has been found to be optimal for combined evaluation of right and left ventricular performance from a single study, because it allows adequate separation of the right atrium and right ventricle and it places the heart closest to the detector, maximizing accumulated radioactive counts. Although the right anterior oblique position separates the right atrium and right ventricle more completely, this position places the left ventricle further away from the detector and increases anatomic overlap of the two ventricles during the left heart phase. The left anterior oblique position often is used in the assessment of left ventricular regional wall motion because the septum and lateral wall are viewed most clearly in this position. Sequential studies in the same position are routinely performed to evaluate interventions, such as drugs or exercise stress, while sequential studies in different positions (anterior and left anterior oblique) are used to determine regional wall motion.

For assessment of right ventricular ejection fraction, a preliminary right ventricular region of interest is chosen, and a corresponding time-activity curve is generated.[5] This curve is used to identify the end-diastolic frame with the greatest counts and its corresponding end-systolic frame. These images are computer-smoothed by expanding the 294 matrix points to 4704 points by a 5-point linear averaging algorithm. Although it is difficult to identify the tricuspid valve plane which provides the anatomic boundary between the right atrium and right ventricle, an anatomically pure right ventricular region of interest can be identified by subtracting the appropriate end-systolic image from the peak end-diastolic image. This approach eliminates the majority of activity originating from the right atrium and pulmonary artery. In the anterior position, the major background activity at the peak of the time-activity curve is from the incoming bolus in the superior vena cava and the right atrium. This background contribution orginates in part from the small degree of anatomic overlap between the right ventricle and right atrium. An empirical background region of interest is chosen as a single row of crystals adjacent to the right ventricle at the right atrial–right ventricular interface. Time-activity curves are generated for the right ventricle and for this background zone. These two curves are out of phase, such that peak activity in the right ventricle occurs at a time when activity in the right atrium is low. Peak background activity is approximately one fourth of peak right ventricular activity and is particularly important in sequential studies and in abnormal patients. The right atrial curve is temporally subtracted from the right ventricular curve, yielding a background-corrected, high-count rate, right ventricular time-activity curve (Fig. 3). The peak end-diastolic frame and its corresponding end-systolic frame are reidentified from this curve; if they differ from those originally chosen, the right ventricular zone is modified and the histograms generated again. Right ventricular ejection fraction (RVEF) is calculated as the average of data obtained from one to four cardiac cycles at the peak of the final background-corrected curve, using the formula:

$$RVEF = \frac{\text{counts at end diastole} - \text{counts at end systole}}{\text{counts at end diastole}} \times 100$$

Based on studies in 50 normal adult subjects without evidence of cardiopulmonary disease, radionuclide right ventricular ejection fraction averaged 55 percent, with a normal range (mean ± 2 standard deviations) of 45 to 65 percent (Fig. 4). To assess the reproducibility of this technique, seven patients were studied sequentially at 20-minute intervals. There was no significant difference between these two studies. Furthermore, to assess the responsiveness of right ventricular ejection fraction to inotropic stimulation, four normal patients were studied in the control state and after isoproterenol infusion. Right ventricular ejection fraction increased significantly in all patients, with a mean increase of 9 percent. Because of the limitations of the contrast angiographic techniques available for calculation of right ventricular volumes, correlative cardiac catheterization data were not obtained as part of the validation of the radionuclide technique.[5] Consideration of the geometric issues outlined earlier would suggest that for a broad spectrum of disease states the first-pass radionuclide technique may be more accurate and reliable than any contrast angiographic approach.

From the original serial analog image display of the angiocardiogram, the left ventricular phase also is identified, allowing determination of indices of left ventricular per-

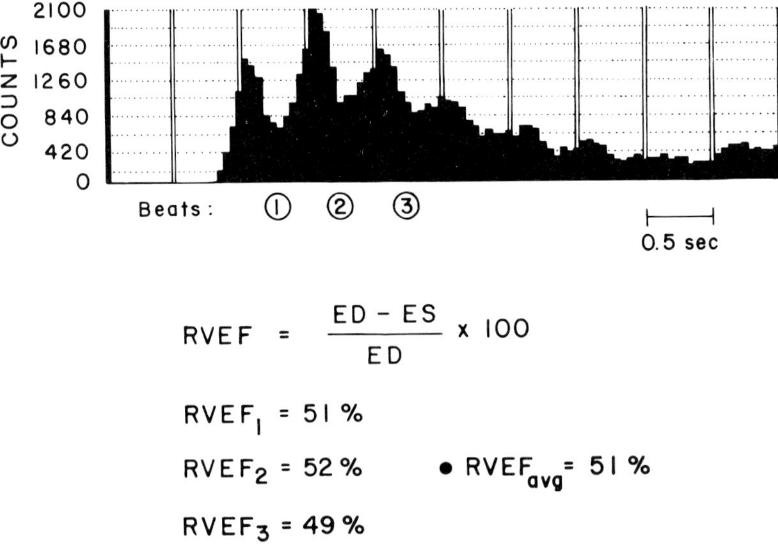

Figure 3. Final background-corrected right ventricular time-activity curve in a normal patient. Right ventricular ejection fraction (RVEF) is calculated on a beat-to-beat basis at the peak of the time-activity curve. ED = counts at end-diastole; ES = counts at end-systole. (Reproduced from Berger et al.[5] with permission of American College of Cardiology.)

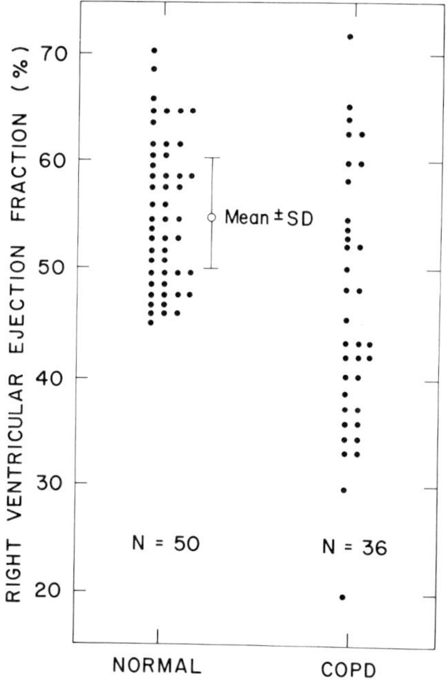

Figure 4. Right ventricular ejection fraction in 50 normal adults without cardiopulmonary disease and in 36 patients with chronic obstructive pulmonary disease (COPD). The normal range, defined as the mean ±2 standard deviations (SD), is 45 to 65 percent. Note the wide range of values in the COPD group. (Reproduced from Berger et al.[5] with permission of American College of Cardiology.)

formance (Fig. 1).[12,26] Because of anatomic and technical considerations the approach used in analysis of the left ventricular time-activity curve is somewhat different from that used on the right. Both are iterative processes involving an empirical background correction. A preliminary high frequency time-activity curve is generated from a left ventricular region of interest obtained from the serial playback. End-diastolic frames are used as starting points to sum together counts at corresponding 50-msec intervals over three to six cardiac cycles at the peak of the time-activity curve. A series of background frames is chosen directly from the left ventricular time-activity curve at a time just prior to the first discernible left ventricular beat. This background represents overlying and scattered radiation from the left atrium and lungs at the time when all activity is in these structures. The same number of background frames as cardiac cycles used are added together. This summed background is subtracted from the summed cardiac cycle, forming a high-count rate, background-corrected "representative" cycle. This is analogous to a relative ventricular volume curve. Using the end-diastolic image generated, the initial left ventricular region of interest is rezoned, and the process reiterated from generation of the time-activity curve through the representative cycle. Left ventricular ejection fraction is calculated directly from end-diastolic and end-systolic counts in the representative cycle. These data also are used to determine mean normalized left ventricular ejection rate and regional wall motion. The latter parameter of regional ventricular performance is obtained by computer-smoothing the end-diastolic and end-systolic analog images generated in the representative cycle. A ring is formed by the computer which represents the perimeter of the end-diastolic image. This ring then is superimposed over the end-systolic image, yielding a composite suitable for regional wall motion analysis.

An excellent correlation has been obtained between left ventricular ejection fraction determined from radionuclide studies and from contrast left ventriculography. In addition, sequential studies in either the same position or in two different positions have been obtained with minimal variation. Regional wall motion analysis evaluated by the radionuclide technique also was found to agree closely with contrast angiographic studies. In

22 patients studied with both methods, 90 percent of abnormal segments were identified correctly by the tracer method.[12]

The multicrystal scintillation camera employed in these studies allows accumulation of much higher count rates than can be attained with conventional single-crystal cameras. The peak end-diastolic count-rate is critical in establishing the reliability of any angiocardiographic technique. For ejection fraction to be calculated with minimal intrinsic variability, high count rates are essential for adequate statistical certainty, especially in patients with abnormally low ejection fractions. In the present studies, up to 250,000 counts per second are obtained in the right heart phase. The average peak end-diastolic count rate in either the background-corrected, right ventricular time-activity curve or the left ventricular representative cycle exceeds 2000/data frame (50 msec). The statistical uncertainty of such count rates is less than ±3 percent.

CLINICAL APPLICATIONS

These first-pass radionuclide techniques have been used to study several subsets of patients in whom the relationship between right and left ventricular performance has important clinical implications. Because the imaging time is extremely short, the techniques described would seem ideally suited for studying acutely ill patients or for evaluating therapeutic or physiologic interventions which involve rapidly changing states.[27,28]

Chronic Obstructive Pulmonary Disease

Studies from this laboratory have evaluated the relationship between left and right ventricular ejection fraction in 36 patients with chronic obstructive pulmonary disease (COPD), as well as the relationship between ventilatory performance and right ventricular function. Right ventricular ejection fraction varied widely in the 36 patients, ranging from 19 to 71 percent (Fig. 4). This parameter of right ventricular function was abnormal ($<45\%$) in 19 patients and normal in 17 others. All 10 patients with clinical and electrocardiographic evidence of cor pulmonale had abnormal right ventricular ejection fraction (mean, $35 \pm 2\%$), which was significantly lower than in the remaining patients with COPD ($p < 0.01$). However, 9 of 26 patients without clinical evidence of cor pulmonale, but with severe ventilatory impairment, also had depressed right ventricular ejection fraction. Within one year of original study, four of these patients had subsequently developed acute respiratory failure and cor pulmonale. In contrast, none of the patients with normal right ventricular ejection has developed evidence of cor pulmonale. This suggests that the radionuclide technique may be sensitive for early detection of right ventricular dysfunction prior to the development of cor pulmonale at a time when it would not be evident by routine clinical assessment.[5]

In the entire group forced expiratory volume (FEV_1) averaged 1.1 ± 0.1 liters. A distinct relationship between pulmonary function and right ventricular performance was noted. Patients were divided into those with severe pulmonary impairment ($FEV_1 < 1$ liter) and those with only moderate impairment ($FEV_1 \geq 1$ liter). Of 19 patients with $FEV_1 < 1$ liter, 14 had depressed right ventricular ejection fraction (mean, $42 \pm 3\%$), which was significantly lower than in the remaining 17 patients ($52 \pm 2\%$) ($p < 0.01$). An alternative interpretation is that FEV_1 was 37 ± 4 percent of the predicted value in those patients with abnormal right ventricular ejection fraction compared to 54 ± 4 percent of the predicted value in the remaining patients ($p < 0.001$). A similar significant difference was found for arterial oxygen tension (Fig. 5). In contrast, left ventricular ejection fraction did not differ significantly in patients with either normal or abnormal right ventricular performance. Left ventricular ejection fraction ranged from 20 to 78 percent and was abnormal in only nine patients, six of whom had documented atherosclerotic or primary myocardial disease.

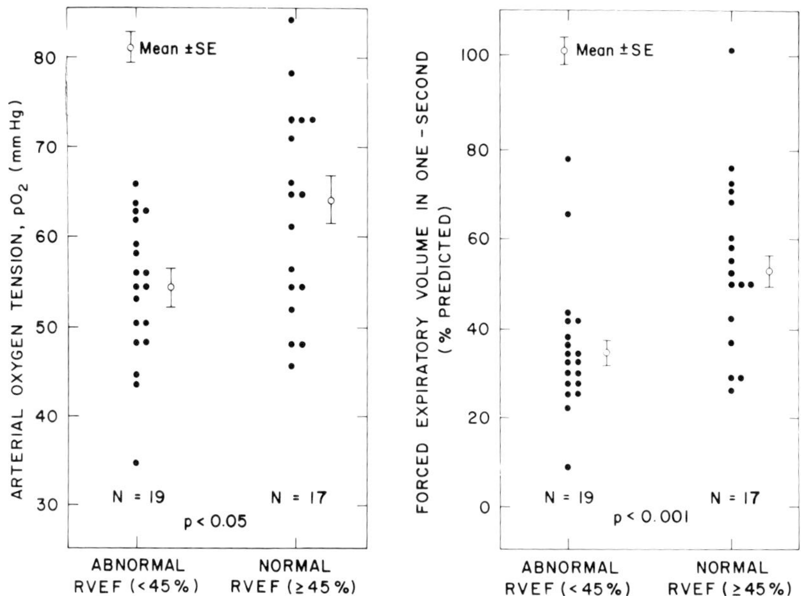

Figure 5. Arterial oxygen tension and forced expiratory volume in one second for 36 patients with chronic obstructive pulmonary disease. Both measures of ventilatory performance were significantly lower in patients with abnormal right ventricular ejection fraction (RVEF) than in those with normal right ventricular function. Open circles with vertical bars at the sides of each panel represent the mean ± standard error. (Reproduced from Berger et al.[5] with permission of American College of Cardiology.)

Previous studies using the scintillation probe or gamma camera also have demonstrated that left ventricular ejection fraction is normal in most patients with COPD and that abnormal left ventricular performance in COPD usually is from concomitant coronary artery disease.[29,30] Matthay and coworkers[31] provided further support for this thesis by studying the effects of dextran volume-loading on radionuclide left ventricular ejection fraction in COPD. They found left ventricular performance to be normal at rest and following dextran infusion, suggesting that left ventricular preload is not diminished in COPD secondary to right ventricular dysfunction. Furthermore, preliminary data on right ventricular ejection fraction from Ellis and coworkers[6] are in agreement with those presented from this laboratory. However, it is important to point out that other clinico-pathologic studies have presented conflicting results concerning the incidence of left and right ventricular dysfunction in COPD.[2]

The development of pulmonary artery hypertension and subsequent right heart failure are well known clinical complications of COPD, and they are usually associated with a poor prognosis. Thus, it would seem appropriate to attempt to modify hemodynamic performance in such individuals. Using the radionuclide techniques described above, the effects of aminophylline, a potent bronchodilator, were assessed.[27] In addition to its actions on bronchial smooth muscle, aminophylline previously has been shown to decrease elevated pulmonary artery pressure in COPD and heart failure. Sequential studies were performed with the patient supine under the camera's detector. Fifteen patients with COPD, including four with documented cor pulmonale and five normal control subjects, were evaluated at baseline and after receiving therapeutic dosages of intravenous amino-phylline (9 mg/kg). In the patients with COPD right ventricular ejection fraction increased significantly from 43 percent at baseline to 52 percent (p < 0.001); while left ventricular ejection fraction increased from 61 to 69 percent (p < 0.001). In six of eight patients with depressed basal right ventricular performance, aminophylline infusion resulted in a return to normal function. Changes in cardiac performance occurred in the presence of only minimal improvement in ventilatory function and no change in arterial oxygen tension. These data demonstrate that in addition to its established value as a bronchodilator, aminophylline is useful for acutely improving global biventricular performance in COPD. Further studies are required to determine whether this hemodynamic improvement is sustained.

Cystic Fibrosis

Of interest to this discussion is a study of a group of 19 young adults with cystic fibrosis.[32] This disease is associated with major anatomic and functional abnormalities of the airways and pulmonary circulation. Despite the fact that most patients with cystic fibrosis die of cardiorespiratory failure, little is known about the physiologic aspects of right and left ventricular performance in this disease.[33] Right ventricular ejection fraction was abnormal in 7 of 19 patients and was normal in the remaining 12. Left ventricular ejection fraction was normal in all patients (Fig. 6). At the time of study none of the patients had evidence of right ventricular hypertrophy on electrocardiogram. However, of two patients with abnormal right ventricular ejection fraction, one died with acute respiratory decompensation and another subsequently developed electrocardiographic evidence of right ventricular hypertrophy within one year of original study.

The clinical severity of cystic fibrosis was determined by the well established Shwachman clinical score, which grades general activity, physical examination, nutritional state, and chest radiograph.[33] All four patients with severely depressed Shwachman score had an abnormal right ventricular ejection fraction. Three of six with mild to moderate Shwachman score had abnormal right ventricular ejection fraction, while none of nine patients with good to excellent scores had abnormal right ventricular performance. These data demonstrate that the degree of clinical impairment is related to abnormalities in right ventricular ejection fraction. In addition, arterial hypoxemia and ventilatory impairment, as judged by arterial oxygen tension and FEV_1, were significantly more severe in patients with abnormal right ventricular function ($p < 0.01$). These findings are similar to those in adults with chronic bronchitis and emphysema and suggest that radionuclide assessment allows

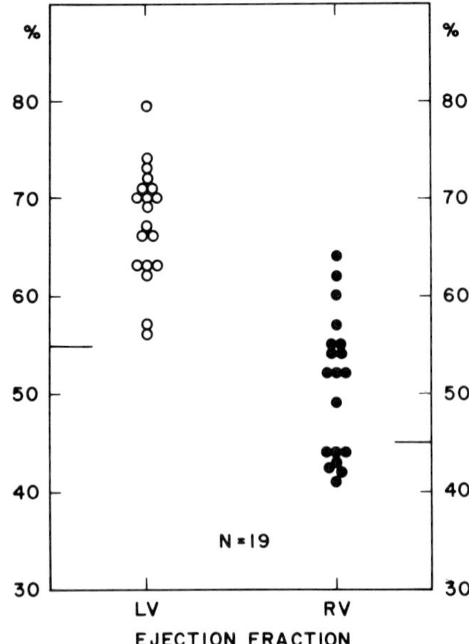

Figure 6. Right ventricular (RV) and left ventricular (LV) ejection fraction in 19 young adults with cystic fibrosis. Note that RV ejection fraction was abnormal (<45%) in 7 patients, while LV ejection fraction was abnormal (<55%) in none.

categorization of patients into hemodynamic subsets. Such categorization may provide the design of more optimal individualized therapeutic regimens. More long-term studies of right ventricular performance and pulmonary function are needed to resolve these questions.

Myocardial Infarction

Functional impairment of the right ventricle in patients with acute myocardial infarction or stable coronary artery disease had previously been considered unusual. Pathologic studies have demonstrated isolated right ventricular infarction in less than 5 percent of autopsies revealing any myocardial damage,[34] while concurrent right and left ventricular infarction has been found to be much more common—approximately 43 percent of patients in one study.[35] A subset of patients with severe hemodynamic abnormalities resulting from right ventricular infarction was first characterized by Cohn and colleagues in 1974.[4] These patients had markedly increased right-sided filling pressures and presented with hypotension and shock. Recent radionuclide studies with ^{99m}Tc-pyrophosphate infarct imaging have shown that 20 to 37 percent of patients with acute inferior wall myocardial infarction have right ventricular involvement, although most do not present in cardiogenic shock.[25,36,37]

To evaluate the relationship of right and left ventricular performance in acute myocardial infarction, 31 patients with their first transmural infarction were studied sequentially using first-pass radionuclide techniques.[38] Radionuclide angiocardiograms were obtained on days 1 (average, 16 hours after admission to the coronary care unit), 3 and 5, and immediately prior to discharge. Of these patients 13 had anterior wall and 18 inferior wall myocardial infarction.

Right ventricular ejection fraction was abnormal in 9 of 18 patients with inferior infarction, but in only 1 of 13 with anterior infarction. On day 1, right ventricular ejection fraction averaged ($\pm$ SEM) 48 $\pm$ 2 percent (range, 38–57%) in inferior wall infarction and 56 $\pm$ 2 percent (range, 42–63%) in anterior wall infarction (Fig. 7). Hypotension and shock were not evident in any of the patients with abnormal right ventricular performance. Mean right ventricular ejection fraction in inferior wall infarction was significantly lower than in anterior infarction on day 1 and at all times throughout the study (p < 0.01). In addition, initial and final right ventricular ejection fractions were not significantly different in either patient group. These data suggest that modest functional impairment of the right ventricle is relatively common in acute inferior myocardial infarction, but not in anterior infarction. However, this dysfunction is not invariably associated with the severe clinically evident hemodynamic compromise first reported by Cohn and associates.[4]

These results are in contrast to the data on left ventricular ejection fraction from the same study. All patients with anterior wall infarction and 13 of 18 with inferior wall infarction had abnormal left ventricular ejection fraction. The magnitude of the dysfunction was significantly greater in anterior wall myocardial infarction (34 $\pm$ 3%) than in inferior wall infarction (50 $\pm$ 3%, p < 0.01). Again, there were no significant sequential changes in ventricular performance.

These results are in agreement with those obtained by gated cardiac blood pool imaging.[36,39] Right ventricular ejection fraction was not measured with this technique; instead, the size of the right and left ventricles was compared. These studies demonstrated disproportionate right ventricular enlargement in acute inferior wall infarction, suggestive of right ventricular dysfunction. These findings are complemented by those in patients with stable coronary artery disease.[24] Right ventricular ejection fraction determined by first-pass technique was well preserved in all patients except in approximately one-third of those with triple-vessel coronary artery disease.

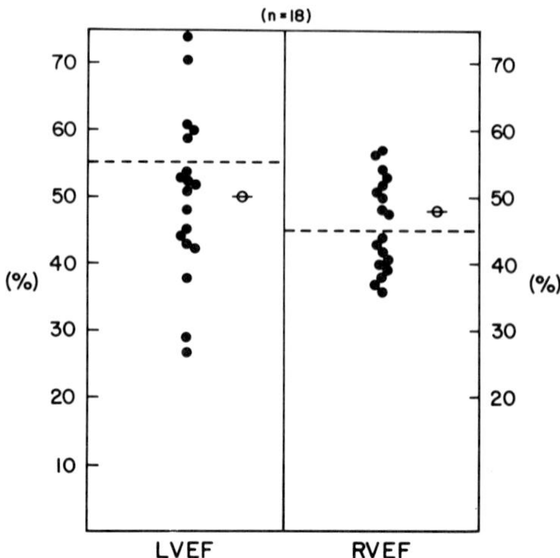

Figure 7. Right ventricular (RV) and left ventricular (LV) ejection fraction in 18 patients with acute inferior wall myocardial infarction studied on the first day of admission to the coronary care unit. RV ejection fraction (EF) was abnormal (<45%) in 9 patients and ranged from 38 to 57 percent. LV ejection fraction was abnormal (<55%) in 13 patients and ranged from 27 to 74 percent. Only one of 13 patients with anterior wall infarction (not shown) had abnormal RV performance. (Reproduced from Reduto et al.[38] with permission of American College of Physicians.)

CONCLUSIONS

First-pass radionuclide angiocardiographic techniques are now suitable for clinical application. They have been validated and standardized and can be relied upon for accurate, reproducible noninvasive assessment of right and left ventricular performance. They already have been applied to the study of patients with cardiopulmonary disease; future directions will include earlier evaluation of myocardial infarction, application to children and adults with a variety of congenital cardiac abnormalities (primarily those affecting the right heart), and assessment of the effects of physiologic stress upon right ventricular function. With this background valuable clinical and investigative insights should become available during the coming years.

REFERENCES

1. STARR, I., JEFFERS, W. A., MEADE, R. H., JR., ET AL.: *The absence of conspicuous increments of venous pressure after severe damage to the right ventricle of the dog, with a discussion of the relation between clinical congestive failure and heart disease.* Am. Heart J. 26:291, 1943.

2. BURROWS, B., KETTEL, L., NIDEN, A., ET AL.: *Patterns of cardiovascular dysfunction in chronic obstructive lung disease.* N. Engl. J. Med. 286:912, 1972.

3. PONTOPPIDAN, H., GELFIN, B., AND LOWENSTEIN, E.: *Acute respiratory failure in the adult.* N. Engl. J. Med. 287:690, 1972.

4. COHN, J., GUIHA, N., BRODER, M., ET AL.: *Right ventricular infarction:clinical and hemodynamic features.* Am. J. Cardiol. 33:209, 1974.

5. BERGER, H., MATTHAY, R., LOKE, J., ET AL.: *Assessment of cardiac performance with quantitative radionuclide angiocardiography: Right ventricular ejection fraction with reference to findings in chronic obstructive pulmonary disease.* Am. J. Cardiol. 41:897, 1978.

6. ELLIS, J., KIRCH, D., AND STEELE, P.: *Right ventricular ejection fraction in severe chronic airway obstruction.* Chest 71 (Suppl.):281, 1977.

7. FISHER, E., DuBROW, I., AND HASTREITER, A.: *Right ventricular volume in congenital heart disease.* Am. J. Cardiol. 36:67, 1975.

8. ARCILLA, R., TSAI, P., THILENIUS, O., ET AL.: *Angiographic method for volume estimation of right and left ventricles.* Chest 60:446, 1971.

9. GRAHAM, T., JARMAKANI, J., ATWOOD, G., ET AL.: *Right ventricular volume determination in children.* Circulation 47:144, 1973.

10. GRAHAM, T., ATWOOD, G., BOUCEK, R., ET AL.: *Right heart volume characteristics in transposition of the great vessels.* Circulation 51:881, 1975.

11. GENTZLER, R., BRISELLI, M., AND GAULT, T.: *Angiographic estimation of right ventricular volume in man.* Circulation 50:324, 1974.

12. MARSHALL, R., BERGER, H., COSTIN, J., ET AL.: *Assessment of cardiac performance with quantitative radionuclide angiocardiography: Sequential left ventricular ejection fraction, normalized left ventricular ejection rate, and regional wall motion.* Circulation 56:820, 1977.

13. RUSHMER, R. F.: *Functional anatomy and control of the heart.* In: Cardiovascular Dynamics, ed.4. Philadelphia, W. B. Saunders Co., 1976, pp. 89–99.

14. FERLINZ, J., GORLIN, R., COHN, P., ET AL.: *Right ventricular performance in patients with coronary artery disease.* Circulation 52:608, 1975.

15. JAFFE, C., AND ELLIS, K.: *Angiographic quantitation of ventricular volume, shape, and mass.* Curr. Probl. Radiol. 4:1, 1974.

16. BLUMGART, H., AND WEISS, S.: *Studies on the velocity of blood flow. VII. The pulmonary circulation time in normal resting subjects.* J. Clin. Invest. 4:399, 1927.

17. PRINZMETAL, M., CORDAY, E., BERGMAN, H., ET AL.: *Radiocardiography: A new method of studying blood flow through the heart in human beings.* Science 108:340, 1948.

18. VAN DYKE, D., ANGER, H., VETTER, W., ET AL.: *Cardiac evaluation from radioisotope dynamics.* J. Nucl. Med. 13:585, 1972.

19. WEBER, P., DOS REMEDIOS, L., JASKO, I., ET AL.: *Quantitative radioisotopic angiocardiography.* J. Nucl. Med. 13:815, 1972.

20. SHELBERT, H., VERBA, J., JOHNSON, A., ET AL.: *Nontraumatic determination of left ventricular ejection fraction by radionuclide angiocardiography.* Circulation 51:902, 1975.

21. STEELE, P., KIRCH, D., MATTHEWS, M., ET AL.: *Measurement of left heart ejection fraction and end-diastolic volume by a computerized, scintigraphic technique using a wedged pulmonary arterial catheter.* Am. J. Cardiol. 34:179, 1974.

22. STEELE, P., LEFREE, M., AND KIRCH, D.: *Measurement of left ventricular mean circumferential fiber shortening velocity and systolic ejection rate by computerized radionuclide angiocardiography.* Am. J. Cardiol. 37:388, 1976.

23. ELLIS, J., AND STEELE, P.: *Quantitative radiocardiography in major pulmonary thromboembolism.* Chest 69:575, 1976.

24. STEELE, P., KIRCH, D., LEFREE, M., ET AL.: *Measurement of right and left ventricular ejection fractions by radionuclide angiocardiography in coronary artery disease.* Chest 70:51, 1976.

25. TOBINICK, E., SCHELBERT, H., HENNING, H., ET AL.: *Right ventricular ejection fraction in acute myocardial infarction assessed by radionuclide angiography.* Circulation 56 (Suppl. III):140, 1977.

26. MARSHALL, R., BERGER, H., REDUTO, L., ET AL.: *Variability in sequential measures of left ventricular performance assessed by radionuclide angiocardiography.* Am. J. Cardiol. 41:531, 1978.

27. BERGER, H., MATTHAY, R., GOTTSCHALK, A., ET AL.: *Effects of aminophylline on right and left ventricular performance in chronic obstructive pulmonary disease: Assessment by quantitative radionuclide angiocardiography.* Clin. Res. 25:208, 1977.

28. BORKOWSKI, H., BERGER, H., LANGOU, R., ET AL.: *Rapid assessment of left ventricular performance during exercise-induced myocardial ischemia: First-pass radionuclide technique.* Circulation 56 (Suppl. III):198, 1977.

29. STEELE, P., ELLIS, J., VAN DYKE, D., ET AL.: *Left ventricular ejection fraction in severe chronic obstructive airways disease.* Am. J. Med. 59:21, 1975.

30. KLINE, L., CRAWFORD, M., MACDONALD, W., ET AL.: *Noninvasive assessment of left ventricular performance in patients with chronic obstructive pulmonary disease.* Chest 72:5, 1977.

31. MATTHAY, R., ELLIS, J., AND STEELE, P.: *Effects of dextran loading on left ventricular performance in chronic obstructive pulmonary disease.* Am. Heart J. 92:730, 1976.

32. MATTHAY, R., BERGER, H., LOKE, J., ET AL.: *Right and left ventricular performance in cystic fibrosis: Assessment by noninvasive radionuclide angiocardiography.* Chest 72:407, 1977.

33. SWACHMAN, H.: *Cystic fibrosis: A new outlook: 70 patients above 25 years of age.* Medicine 56:129, 1977.

34. YATER, W., TRAUM, A., BROWN, W., ET AL.: *Coronary artery disease in men eighteen to thirty-nine years of age. Report of eight hundred sixty-six cases, four hundred fifty with necropsy examinations.* Am. Heart J. 36:683, 1948.

35. ERHARDT, L.: *Clinical and pathological observations in different types of acute myocardial infarction A study of 84 patients deceased after treatment in a coronary care unit.* Acta Med. Scand. (Suppl.) 560:1, 1974.

36. SHARPE, N., NUTVINICK, E., SHAMES, D., ET AL.: *Noninvasive diagnosis of right ventricular infarction: A common clinical entity.* Circulation 54 (Suppl. II):76, 1976.

37. BUSEMAN SOKOLE, E., WACKERS, F., RES, J., ET AL.: *Incidence of right ventricular involvement in acute inferior infarction assessed by* [201]*thallium and* [99m]*Tc-pyrophosphate imaging.* Circulation 56 (Suppl. III):62, 1977.

38. REDUTO, L., BERGER, H., COHEN, L., ET AL.: *Sequential radionuclide assessment of left and right ventricular performance after acute transmural myocardial infarction.* Ann. Intern. Med. 89:441, 1978.

39. RIGO, P., MURRAY, M., TAYLOR, D., ET AL.: *Right ventricular dysfunction detected by gated scintiphotography in patients with acute inferior myocardial infarction.* Circulation 52:268, 1975.

Pathophysiology of "Cold Spot" and "Hot Spot" Myocardial Imaging Agents Used to Detect Ischemia or Infarction*

L. Maximilian Buja, M.D., Robert W. Parkey, M.D., Frederick J. Bonte, M.D., and James T. Willerson, M.D.

The purpose of this chapter is to summarize available information regarding the pathophysiological basis for the scintigraphic detection of ischemic and infarcted myocardium. Two general approaches have been developed: (1) negative ("cold spot") imaging with agents which concentrate in normal myocardium, and (2) positive ("hot spot") imaging with agents which concentrate in acutely infarcted myocardium (Table 1). These two techniques for the detection of myocardial lesions are complimented by scintigraphic techniques which evaluate ventricular function and wall motion with agents which label the cardiac blood pool.

EXPERIMENTAL MODELS

A variety of models have been used to evaluate mechanisms of uptake of scintigraphic agents, including experimental myocardial infarction,[1-4] thermal injury to the heart,[5,6] electrical injury (cardioversion),[7-9] myocardial contusion,[10] and in vitro tissue and organ culture systems.[11-13] Since most studies have employed models involving some form of coronary occlusion, this discussion begins with a review of the pathophysiology of experimental myocardial infarction as a basis for interpretation of the correlations with myocardial scintigraphy. Following coronary occlusion, myocardial necrosis initially develops in maximally ischemic subendocardial regions followed by progressive spread of irreversible injury into less severely ischemic regions which are located predominantly in the subepicardium.[14] Prolonged coronary occlusion also is accompanied by a progressive increase in collateral blood flow into the ischemic region.[15-17] Individual variation in coronary anatomy and collateral flow has a major influence on ultimate size and transmural distribution of infarction. The exact nature and distribution of collateral perfusion and related border zone phenomena in the ischemic region has been the subject of debate and active investigation.[14,18-20] However, histologic study has shown that the established transmural infarct in the dog has histopathologically distinct zones which appear to correlate with differences in regional perfusion within the ischemic region (Figs. 1 through 4, Table 2).[1-4] Experimental and human infarcts share many similar histologic features,[21,22] although human infarcts tend to have more irregular and complicated zonal development, perhaps as a result of more complex patterns of collateral perfusion.

*Work from Southwestern Medical School included in this chapter was supported by NIH Ischemic Heart Disease Specialized Center of Research (SCOR) Grant HL 17669, NIH Grant HL 17777 and by the Harry S. Moss Heart Fund.

Table 1. Pathophysiologic classification of cardiac scintigraphic techniques

I. Myocardial Perfusion (Cold Spot) Scintigraphy
   A. Radiolabeled particulate agents (macroaggregated albumin, microspheres)
   B. Inert gases
      1. krypton-85
      2. xenon-133
   C. Radionuclides of potassium and its analogues (cesium, rubidium, thallium)
   D. Metabolic substrates
      1. radiolabeled ammonia
      2. radiolabeled fatty acids
II. Infarct-Avid (Hot Spot) Scintigraphy
   A. Nonimmunologic complexing with damaged cellular constituents
      1. ^{99m}Tc-phosphates and diphosphonates
      2. ^{99m}Tc-tetracycline
      3. ^{99m}Tc-glucoheptonate
      4. ^{99m}Tc-heparin
      5. [203]Hg-mercurials
   B. Immunologic complexing with damaged cellular constituents
      1. ^{131}I-antibody to myosin
      2. ^{131}I-antibody to myoglobin
   C. Labeling of inflammatory infiltrate
      1. gallium-67 citrate
      2. indium-111 leukocytes
III. Ventricular Function and Wall Motion Scintigraphy

The typical acute transmural infarct has a subendocardially located central zone which is characterized by necrotic myocardium without neutrophilic infiltrate.[1-4,21,22] The necrotic muscle cells are arrested in a relaxed state and do not exhibit calcification; however, calcification may be observed in necrotic blood vessels. Virtual absence of blood flow in this zone appears to explain the lack of muscle cell calcification and the inability of neutrophils to gain access to this region. The central zone of severe ischemia is the last area to undergo organization and may still be identified two or more weeks after onset of infarction.

The central infarct region is surrounded by a peripheral zone of necrotic myocardium which is heavily infiltrated by neutrophils.[1-4,21,22] Collateral blood flow in this zone results in influx of blood constituents, including the entry of plasma proteins and calcium into muscle cells with membrane damage. Calcium accumulation is associated with widespread hypercontraction damage with or without demonstrable mitochondrial calcification.[3,23,24] Some retained mitochondrial function during initial stages of calcium influx is apparently required for initial formation of calcium phosphate precipitates which serve as foci for progressive calcification after mitochondrial function ceases (Fig. 1).[3,25] However, available evidence suggests that pathological calcium accumulation includes complexing of calcium with organic macromolecules and formation of soluble amorphous calcium phosphate precipitates in addition to growth of insoluble hydroxyapatite-like deposits.[3] Heavily calcified necrotic muscle cells are located primarily in the outer region of the peripheral infarct zone. These cells also frequently exhibit prominent accumulation of neutral lipid droplets.[4]

The peripheral zone of necrotic myocardium interdigitates with a narrow border zone of muscle cells which exhibit significant histologic damage but do not show evidence of advanced necrosis.[3,4] These cells are characterized by prominent accumulation of neutral lipid droplets, variable decrease in glycogen granules, focal lysis of myofibrils, and focal mitochondrial alterations, including small calcific deposits observed only at the electron

106

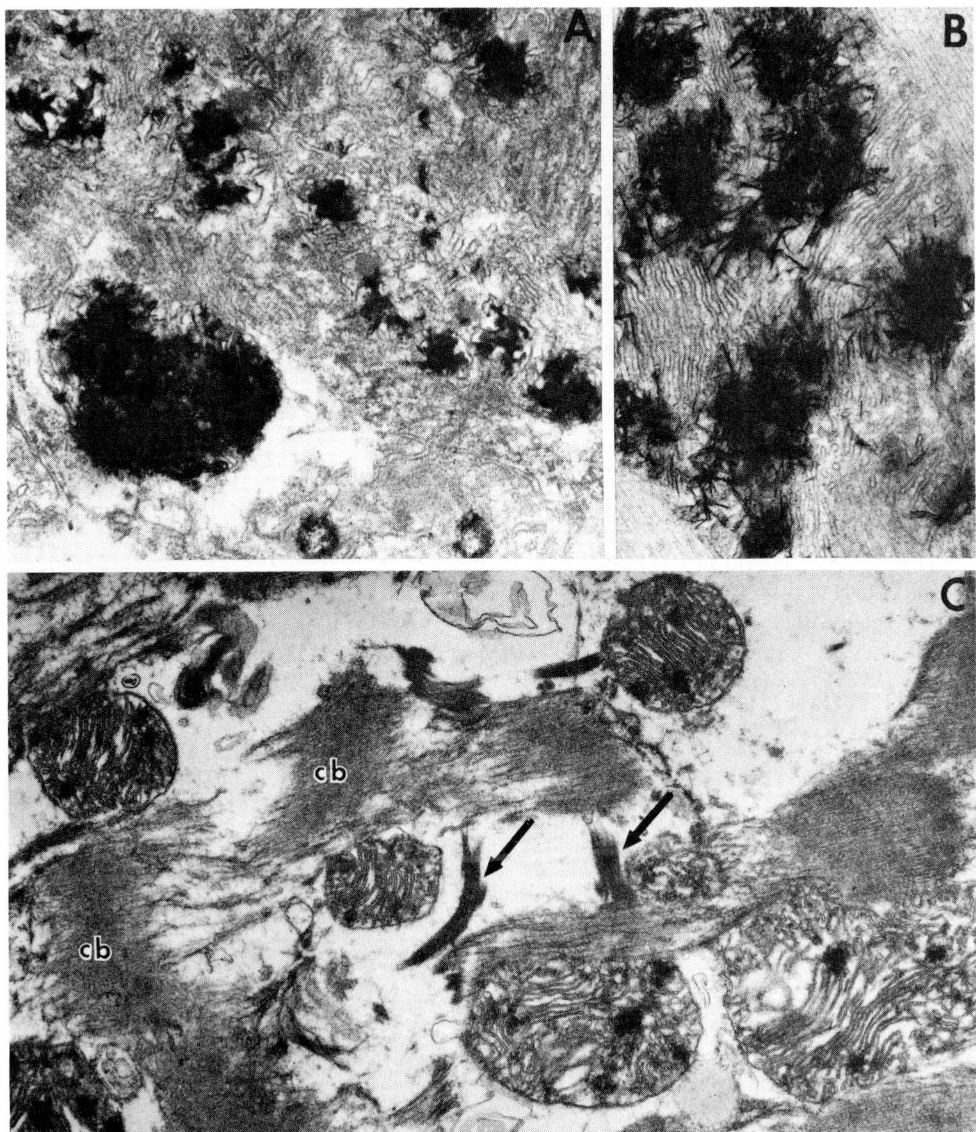

Figure 1. Electron micrographs of necrotic muscle cells of the infarct periphery. A, Dense calcific deposits are localized to the mitochondria of a heavily calcified necrotic muscle cell ($\times$ 34,000). B, Mitochondrial calcium deposits frequently exhibit the spicular appearance of hydroxyapatite ($\times$ 36,000). C, Another necrotic muscle cell exhibits contraction bands (cb) and fibrin-like deposits (arrows). The mitochondria of this cell contain amorphous matrix (flocculent) densities composed of organic material, but they are devoid of calcific deposits ($\times$ 15,000).

microscopic level (Fig. 2).[3,4,24] Hydropic degeneration (colliquative myocytolysis) is not as prominent as in human infarcts.[21,22]

The characteristic infarct zones are shown in a simplified diagrammatic form in Figure 3. These zones tend to have irregular margins and to interdigitate with one another. In addition, the relative size of the regions varies for different infarcts. A central infarct zone may not be identified with small, patchy and/or subendocardial infarcts. At the lateral

Table 2. Distribution of ^{99m}Tc-PYP, ^{201}Tl and ^{125}I microspheres in nine canine myocardial infarcts after one to two days of coronary occlusion*

Infarct Region	$^{99m}Tc\text{-}PYP$ (n = 9)	^{201}Tl (n = 5)	$^{125}I\text{-}Microspheres$ (n = 4)
Outer peripheral and border zones with 1-2+ necrosis	21.5 ± 8.3	0.74 ± 0.02	0.81 ± 0.10
Outer periphery with 3-4+ necrosis	24.5 ± 6.2	0.27 ± 0.03	0.26 ± 0.04
Inner periphery with 4+ necrosis	7.8 ± 1.9	0.08 ± 0.01	0.06 ± 0.01
Center with 4+ necrosis	1.7 ± 0.5	0.03 ± 0.00	0.04 ± 0.01

*Data summarized from Table 1 of Buja et al.[2] and reproduced with permission of the Journal of Clinical Investigation. Values represent ratios to normal myocardium.

margins of the infarcts, the greatest extent of the histologically defined border and peripheral zones is located in the subepicardium; however, the central infarct zone is typically surrounded laterally—even in the subendocardium—by narrow peripheral and border zones (Fig. 4). The lateral extent of these zones is variable but can be as wide as 5 to 10 mm. Furthermore, the border zone of histologic damage with admixed foci of advanced necrosis is usually limited to a very narrow lateral and subepicardial regional of approximately 1 or 2 mm, although lipid accumulation can be observed deeper into the peripheral infarct zone.[3,4]

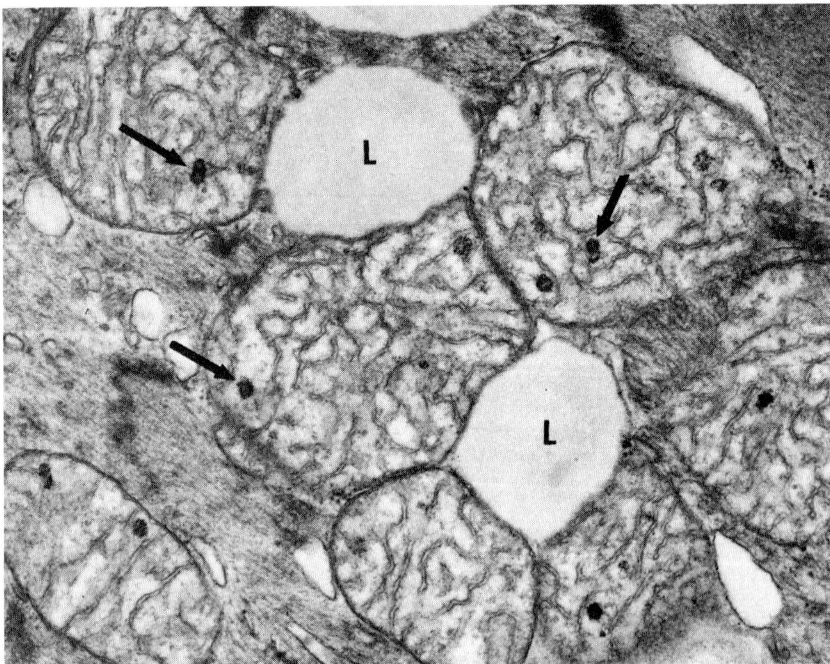

Figure 2. Electron micrograph of injured border-zone muscle cell. This cell has prominent lipid droplets (L) and the mitochondria are swollen and contain small, electron-dense deposits indicative of early calcification (arrows) (× 28,500).

108

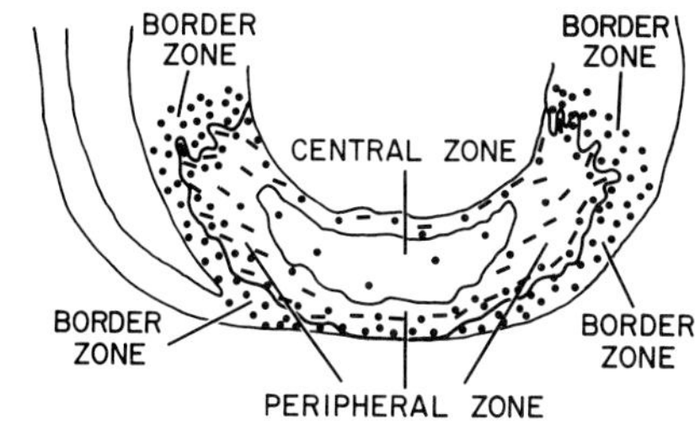

Figure 3. Diagram of the histopathological zones of the typical acute transmural myocardial infarct produced by permanent coronary occlusion in the dog. (From Bilheimer et al.,[4] reproduced with permission of the Journal of Nuclear Medicine.)

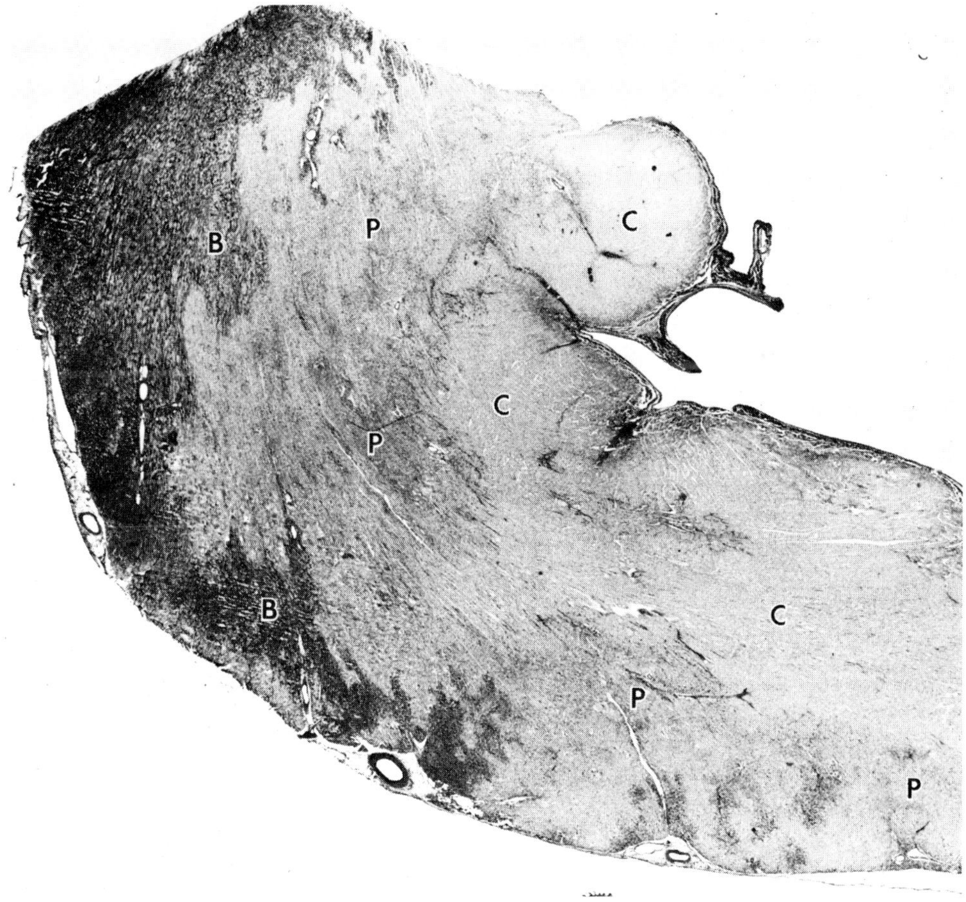

Figure 4. The different infarct zones are shown in this low magnification light micrograph of a histological section stained by the periodic acid-Schiff (PAS) method for carbohydrates. Necrotic myocardium is palely stained due to total glycogen depletion. The central zone (C) is surrounded by a peripheral zone (P) extensively infiltrated with PAS-positive neutrophils. A narrow border zone (B) of damaged muscle cells with partial glycogen depletion interdigitates with necrotic muscle cells of the outer periphery (PAS stain, × 13).

109

COLD SPOT SCINTIGRAPHY

Primary Myocardial Perfusion Techniques

Two general classes of perfusion imaging agents consist of radiolabeled particulate materials that become lodged in the microvasculature on initial passage[26] and soluble radionuclides that rapidly enter the intracellular space (Table 1). The major groups of soluble radionuclides are the inert gases, such as ^{85}Kr and ^{133}Xe, and the potassium analogues, including radionuclides of potassium, cesium, rubidium and thallium. The inert gases diffuse very rapidly into and out of tissues, and distribution of these agents is entirely dependent upon blood flow.[27] Tissue concentration of the potassium analogues is dependent not only on blood flow but also on active membrane transport involving the sodium, potassium ATPase emzyme system.[28] With normal or reduced levels of blood flow, a direct linear relationship is observed between perfusion level and ^{201}Tl or ^{43}K uptake, although tracer uptake may exceed blood flow in areas of severe ischemia.[29-33] With progressive hyperemia, blunting of the uptake of the agents occurs; however, ^{201}Tl exhibits a better correlation with hyperemic flow levels than does ^{43}K.[29-33] Impairment of active membrane function by hypoxia, even with normal perfusion, also may result in reduced uptake of the agents.[34,35]

Although tissue concentration of potassium analogues is a multifactoral phenomenon, the major role of myocardial perfusion in the distribution of these agents has been demonstrated in experimental studies of myocardial infarction (Fig. 5, Table 2).[2,32,33] These studies compared regional distribution of ^{201}Tl or ^{43}K with regional myocardial blood flow determined by the radiolabeled microsphere technique, and showed progressive reduction in radionuclide levels in parallel with reduction in blood flow extending from border and peripheral regions into the centers of the infarcts (Table 2). Relatively high ^{201}Tl or ^{43}K uptake and blood flow in the most peripheral infarct samples may be partially explained by admixtures of normally perfused areas supplied by the left circumflex coronary artery as well as ischemic areas supplied by the occluded left anterior descending artery (LAD).[18,19] Significant radionuclide activity occurred, however, not only in border zone regions with minimal blood flow reduction but also in areas of extensive necrosis where residual perfusion was reduced to 15 to 40 percent of normal.[2,32,33]

The various potassium analogues have different kinetics of uptake and release from myocardium and other tissues; these are based on individual chemical and biological properties of the agents.[36-38] The myocardial clearance rates of potassium analogues also can be influenced by various pharmacologic agents, such as digoxin, isoproterenol and propranolol.[37,39] The duration of tissue retention and release of a given agent is important in the clinical timing of myocardial imaging. Delayed imaging studies have revealed a filling-in of perfusion defects initially seen on scintigrams obtained shortly after injection of radionuclide during transient myocardial ischemia.[40,41] This phenomenon has been attributed to progressive redistribution of radionuclide from normal to ischemic areas. Apparent redistribution of radionuclide may also result in part from faster washout of radioiosotope from normal regions with higher perfusion levels than in diseased myocardium.

Clinical uses of myocardial perfusion imaging with potassium analogues have included diagnosis and sizing of acute myocardial infarction and diagnosis of coronary heart disease on the basis of perfusion defects which appear during resting studies or following exercise. Clinical interpretation of potassium analogue scintigraphy requires recognition that the only direct information provided by the method concerns the status of relative perfusion in different regions of the myocardium. Filling defects may be associated with transiently ischemic tissue, acutely necrotic tissue, severely ischemic tissue adjacent to regions of acute necrosis as well as from areas of old myocardial infarction. However, clinical studies sug-

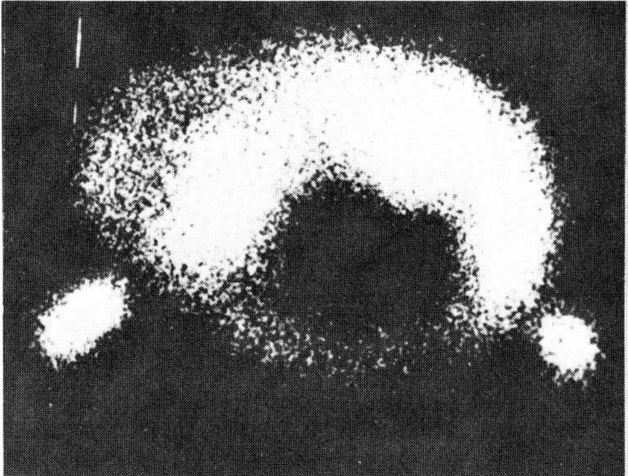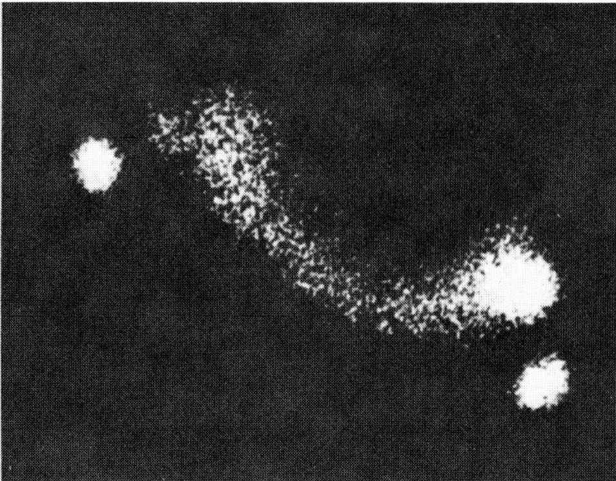

Figure 5. Scintigraphy of transverse ventricular slices from two dogs with one-day-old myocardial infarcts. The external spots are markers for the gross infarct boundaries. The ^{201}Tl scintigram (left) shows intense activity in normal myocardium and decreased but significant activity in the infarct. The ^{99m}Tc-PYP scintigram (right) demonstrates activity which is most intense at the infarct periphery. (From Buja et al.,[2] reproduced with permission of the Journal of Clinical Investigation.)

gest that a large filling defect on a resting study is usually indicative of myocardial infarction rather than transient ischemia alone.[42-45] Evolution of collateral coronary perfusion plays an important role in results obtained with potassium analogue scintigraphy. Thus, Wackers and coworkers reported reduced diagnostic accuracy for acute myocardial infarction when ^{201}Tl scintigraphy was delayed past the first few hours after onset of chest pain.[42] This phenomenon may be caused by progressive increase of collateral flow into the ischemic and infarcted area. The dynamics of the flow-related nature of potassium analogue scintigraphy also have important implications for infarct sizing with these agents.

Metabolic Imaging Techniques

These techniques employ radioisotopically labeled compounds which become incorporated into various metabolic pathways of the myocardium after being distributed according to regional myocardial perfusion. The goal of metabolic imaging is to provide information regarding tissue integrity and viability rather than simply information regarding regional perfusion. Major work in this area has been performed using positron-emitting ^{13}N-ammonia[46] and fatty acids labeled with various radionuclides of iodine (gamma emitters)[47-49] or with positron-emitting ^{11}C.[50-52] Imaging with ^{13}N-ammonia has been performed with modified scintillation cameras, whereas positron emission transaxial tomography (PETT) has been used for imaging with ^{11}C-palmitate.

Although the exact mechanism of ^{13}N-ammonia uptake into myocardium has not been proven, circumstantial evidence suggests a role for active membrane transport mechanisms similar to those involved in uptake of potassium analogues. After entry into the muscle cell, ammonia becomes incorporated into pathways of glutamine and urea metabolism. Imaging results obtained with ^{13}N-ammonia suggest that this agent provides information regarding regional myocardial perfusion very similar to that provided by the potassium analogues.[46]

Myocardial imaging with radiolabeled fatty acids is based on biochemical studies which have demonstrated that free fatty acids represent a major substrate for myocardial metabolism and that hypoxic or ischemic injury results in decreased uptake and metabolism of free fatty acids in the heart.[53] Weiss, Sobel, Ter-Pogossian and associates have shown that areas of myocardial ischemia and infarction can be detected as filling

111

defects on ^{11}C-palmitate positron emission transaxial tomograms.[50-52] This group has also demonstrated close correlation between defect size in tomographic cuts and infarct size subsequently determined in transverse sections of hearts.[51]

Pathological studies have shown a seemingly paradoxical, focally increased accumulation of neutral lipid droplets which begins as early as six hours after coronary occlusion and is more prominent at one to two days after occlusion (Fig. 1).[4,54] In addition, Bilheimer and colleagues have demonstrated increased incorporation of circulating fatty acids into peripheral infarct regions.[4] Accumulation of fatty acids in lipid deposits in damaged muscle cells may be explained by failure of mitochondrial metabolism of fatty acids derived from internal metabolic derangements, as well as from continued extraction from the circulating pool.[4,55] With time, many damaged cells with accumulated lipid progress to a stage of advanced necrosis, as indicated by the finding of prominent lipid deposits in damaged border zone muscle cells and necrotic muscle cells of the infarct periphery.[3,4]

The work of Weiss and associates shows that accurate detection and sizing of myocardial lesions as filling defects with radiolabeled fatty acids can be obtained if the relative proportions of peripheral regions with increased fat accumulation and central regions with decreased fatty acid uptake remain relatively constant in the population examined.[50-52] However, pathophysiologic studies indicate that metabolic imaging with fatty acids does not provide clean separation of cell populations with normal and significantly deranged metabolism.[4,54,55] Future work in the area of metabolic imaging should provide important advances in clinical nuclear cardiology as well as in basic understanding of myocardial pathophysiology.

HOT SPOT SCINTIGRAPHY

General Principles

A useful classification of infarct-avid (hot spot) scintigraphic techniques is based on general mechanisms of concentration of the various radiopharmaceuticals in acutely infarcted myocardium (Table 1). Agents in one group concentrate in damaged myocardium as a result of relatively nonspecific biochemical reactions with various constituents of damaged myocardial cells. This group includes ^{99m}Tc-stannous pyrophosphate (PYP) and other ^{99m}Tc-labeled phosphates and diphosphonates (designated as a class ^{99m}Tc-P);[56-58] ^{99m}Tc-tetracycline;[59-62] ^{99m}Tc-glucoheptonate;[60,63-65] ^{99m}Tc-heparin;[66] and [203]Hg-organic mercurials.[6,62] A second group of agents concentrate in infarcted myocardium on the basis of immunologic complexing of radiolabeled specific antibodies to various myocardial constituents. The two agents in this category are radiolabeled antibodies to myosin[67] and myoglobin.[68] A third group of radiopharmaceuticals concentrate in acutely infarcted myocardium as a result of labeling of the inflammatory infiltrate of infarcted tissue. This group includes an in vivo method for labeling leukocytes with ^{67}Ga-citrate[60,69] and injection of leukocytes labeled in vitro with [111]In.[70]

Results obtained with the various infarct-avid scintigraphic techniques are dependent upon the interaction of several key factors (Table 3): (1) physicochemical properties of the radiopharmaceutical, including properties of the radionuclide and the organic molecule; (2) timing of scintigraphic study after onset of myocardial injury; (3) initial levels of concentration in blood which influence the amount of agent delivered to the lesion; (4) specific mechanism(s) of concentration of the agent in damaged myocardium; (5) types, severity, and duration of injury associated with concentration of the agent; (6) characteristics of the lesion, including size, location and level of residual blood flow; and (7) rate of clearance of the agent from normal structures and blood which influences background radioactivity and target-to-background ratio.

112

Table 3. Factors influencing results obtained with various infarct-avid scintigraphic techniques

1. Physicochemical properties of radiopharmaceutical (radionuclide and organic molecule)
2. Timing of scintigraphic study after onset of myocardial injury
3. Initial level of concentration in blood
4. Specific mechanisms(s) of concentration in damaged myocardium
5. Types, severity and duration of injury associated with concentration
6. Characteristics of myocardial lesion—size, location, level of residual blood flow
7. Rate of clearance of agent from normal structures and blood

Pathophysiology of ^{99m}Tc-P Scintigraphy

Experimental studies have shown that the pathophysiological basis for a positive ^{99m}Tc-P myocardial scintigram is selective concentration of the agent in severely damaged and necrotic myocardium with calcium accumulation and residual blood flow (Figs. 1 through 6, Table 2).[1-3,7-9,32,35,62,67,71-77,88] Many aspects of ^{99m}Tc-P myocardial scintigraphy can be related to the temporal and topographical relationships which have been demonstrated between ^{99m}Tc-P uptake and pathologic calcium accumulation in damaged muscle (Fig. 6).[1-3,76,88]

In animals ^{99m}Tc-P myocardial scintigraphy has a characteristic time course following permanent coronary occlusion; the scintigrams become consistently positive approximately 12 hours after coronary occlusion, increase in intensity over the next 48 to 72 hours, decrease in intensity thereafter, and are usually negative by one to two weeks after coronary

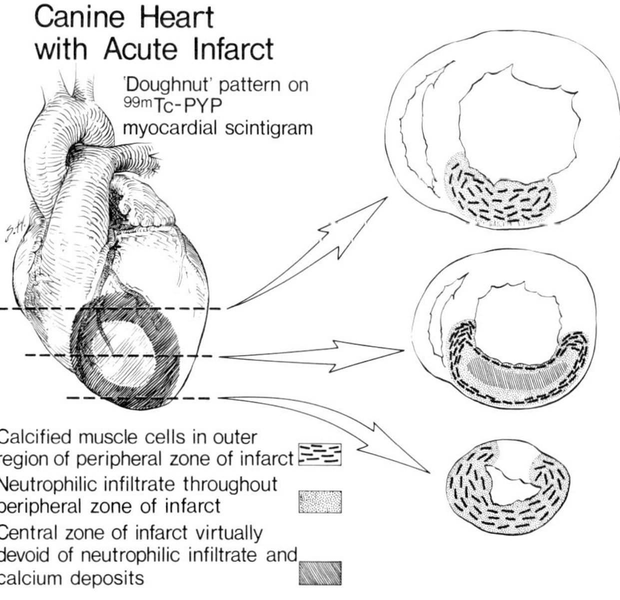

Figure 6. Correlation of scintigraphic and histopathologic findings in dogs with large transmural anterior infarcts produced by proximal LAD ligation. The scintigraphic "doughnut" pattern results from selective concentration of ^{99m}Tc-PYP in peripheral infarct regions with residual blood flow. Small, patchy, and/or subendocardial infarcts have diminutive or absent central zones of no perfusion and also show more homogeneous patterns on ^{99m}Tc-PYP myocardial scintigrams. (From Buja et al.,[1] reproduced with permission of the American Heart Association, Inc.)

occlusion.[1-3,56] A similar time course has been observed in most humans with myocardial infarction.[58,78] Available data suggest that the onset of [99mTc-P] concentration occurs relatively early in the course of myocardial infarction, but that the exact timing of the onset of positive [99mTc-P] scintigraphy is influenced by the rate of evolution of infarction and by variation in the level of perfusion in the ischemic region. Thus, severe temporary ischemia followed by reflow can result in a positive scintigram as early as a few hours after onset of damage.[72,76] In contrast, the usual threshold of 12 hours or longer for positive [99mTc-P] scintigraphy with permanent coronary occlusion appears related to slower evolution of phenomena important for [99mTc-P] concentration, such as calcium accumulation and collateral perfusion.[1-3] Holman and associates, however, have shown that positive [99mTc-P] myocardial scintigrams may develop as early as 4 hours after onset of symptoms in some patients, possibly related to relatively abundant collateral blood flow in these subjects.[89]

Analytical chemical studies have demonstrated that selective in vivo concentration of [99mTc-P] in altered soft tissues as well as bone is invariably associated with elevated tissue calcium content, but that a linear relationship between [99mTc-P] and calcium levels does not occur (Fig. 7).[3] In considering this analytical data it is important to realize that tissue calcium content represents the total calcium sequestered since the onset of the process responsible for calcium accumulation, whereas the [99mTc-P] content reflects [99mTc-P] uptake over a relatively short period after injection. Thus, the lack of a linear relationship can be adequately explained by different affinity of [99mTc-P] for various types of tissue calcium stores in addition to regional differences in delivery of [99mTc-P] attributed to local vascularity and blood flow variations.[1-3]

Concentration of [99mTc-P] may result from complexing with organic molecules in addition to soluble and insoluble inorganic calcific deposits (Fig. 8). The process of polynuclear complexing of [99mTc-P] to organic macromolecules may result from (1) binding of the entire [99mTc-P] complex through a reaction between the phosphate moiety and calcium associated with the macromolecules; (2) complexing of the [99mTc-P] via the [99mTc] moiety, or, possibly,

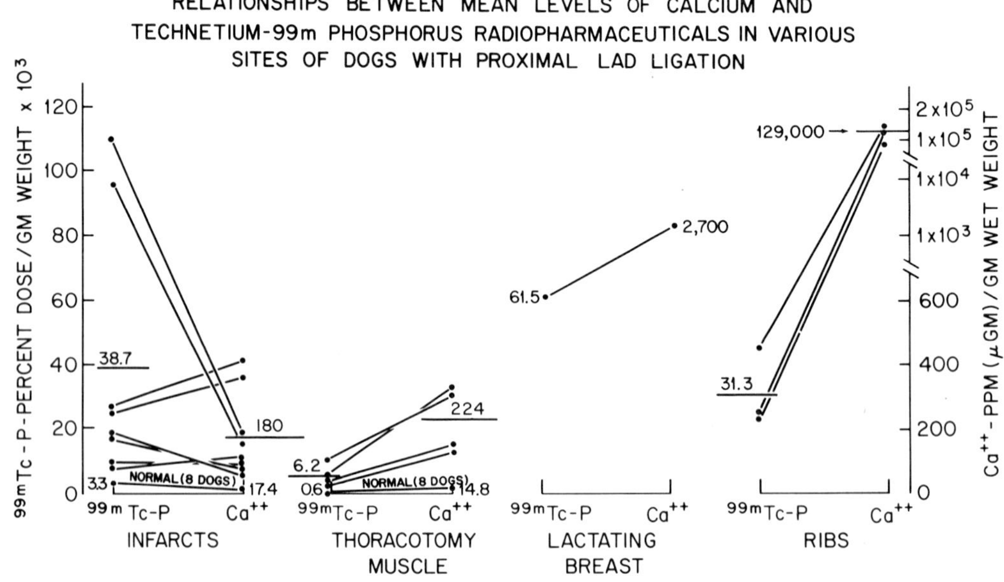

Figure 7. Analytical chemical data show elevated calcium levels in all sites of increased [99mTc-P] visualized scintigraphically. (From Buja et al.,[3] reproduced with permission of the Journal of Clinical Investigation.)

(3) transchelation of reduced ^{99m}Tc from the ^{99m}Tc-P to the macromolecules.[3,11,12,49,74] Significant ^{99m}Tc-P uptake does not require extensive deposition of insoluble hydroxyapatite and can be associated with calcium accumulation in the form of relatively soluble tissue calcium stores which may not be preserved by routine histochemical methods.[3] The studies of Schelbert and coworkers also have shown that increased ^{99m}Tc-P uptake can occur in vitro when necrosis is induced in fetal mouse myocardium and pathological calcium accumulation is prevented by incubation in calcium-free medium.[13] However, these data do not exclude the possibility that native stores of tissue calcium participated in the modest uptake (2.8 times normal) observed in necrotic myocardium in this in vitro system.[13]

Dewanjee has suggested that ^{99m}Tc-P uptake in acute myocardial infarcts results primarily from polynuclear complexing with denatured macromolecules; this is based on the observations that complexing of pyrophosphate with ^{99m}Tc imparted an affinity for protein binding and resulted in higher infarct-to-normal ratios for ^{99m}Tc-P than for free pyrophosphate.[74] But other studies have shown that the presence or absence of ^{99m}Tc does not influence the absolute levels of concentration of phosphates and diphosphonates in infarcted myocardium.[3,73] It has also been shown that partial complexing of ^{99m}Tc-PYP with serum proteins is associated with only minimal reduction in complexing affinity of serum ^{99m}Tc-PYP with hydroxyapatite or amorphous calcium phosphate in vitro.[3] Furthermore, cell fraction data provide an artifactually low estimate of the amount of ^{99m}Tc-P uptake in mitochondiral calcific deposits of infarcted myocardium.[3,25] Thus, available data suggest multiple mechanisms for in vivo concentration of ^{99m}Tc-P, including complexing with soluble and insoluble inorganic calcium deposits as well as organic macromolecules (Fig. 8).

Experimental studies of ^{99m}Tc-P scintigraphy also have provided evidence which suggests that significant cardiac uptake of ^{99m}Tc-P is dependent upon the presence of irreversible myocardial damage. Further evidence has been provided by autoradiographic studies which have shown extensive radiopharmaceutical uptake in frankly necrotic muscle cells as well as significant uptake in a small population of damaged, lipid-laden muscle cells admixed with frankly necrotic myocardium (Table 4).[3] Further studies are needed to determine the ultimate fate of damaged border-zone muscle cells with increased ^{99m}Tc-P uptake; however, the finding of focal, early mitochondrial calcification suggests that at least some

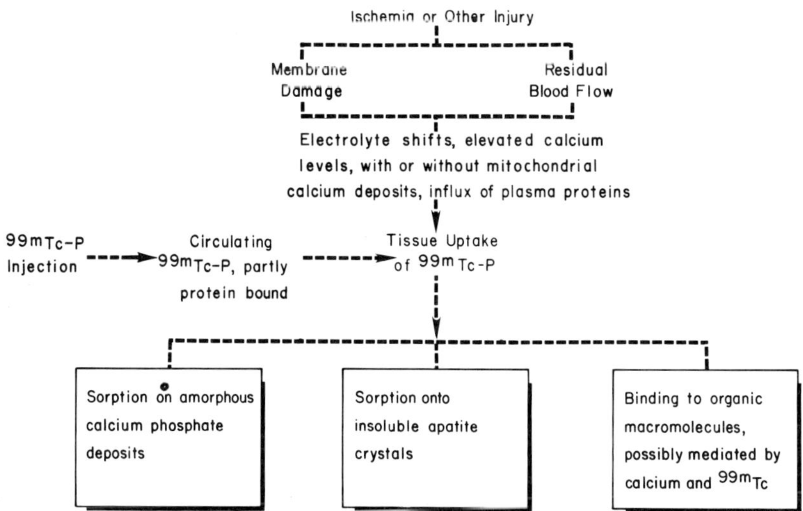

Figure 8. Pathophysiologic factors involved in the concentration of ^{99m}Tc-P agents in damaged myocardium and other soft tissues.

of these cells show evidence of an early stage of irreversible damage at the time of study (Fig. 2).[3] It also seems unlikely that cellular injury of the type manifested by autoradiographically labeled border-zone myocardium occurs in the absence of associated cellular necrosis.

The experimental findings have been supported by clinicopathologic studies in more than 50 patients.[79-81] These studies have shown that ^{99m}Tc-PYP myocardial scintigraphy accurately detects subendocardial as well as transmural myocardial infarcts (Figs. 9 and 10). In addition, the significance of relatively low grade positive ^{99m}Tc-PYP myocardial scintigrams in the clinical settings of unstable angina pectoris and chronic ischemic heart disease following myocardial infarction (persistently positive myocardial scintigram) was evaluated.[58, 78, 80] Positive ^{99m}Tc-PYP scintigrams in these clinical settings were associated with evidence of multifocal, irreversible myocardial damage in the form of coagulation necrosis, myocytolysis, and/or fibrosis, and the histologic age of the lesions was consistent with acute injury at the time of positive scintigraphic examination in each patient studied (Figs. 11 and 12).[80, 81] Positive ^{99m}Tc-PYP myocardial scintigrams also can be associated with myocardial necrosis of varied pathogenesis, including necrosis resulting from repeated cardioversion,[7-9, 82] heat injury,[5, 6] contusion,[10] metastatic cardiac tumor,[83] and metastatic myocardial calcification.[58, 84, 85]

Thus, a large body of experimental and clinicopathological evidence supports the conclusion that a positive ^{99m}Tc-PYP myocardial scintigram is a sensitive indicator of myocardial necrosis of varied pathogenesis and that a positive ^{99m}Tc-PYP myocardial scintigram does not occur when the severity of myocardial damage is limited to reversible injury unassociated with myocardial necrosis.

Comparisons of ^{99m}Tc-P with Other Infarct Avid Agents

Scintigraphic findings obtained with different infarct avid agents are determined by the general pathophysiologic factors listed in Table 3. Clear differences in the rate of clearance of the various infarct avid agents have been demonstrated. The ^{99m}Tc-P agents exhibit rapid concentration in infarcted myocardium and rapid clearance from blood and normal tissues,[86] so that ^{99m}Tc-P scintigraphy can be performed as early as one hour after injection.[78] Other agents, such as ^{99m}Tc-glucoheptonate (a seven-carbon sugar acid) and ^{99m}Tc-heparin (a mucopolysaccharide), exhibit moderately rapid clearance so that scin-

Table 4. Autoradiographic localization of tritiated diphosphonate in two canine hearts with LAD occlusion for two days*

Area	Labeling Index Per Unit Area	Ratio with Normal Myocardium
Normal posterior myocardium	6.3 ± 2.1	1.0
Myocardium without histologic damage adjacent to infarct	12.3 ± 2.6†	2.0
Narrow border region with severely damaged muscle cells	143.4 ± 20.6‡	22.8
Peripheral zone of infarct with advanced necrosis	167.1 ± 35.9‡	26.5
Central zone of infarct with advanced necrosis	1.0 ± 0.6†	0.2

*Adapted from Table 1 of Buja et al.[3] and reproduced with permission of the Journal of Clinical Investigation.

†P < 0.05 compared with normal posterior myocardium (Student's t test).

‡P < 0.0005 compared with normal posterior myocardium (Student's t test).

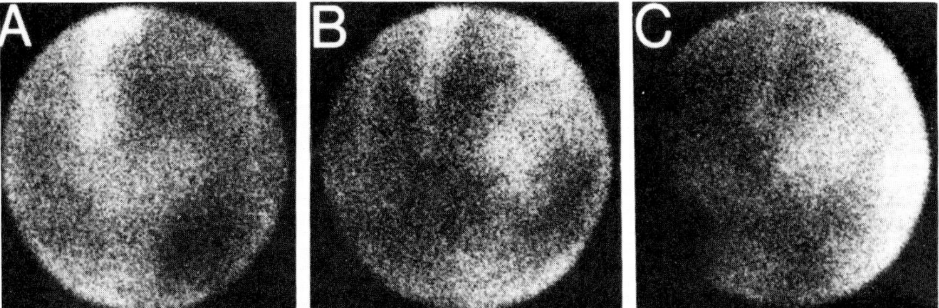

Figure 9. Patient with an acute lateral subendocardial myocardial infarct and a positive ^{99m}Tc-PYP myocardial scintigram. Anterior (A), left anterior oblique (B), and lateral (C) scintigraphic projections show prominent myocardial activity.

tigraphy with these agents can be performed within the first few hours after injection.[60, 63-66] Myocardial scintigraphy with ^{99m}Tc-tetracycline must be delayed for approximately 24 hours after injection, and with the antibodies to myosin and myoglobin a clearance time of 2 to 3 days is optimal.[59-62, 67, 68]

Scintigraphy with ^{99m}Tc-heparin has been found to be particularly successful in the experimental setting of temporary coronary occlusion followed by reflow,[66] and ^{99m}Tc-glucoheptonate has unusual properties as an imaging agent since myocardial scintigrams with this agent exhibit maximal positivity during initial stages of myocardial infarction (as early as five hours) but become less satisfactory later in the course of infarction.[60, 63-65]

Differences in the rate of clearance and concentration of various infarct-avid agents appear to influence the distribution of the agents in infarcted myocardium. Rapid clearance and concentration of ^{99m}Tc-P agents is associated with maximal concentration of these agents in infarct regions with blood flows in the range of 15 to 40 percent of normal.[2, 32, 67, 77] Other agents, including ^{99m}Tc-glucoheptonate, ^{99m}Tc-tetracycline, and the radiolabeled an-

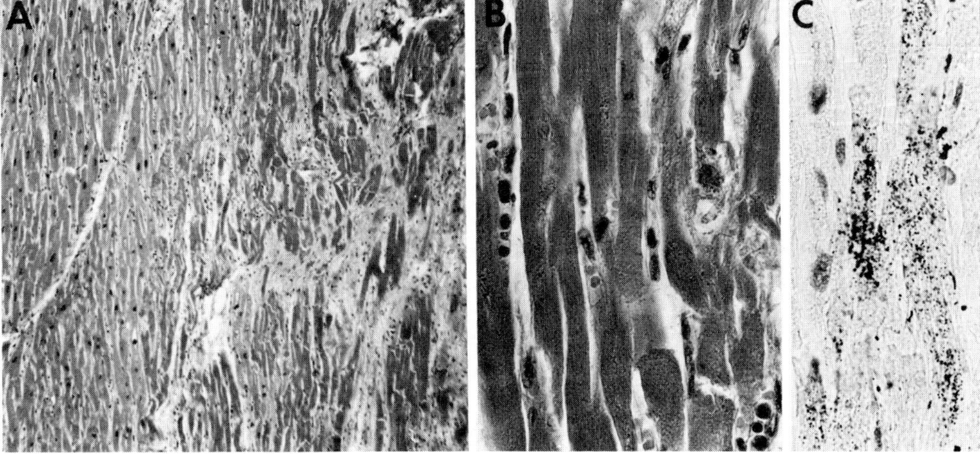

Figure 10. Histopathologic findings in the patient with subendocardial infarction (Fig. 9). A, low magnification view of lateral left ventricle. The subendocardial myocardium (right) shows pale amorphous areas of fibrosis as well as densely stained, necrotic muscle cells (Hematoxylin and eosin stain, × 54) B, The necrotic muscle cells are devoid of nuclei and have densely stained, homogeneous sarcoplasm. Neutrophilic infiltrate is present (Hematoxylin and eosin stain, × 340). C, Foci of muscle cell calcification also are present (Von Kossa stain for calcium salts, × 410). Figure reduced 16 percent.

117

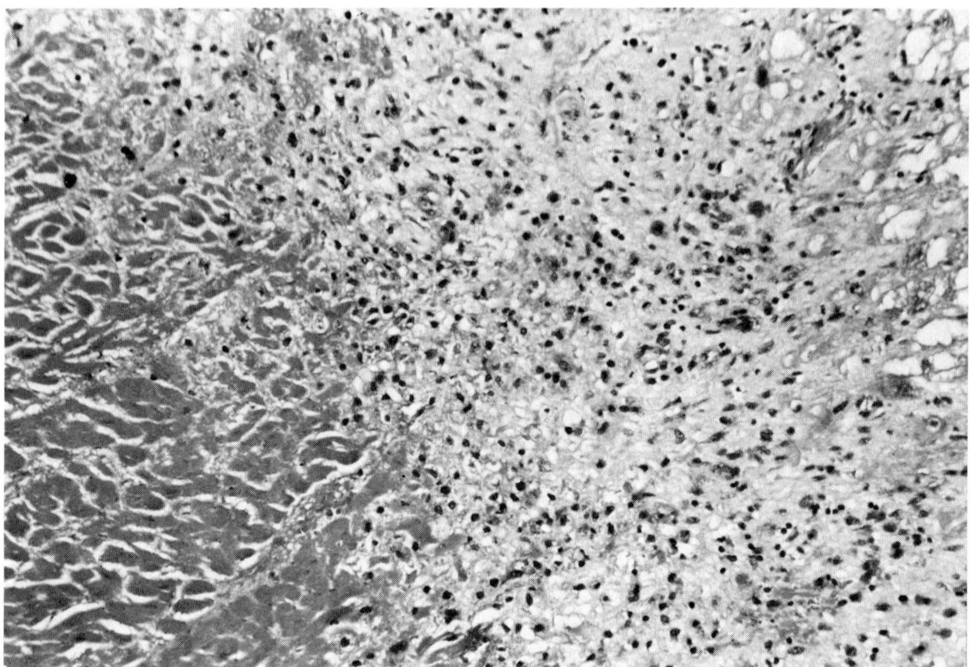

Figure 11. Healing microinfarct from a patient with unstable angina pectoris. The patient had a positive ^{99m}Tc-PYP myocardial scintigram without definitive electrocardiographic or serum enzyme changes of acute myocardial infarction three weeks prior to coronary bypass surgery and operative resection of a severely dyskinetic segment of anterior myocardium (Hematoxylin and eosin stain, × 150).

tibodies, have slower clearance rates, and are therefore exposed in high concentration to infarcted tissue for longer intervals than the ^{99m}Tc-P agents. Longer exposure appears to contribute to the typical infarct distribution of relatively slow clearing agents characterized by an inverse linear relationship between perfusion and uptake with maximal concentration in the infarcted regions which have the most severe blood flow reduction.[59-62,67,68] There also appears to be a general relationship between molecular size of radiopharmaceutical, clearance rate, and maximal target-to-normal ratios, with the highest ratios generally observed with ^{99m}Tc-P agents.

In general, concentration of an infarct-avid agent appears dependent upon the development of membrane damage and its consequences and the presence of at least some residual perfusion (Fig. 8). The theoretical basis for concentration of radiolabeled antibodies in acutely infarcted myocardium is immunologic reaction with specific subcellular constituents of muscle cells which have sufficient membrane damage to permit exposure of the antibodies—which are large molecules—to the cellular interior.[67,68] With [203]Hg-organomercurials the degree of concentration in damaged myocardium is influenced both by the presence of mercury and by the nature of the organic molecule.[6,62] Electrochemical properties of infarcted tissue also may play a role in concentration of some agents as suggested by uptake of highly negatively charged ^{99m}Tc-heparin in damaged myocardium.[66] Transchelation of ^{99m}Tc from radiopharmaceutical to damaged tissue components has been proposed as a general mechanism for uptake of various ^{99m}Tc-chelates in damaged tissues.[49] It also is possible that calcium-mediated concentrating mechanisms, similar to those involved in ^{99m}Tc-P uptake, may have a general role in the uptake of many infarct-avid agents. Indeed, Malek and associates using fluorescence and histochemical methods

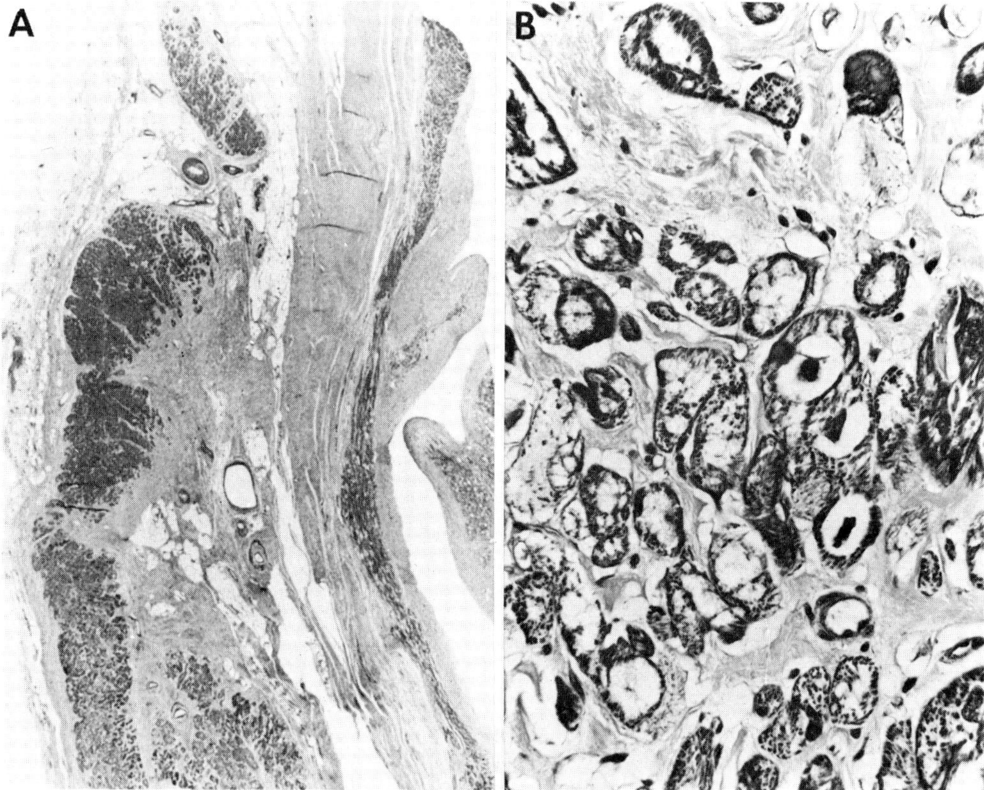

Figure 12. Patient with persistently positive ^{99m}Tc-PYP myocardial scintigram, recurrent chest pain, and a ventricular aneurysm. A, The operatively resected aneurysm wall exhibits extensive replacement fibrosis but contains foci of myocardium (darkly stained areas) which have survived the initial infarction. B, These foci of myocardium show interstitial fibrosis as well as myocytolytic degeneration of muscle cells ranging from edema to complete loss of stained cytoplasmic constituents (Masson trichome stains: A, × 15; B, × 400; figure reduced 17 percent). (From Buja et al.,[80] reproduced with permission of the American Heart Association, Inc.)

showed prominent accumulation of tetracycline analogues in areas of intense muscle cell calcification.[87] It is obvious that additional information is needed to clarify specific subcellular concentrating mechanisms for the various infarct-avid agents; such work also should lead to new and varied methods for infarct-avid scintigraphy.

REFERENCES

1. BUJA, L. M., PARKEY, R. W., DEES, J. H., ET AL.: *Morphologic correlates of technetium-99m stannous pyrophosphate imaging of acute myocardial infarcts in dogs.* Circulation 52:596, 1975.

2. BUJA, L. M., PARKEY, R. W., STOKELY, E. M., ET AL.: *Pathophysiology of technetium-99m stannous pyrophosphate and thallium-201 scintigraphy of acute anterior myocardial infarcts in dogs.* J. Clin. Invest. 57:1508, 1976.

3. BUJA, L. M., TOFE, A. J., KULKARNI, P. V., ET AL.: *Sites and mechanisms of localization of technetium-99m phosphorus radiopharmaceuticals in acute myocardial infarcts and other tissues.* J. Clin. Invest. 60:724, 1977.

4. BILHEIMER, D. W., BUJA, L. M., PARKEY, R. W., ET AL.: *Fatty acid accumulation and abnormal lipid deposition: A characteristic feature of the peripheral and border zones of experimental myocardial infarcts.* J. Nucl. Med. 19:276, 1978.

5. ADLER, N., CAMIN, L. L., AND SCHULKIN, P.: *Rat model for acute myocardial infarction. Application to technetium-labeled glucoheptonate, tetracycline and polyphosphate.* J. Nucl. Med. 17:203, 1976.

6. DAVIS, M. A., HOLMAN, B. L., AND CARMEL, A.N.: *Evaluation of radiopharmaceuticals sequestered by acutely damaged myocardium.* J. Nucl. Med. 17:911, 1976.

7. PUGH, B. R., BUJA, L. M., PARKEY, R. W., ET AL.: *Cardioversion and "false positive" technetium-99m stannous pyrophosphate myocardial scintigrams.* Circulation 54:399, 1976.

8. DICOLA, V. C., FREEDMAN, G. S., DOWNING, S. E., ET AL.: *Myocardial uptake of technetium-99m stannous pyrophosphate following direct current transthoracic countershock.* Circulation 54:980, 1976.

9. SCHNEIDER, R. M., HAYSLETT, J. P., DOWNING, S. E., ET AL.: *Effect of methylprednisolone upon technetium-99m pyrophosphate assessment of myocardial necrosis in the canine countershock model.* Circulation 56:1029, 1977.

10. GO, R. T., CHIU, C. L., DOTY, D. B., ET AL.: *Radionuclide imaging of experimental myocardial contusion.* J. Nucl. Med. 15:1174, 1974.

11. DEWANJEE, M. K., AND PRINCE, E. W.: *Cellular necrosis model in tissue culture: Uptake of ^{99m}Tc-tetracycline and the pertechnetate ion.* J. Nucl. Med. 15:577, 1974.

12. DEWANJEE, M. K.: *Localization of skeletal-imaging ^{99m}Tc chelates in dead cells in tissue culture: Concise communication.* J. Nucl. Med. 17:993, 1976.

13. SCHELBERT, H. R., INGWALL, J. S., SYBERS, H. D., ET AL.: *Uptake of infarct-imaging agents in reversibly and irreversibly injured myocardium in cultured fetal mouse heart.* Circ. Res. 39:860, 1976.

14. REIMER, K. A., LOWE, J. E., RASMUSSEN, M. M., ET AL.: *The wavefront phenomenon of ischemic cell death. I. Myocardial infarct size vs. duration of coronary occlusion in dogs.* Circulation 56:786, 1977.

15. GREGG, D. E.: *The natural history of coronary collateral development.* Circ. Res. 35:335, 1974.

16. BISHOP, S. P., WHITE, F. C., AND BLOOR, C. M.: *Regional myocardial blood flow during acute myocardial infarction in the conscious dog.* Circ. Res. 38:429, 1976.

17. RIVAS, F., COBB, F. R., BACHE, R. J., ET AL.: *Relationship between blood flow to ischemic regions and extent of myocardial infarction. Serial measurement of blood flow to ischemic regions in dogs.* Circ. Res. 38:439, 1976.

18. HIRZEL, H. O., NELSON, G. R., SONNENBLICK, E. H., ET AL.: *Redistribution of collateral blood flow from necrotic to surviving myocardium following coronary occlusion in the dog.* Circ. Res. 39:214, 1976.

19. HIRZEL, H. O., SONNENBLICK, E. H., AND KIRK, E. S.: *Absence of a lateral border zone of intermediate creatine phosphokinase depletion surrounding a central infarct 24 hours after acute coronary occlusion in the dog.* Circ. Res. 41:673, 1977.

20. HEARSE, D. J., OPIE, L. H., KATZEFF, I. E., ET AL.: *Characterization of the "border zone" in acute regional ischemia in the dog.* Am. J. Cardiol. 40:716, 1977.

21. MALLORY, G. K., WHITE, P. D., AND SALCEDO-SALGAR, J.: *The speed of healing of myocardial infarction. A study of the pathologic anatomy in seventy-two cases.* Am. Heart J. 18:647, 1939.

22. BAROLDI, G.: *Different types of myocardial necrosis in coronary heart disease: A pathophysiologic review of their functional significance.* Am. Heart J. 89:742, 1975.

23. BUJA, L. M., DEES, J. H., HARLING, D. F., ET AL.: *Analytical electron microscopic study of mitochondrial inclusions in canine myocardial infarcts.* J. Histochem. Cytochem. 24:508, 1976.

24. HAGLER, H. K., BURTON, K. P., BROWNE, R. H., ET AL.: *Energy dispersive x-ray spectroscopic (EDS) analysis of small particulate inclusions in hypoxic and ischemic myocardium.* In: Johari, O. (ed): *Proceedings of the Illinois Institute of Technology (IIT) Research Institute Scanning Electron Microscopy Symposium 1977, vol. II.* Chicago Press Corporation, Chicago, 1977, p. 145.

25. JENNINGS, R. B., AND GANOTE, C. E.: *Mitochondrial structure and function in acute myocardial ischemic injury.* Circ. Res. 38 (Suppl. I):80, 1976.

26. KIRK, G. A., ADAMS, R., JANSEN, C., ET AL.: *Particulate myocardial perfusion scintigraphy: Its clinical usefulness in evaluation of coronary artery disease.* Semin. Nucl. Med. 7:67, 1977.

27. BONTE, F. J., PARKEY, R. W., STOKELY, E. M., ET AL.: *Radionuclide determination of myocardial blood flow.* Semin. Nucl. Med. 3:153, 1973.

28. BRITTEN, J. S., AND BLANK, M.: *Thallium activation of the (Na^+-K^+)-activated ATPase of rabbit kidney.* Biochim. Biophys. Acta 159:160, 1968.

29. PROKOP, E. K., STRAUSS, H. W., SHAW, J., ET AL.: *Comparison of regional myocardial perfusion determined by ionic potassium-43 to that determined by microspheres.* Circulation 50:978, 1974.

30. STRAUSS, H. W., HARRISON, K., LANGAN, J. K., ET AL.: *Thallium-201 for myocardial imaging. Relation of thallium-201 to regional myocardial perfusion.* Circulation 51:641, 1975.

31. WEICH, H. F., STRAUSS, H. W., AND PITT, B.: *Myocardial extraction fraction of thallium-201.* Circulation 54(Suppl. II):218, 1976.

120

32. ZARET, B. L., DiCOLA, V. C., DONABEDIAN, R. K., ET AL.: *Dual radionuclide study of myocardial infarction. Relationships between myocardial uptake of potassium-43, technetium-99m stannous pyrophosphate, regional myocardial blood flow and creatine phosphokinase depletion.* Circulation 53:422, 1976.

33. DiCOLA, V., DOWNING, S. E., DONABEDIAN, R. K., ET AL.: *Pathophysiological correlates of thallium-201 myocardial uptake in experimental infarction.* Cardiovasc. Res. 11:141, 1977.

34. LEVENSON, N. I., ADOLPH, R. J., ROMHILT, D. W., ET AL.: *Effects of myocardial hypoxia and ischemia on myocardial scintigraphy.* Am. J. Cardiol. 35:251, 1975.

35. ADOLPH, R., ROMHILT, D., NISHIYAMA, H., ET AL.: *Use of positive and negative imaging agents to visualize myocardial ischemia.* Circulation 54(Suppl. II):220, 1976.

36. POE, N. D.: *Comparative myocardial uptake and clearance characteristics of potassium and cesium.* J. Nucl. Med. 13:557, 1972.

37. SCHELBERT, H. R., ASHBURN, M. L., CHAUNCEY, D. M., ET AL.: *Comparative myocardial uptake of intravenously administered radionuclides.* J. Nucl. Med. 15:1092, 1974.

38. ATKINS, H. L., BUDINGER, T. F., LEBOWITZ, E., ET AL.: *Thallium-201 for medical use. III. Human distribution and physical imaging properties.* J. Nucl. Med. 18:133, 1977.

39. SCHELBERT, H., INGWALL, J., WATSON, R., ET AL.: *Factors influencing the myocardial uptake of thallium-201.* J. Nucl. Med. 18:598, 1977.

40. POHOST, G. M., ZIR, L. M., MOORE, R. H., ET AL.: *Differentiation of transiently ischemic from infarcted myocardium by serial imaging after a single dose of thallium-201.* Circulation 55:294, 1977.

41. SCHELBERT, H., SCHULER, G., ASHBURN, W., ET AL.: *Time course redistribution of Tl-201 after transient ischemia.* J. Nucl. Med. 18:598, 1977.

42. WACKERS, F. J. TH., SOKOLE, E. B., SAMSON, G., ET AL.: *Value and limitations of thallium-201 scintigraphy in the acute phase of myocardial infarction.* N. Engl. J. Med. 295:1, 1976.

43. WACKERS, F. J. TH., BECKER, A. E., SAMSON, G., ET AL.: *Location and size of acute transmural myocardial infarction estimated from thallium-201 scintiscans: A clinicopathologic study.* Circulation 56:72, 1977.

44. BULKLEY, B. H., HUTCHINS, G. M., BAILEY, I., ET AL.: *Thallium-201 imaging and gated cardiac blood pool scans in patients with ischemic and idiopathic congestive cardiomyopathy: A clinical and pathologic study.* Circulation 55:753, 1977.

45. HAMILTON, G. W., TROBAUGH, G. B., RITCHIE, J. L., ET AL.: *Myocardial imaging with intravenously injected thallium-201 in patients with suspected coronary artery disease. Analysis of technique and correlation with electrocardiographic, coronary anatomic and ventriculographic findings.* Am. J. Cardiol. 39:347, 1977.

46. WALSH, W. F., FILL, H. R., AND HARPER, P. V.: *Nitrogen-13-labeled ammonia for myocardial imaging.* Semin. Nucl. Med. 7:59, 1977.

47. EVANS, J. R., GUNTON, R. W., BAKER, R. G., ET AL.: *Use of radioiodinated fatty acid for photoscans of the heart.* Circ. Res. 16:1, 1965.

48. BONTE, F. J., GRAHAM, K., AND MOORE, J.: *Experimental myocardial imaging with ^{131}I-labeled oleic acid.* Radiology 108:195, 1973.

49. POE, N. D.: *Rationale and radiopharmaceuticals for myocardial imaging.* Semin. Nucl. Med. 7:7, 1977.

50. WEISS, E. S., HOFFMAN, E. J., PHELPS, M. E., ET AL.: *External detection and visualization of myocardial ischemia with ^{11}C-substrates in vitro and in vivo.* Circ. Res. 39:24, 1976.

51. WEISS, E. S., AHMED, S. A., WELCH, M. J., ET AL.: *Quantification of infarction in cross sections of canine myocardium in vivo with positron emission transaxial tomography and ^{11}C-palmitate.* Circulation 55:66, 1977.

52. SOBEL, B. E., WEISS, E. S., WELCH, M. J., ET AL.: *Detection of remote myocardial infarction in patients with positron emission transaxial tomography and intravenous ^{11}C-palmitate.* Circulation 55:853, 1977.

53. SCHEUER, J., AND BRACHFELD, N.: *Myocardial uptake and functional distribution of palmitate-1-C^{14} by the ischemic dog heart.* Metabolism 15:945, 1966.

54. WARTMAN, W. B., JENNINGS, R. B., YOKOYAMA, H. O., ET AL.: *Fatty change of the myocardium in early experimental myocardial infarction.* Arch. Pathol. 62:318, 1956.

55. WOOD, J. M., SORDAHL, L. A., LEWIS, R. M., ET AL.: *Effect of chronic myocardial ischemia on the activity of carnitine palmityl coenzyme A transferase of isolated canine heart mitochondria.* Circ. Res. 32:340, 1973.

56. BONTE, F. J., PARKEY, R. W., GRAHAM, K. D., ET AL.: *A new method of radionuclide imaging of myocardial infarcts.* Radiology 110:473, 1974.

57. BONTE, F. J., PARKEY, R. W., GRAHAM, K. D., ET AL.: *Distribution of several agents useful in imaging myocardial infarcts.* J. Nucl. Med. 16:132, 1975.

58. WILLERSON, J. T., PARKEY, R. W., BUJA, L. M., ET AL.: *Are ^{99m}Tc-stannous pyrophosphate myocardial scintigrams clinically useful?* Clin. Nucl. Med. 2:137, 1977.

59. Holman, B. L., Dewanjee, M. K., Idoine, J., et al.: *Detection and localization of experimental myocardial infarction with ^{99m}Tc-tetracycline.* J. Nucl. Med. 14:595, 1973.

60. Zweiman, F. G., Holman, B. L., O'Keefe, A., et al.: *Selective uptake of ^{99m}Tc-complexes and ^{67}Ga in acutely infarcted myocardium.* J. Nucl. Med. 16:975, 1975.

61. Holman, B. L., and Zweiman, F. G.: *Time course of ^{99m}Tc(Sn)-tetracycline uptake in experimental acute myocardial infarction.* J. Nucl. Med. 16:1144, 1975.

62. Holman, B. L., Davis, M. A., and Hanson, R. N.: *Myocardial infarct imaging with technetium-labeled complexes.* Semin. Nucl. Med. 7:29, 1977.

63. Rossman, D. J., Rouleau, J., Strauss, H. W., et al.: *Detection and size estimation of acute myocardial infarction using ^{99m}Tc-glucoheptonate.* J. Nucl. Med. 16:980, 1975.

64. Combes, J. R., Jacobstein, J. G., Post, M. R., et al.: *^{99m}Tc-glucoheptonate imaging for the early diagnosis of acute myocardial infarction.* Am. J. Cardiol. 39:315, 1977.

65. Roberts, A. J., Cipriano, P. R., Alonso, D. R., et al.: *Evaluation of methods for the quantification of experimental myocardial infarction.* Circulation 57:35, 1978.

66. Kulkarni, P., Parkey, R., Buja, L., et al.: *Technetium-99m heparin: A new radiopharmaceutical to identify damaged coronary endothelium and damaged myocardium.* J. Nucl. Med. 19:810, 1978.

67. Beller, G. A., AnKhaw, B., Haber, E., et al.: *Localization of radiolabeled cardiac myosin-specific antibody in myocardial infarcts. Comparison with technetium-99m stannous pyrophosphate.* Circulation 55:74, 1977.

68. Parkey, R. W., Buja, L. M., Kulkarni, P., et al.: *Localization of a specific I-131 antibody to myoglobin in myocardial tissue and factors which influence myoglobin release from cardiac cells.* J. Nucl. Med. 18:611, 1977.

69. Kramer, R. J., Golstein, R. E., Hirshfield, J. W., Jr., et al.: *Accumulation of gallium-67 in regions of acute myocardial infarction.* Am. J. Cardiol. 33:861, 1974.

70. Weiss, E. S., Ahmed, S. A., Thakur, M. L., et al.: *Imaging of the inflammatory response in ischemic canine myocardium with ^{111}indium-labeled leukocytes.* Am. J. Cardiol. 40:195, 1977.

71. Botvinick, E. H., Shames, D., Lappin, H., et al.: *Noninvasive quantitation of myocardial infarction with technetium-99m pyrophosphate.* Circulation 52:909, 1975.

72. Bruno, F. P., Cobb, F. R., Rivas, F., et al.: *Evaluation of technetium stannous pyrophosphate as imaging agent in acute myocardial infarction.* Circulation 54:71, 1976.

73. Coleman, R. E., Klein, M. S., Ahmed, S. A., et al.: *Mechanisms contributing to myocardial accumulation of technetium-99m stannous pyrophosphate after coronary occlusion.* Am. J. Cardiol. 39:55, 1977.

74. Dewanjee, M. K., and Kahn, P. C.: *Mechanisms of localization of ^{99m}Tc-labeled pyrophosphate and tetracycline in infarcted myocardium.* J. Nucl. Med. 17:639, 1976.

75. Ettinger, U., Kronenberg, M., Wilson, G., et al.: *Localization of ^{99m}Tc-pyrophosphate uptake and CPK depletion in acutely infarcted dog hearts: A function of variable tissue pathology.* Circulation 52(Suppl. II):179, 1975.

76. Reimer, K. A., Martonffy, K., Schumacher, B. L., et al.: *Localization of ^{99m}Tc labeled pyrophosphate and calcium in myocardial infarcts after temporary coronary occlusion in dogs.* Proc. Soc. Exper. Biol. Med. 156:272, 1977.

77. Marcus, M. L., Tomanek, R. J., Ehrhardt, J. C., et al.: *Relationships between myocardial perfusion, myocardial necrosis, and technetium-99m pyrophosphate uptake in dogs subjected to sudden coronary occlusion.* Circulation 54:647, 1976.

78. Parkey, R. W., Bonte, F. J., Buja, L. M., et al.: *Myocardial infarct imaging with technetium-99m phosphates.* Semin. Nucl. Med. 7:15, 1977.

79. Holman, B. L., Ehrie, M., and Lesch, M.: *Correlation of acute myocardial infarct scintigraphy with post-mortem studies.* Am. J. Cardiol. 37:311, 1976.

80. Buja, L. M., Poliner, L. R., Parkey, R. W., et al.: *Clinicopathologic study of persistently positive technetium-99m stannous pyrophosphate myocardial scintigrams and myocytolytic degeneration after acute myocardial infarction.* Circulation 56:1016, 1977.

81. Poliner, L. R., Buja, L. M., Parkey, R. W., et al.: *Clinicopathologic correlations in 52 patients studied by technetium-99m stannous pyrophosphate myocardial scintigraphy.* Circulation. In press, 1979.

82. McDaniel, M. M., and Morton, M. E.: *^{99m}Tc-pyrophosphate imaging demonstrating skeletal muscle and myocardial activity following cardioversion.* Clin. Nucl. Med. 2:57, 1977.

83. Harford, W., Weinberg, M. N., Buja, L. M., et al.: *Positive technetium-99m stannous pyrophosphate myocardial scintigram in a patient with carcinoma of the lung.* Radiology 122:747, 1977.

84. Epstein, D. A., Solar, M., and Levin, E. J.: *Demonstration of long-standing metastatic soft tissue calcification by ^{99m}Tc-diphosphonate.* Am. J. Roentgenol. 128:145, 1977.

85. JANOWITZ, W. R., AND SERAFINI, A. N.: *Intense myocardial uptake of ^{99m}Tc-diphosphonate in a uremic patient with secondary hyperparathyroidism and pericarditis: Case Report.* J. Nucl. Med. 17:896, 1976.
86. KRISHNAMURTHY, G. T., HUEBOTTER, R. J., WALSH, C. F., ET AL.: *Kinetics of ^{99m}Tc-labeled pyrophosphate and polyphosphate in man.* J. Nucl. Med. 16:109, 1975.
87. MALEK, P., KOLC, J., ZASTAVA, V., ET AL.: *Fluorescence of tetracycline analogues fixed in myocardial infarction.* Cardiologia 42:303, 1963.
88. SIEGAL, B. A., ENGEL, W. K., AND DERRER, E. C.: *Localization of technetium-99m diphosphonate in acutely injured muscle. Relationship to muscle calcium deposition.* Neurology 27:230, 1977.
89. HOLMAN, B. L., LESCH, M., AND ALPERT, J. S.: *Myocardial scintigraphy with technetium-99m pyrophosphate during the early phase of acute infarction.* Am. J. Cardiol. 41:39, 1978.

Clinical Application of Myocardial Imaging with Thallium-201

Bertram Pitt, M.D., and H. William Strauss, M.D.

Myocardial imaging with thallium-201 (^{201}Tl) is finding wide application in the clinical evaluation of patients with both ischemic and nonischemic cardiac disease.[1,2]

A monovalent cationic radionuclide, ^{201}Tl has biologic properties similar to potassium and has been used as a surrogate for K^+ in studies of membrane transport. The uptake of ^{201}Tl by the myocardium has been shown to be dependent upon the presence of an active Na^+-K^+ ATPase transport system, intracellular cation concentration, and regional myocardial blood flow. Over a wide physiologic range of myocardial blood flow, the uptake of ^{201}Tl by the heart is proportional to myocardial blood flow.[3] The extraction rate for ^{201}Tl within the normal range of myocardial blood flow is approximately 88 percent.[4] At the extremes of myocardial flow the extraction of ^{201}Tl by the myocardium is altered such that at low flow rates the extraction is greater and at high flow rates less. The relationship between myocardial flow and ^{201}Tl uptake may also be altered by a number of pharmacological and physiological interventions which affect Na^+-K^+ ATPase.[4,5]

PHYSICAL PROPERTIES OF ^{201}TL

The physical properties of ^{201}Tl are different from previously used cationic tracers in that a large percentage of the energy emitted by ^{201}Tl is an x-ray of between 69 and 80 keV. Potassium-43 and ^{81}Rb emit relatively high energy gamma rays which require special shielding when used with commercially available scintillation cameras with a 1.25-cm-thick crystal. The shielding necessary for imaging of the high energy gamma rays results in a loss of image resolution. Although the 80-keV energy emitted by ^{201}Tl is below the optimum for imaging with the Anger scintillation camera, the image resolution is considerably greater than with ^{43}K and ^{81}Rb. The half-life of ^{201}Tl (approximately 72 hours) is an advantage for clinical myocardial imaging since it permits delivery to clinical facilities on a once- or twice-a-week basis in comparison to several times a week for ^{43}K. The disadvantage of the longer half-life of ^{201}Tl is an inability to rapidly assess the effects of physiologic and therapeutic interventions on the distribution of myocardial blood flow.

In normal individuals ^{201}Tl is homogeneously distributed throughout the myocardium. There is often a relative diminution of tracer uptake at the apex in the anterior view since the apex is physiologically thinner than the left ventricular free wall.[6] Uptake of ^{201}Tl by the right ventricle is usually not appreciated in the left anterior oblique (LAO) view; even though ^{201}Tl is taken up by the right ventricular free wall, the mass of the right ventricular free wall and blood flow per gram are considerably less than in the left ventricle. However, during tachycardia, exercise, and conditions resulting in increased right ventricular mass the right ventricular free wall is easily detectable on the LAO ^{201}Tl image.[7]

^{201}TL IMAGE DISPLAY

There is at present no general agreement as to the best way to display ^{201}Tl myocardial images for analysis. Adequate results have been obtained by viewing the unprocessed image recorded on Polaroid or 35-mm film from the CRT of the scintillation camera. More recently the ^{201}Tl image has been subjected to various forms of computer processing and enhancement. Programs have been developed to extract the myocardium from surrounding background areas such as the lung and hepatic-gastric areas. The extracted myocardial image can then be further processed by simple background subtraction or more complex computer programs. Attempts have also been made to analyze the distribution of ^{201}Tl within the myocardium by computer processing. An early attempt was made to relate relative myocardial ^{201}Tl activity to background. The ratio of myocardial ^{201}Tl to background (RMTA) in normal individuals and in those with infarction or ischemia was then compared.[8] Although initial studies showed good separation of values obtained in normal individuals and those with ischemia or infarction, further experience with a larger number of patients pointed to difficulties in finding a standard area to select background. Background activity within the region of the lungs is heterogeneous and varies considerably in normal patients, smokers, and patients with pulmonary congestion.

More recently, computer programs have been generated to detect the center of activity within the myocardial image.[9] Radii are constructed from the center of activity to the edge of the myocardial image and activity along each radii is determined. The activity for the entire 360 degrees of the outlined edge of the myocardial images is then displayed on a linear graph. Patients with myocardial ischemia or infarction will have radii with decreased activity as compared to normal patients. Although they are attractive and of possible clinical value, caution should be used in the application of currently available computer-assisted programs for analysis and interpretation of ^{201}Tl activity: the nature of the background activity is heterogeneous and variable and there are always differences in individual attenuation coefficients.

Although many of the difficulties may be overcome with further experience and tomographic imaging, it is essential that for the immediate future each institution establish criteria for interpretation of normal and abnormal images on their own equipment. If computer facilities are available, normal volunteers should be imaged and their data stored on magnetic tape or disc. The images should be subject to whatever processing is decided upon (e.g., 20 percent background subtraction) and all subsequent patients should be interpreted relative to the normal studies. It would be dangerous to view patient studies at high contrast or with 30 to 50 percent background subtraction and to diagnose the presence of myocardial ischemia or infarction if normal images had not been processed in an identical manner and found to have uniform ^{201}Tl uptake. It is relatively easy with current computer and image display systems to obtain false positive defects in tracer uptake. This is especially true if the images are viewed on a color display. Color tables can be constructed in which a 5 percent reduction in activity appears as a major defect in tracer uptake. For example, by choosing a color scale such that 5 percent increments in activity result in a difference between red and white, a 5 percent difference may appear to be a defect in tracer uptake leading to a false diagnosis of myocardial ischemia.

As already mentioned, the right ventricle is not appreciated on the LAO ^{201}Tl myocardial image; however, it is easily seen in conditions with increased right ventricular myocardial flow such as exercise or excitement, or in conditions leading to right ventricular hypertrophy. Patients with right ventricular hypertrophy are often difficult to detect by standard electrocardiography, particularly if left ventricular hypertrophy is also present. The LAO ^{201}Tl image, however, allows independent assessment of both the right and left ventricular myocardium. In patients with pulmonary hypertension and resultant right ventricular hypertrophy, the ventricular free wall is easily visualized on the resting LAO ^{201}Tl myocardial image.[7] The ability of the ^{201}Tl myocardial image to independently examine the intraventricular septum as well as the left and right ventricular free walls has been found to be useful in the evaluation of patients with cyanotic congenital heart disease.[10] The detection of patients with single ventricles is often difficult; even at contrast ventriculography it is sometimes difficult to determine the exact diagnosis and anatomic relationships. In patients with cyanotic congenital heart disease with a ventricular septal defect (e.g., tetralogy of Fallot), the LAO ^{201}Tl image reveals the presence of an intraventricular septum with clear separation of both the right and left ventricular free walls. In patients with single ventricles, however, a single ventricular cavity or a large chamber with a small outlet chamber is seen—no matter how the patient is rotated—without the presence of an interventricular septum. The right atrium is also detected occasionally in patients with cyanotic congenital heart disease, reflecting right atrial hypertrophy.[11]

EXERCISE MYOCARDIAL IMAGING

Thallium-201 myocardial imaging has found its greatest use in the evaluation of patients with suspected ischemic heart disease.[12] Results from several centers and a recent multicenter trial have shown that the sensitivity and specificity of exercise ^{201}Tl myocardial imaging for the detection of patients with suspected myocardial ischemia are greater than for conventional exercise electrocardiography.[12-14] Although exact figures vary from center to center depending upon patient selection and technique, the sensitivity and specificity for exercise ^{201}Tl myocardial imaging appears to be in the range of 85 to 90 percent, compared to 65 to 70 percent for exercise electrocardiography.

Exercise ^{201}Tl myocardial imaging is normally performed in conjunction with exercise electrocardiography. Resting electrocardiograms, supine and upright, are obtained first. The patient is then exercised to his maximum capacity on a bicycle ergometer or treadmill. At that point 1½ to 2 mCi of ^{201}Tl is injected through an indwelling intravenous catheter and the patient is asked to continue exercising for an additional 20 to 30 seconds at a lower level of exercise. This allows the tracer to be taken up by the myocardium over several cardiac cycles. Approximately five minutes after injection of the tracer the patient is placed beneath the detector of the scintillation camera and imaging is begun in the anterior, LAO, and left lateral positions. It is important to begin imaging within 5 to 10 minutes after injecting the tracer since redistribution of ^{201}Tl into an ischemic area is relatively rapid and can result in a loss of definition and sensitivity for detection of ischemia (Fig. 1).

A focal defect of tracer uptake seen on the postexercise myocardial image suggests the presence of myocardial ischemia or infarction. Ischemia can be diagnosed if the initial defect found at exercise is no longer seen (1) on reimaging after several hours to observe the redistribution of ^{201}Tl,[15] or (2) on imaging at rest following a second injection of tracer several days later.[12] Myocardial infarction can be diagnosed if the tracer defect seen on the postexercise image persists at rest (Fig. 2). The ability to reimage ^{201}Tl with the same dose of tracer several hours later has the advantage of considerable savings in cost and radiation exposure. If, however, the initial defect is smaller but persistent on reimaging several hours later, the distinction between ischemia and infarction becomes uncertain and the patient should either be reimaged sometime within the next 12 hours, or given a second dose of ^{201}Tl at rest several days later. The variability in the time of redistribution for ^{201}Tl has diminished the clinical utility of redistribution ^{201}Tl imaging. However, in a patient being evaluated for suspected ischemic heart disease it may be sufficient to detect

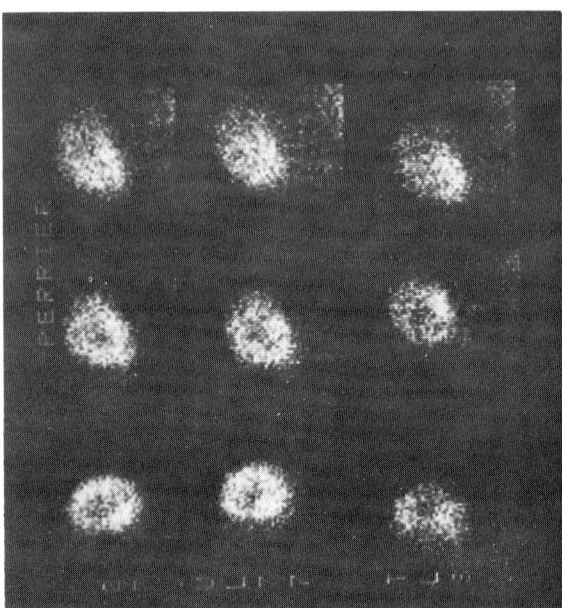

Figure 1. Thallium scan in the anterior position (left panel), 45 degree LAO position (middle panel), and 70 degree LAO position (right panel) performed immediately following injection at peak stress (top row), 30 minutes later (middle row) and at three hours (bottom row). The examination shows a homogeneous distribution of thallium in all views (the thinning noted at the apex is a normal variant). On the delayed images, however, there is marked decrease of thallium in the anterolateral and septal surfaces. This has not been correlated with coronary artery disease and is observed in approximately 10 percent of patients. Its importance is uncertain at this time. On coronary arteriography performed subsequently, no lesions were demonstrable.

a defect in tracer uptake during stress and thereby establish the presence of ischemic heart disease (Figs. 3, 4 and 5).

Redistribution of ^{201}Tl into an ischemic region over time may occur independent of any change in myocardial blood flow. In a patient with a persistent area of myocardial ischemia the initial ^{201}Tl myocardial image will show a defect in tracer uptake; this is largely dependent upon the initial delivery of ^{201}Tl. Since myocardial flow to the ischemic region is less than normal, regional tracer uptake will be less and a tracer defect will be appreciated on imaging. Over time, however, the ischemic area continues to extract ^{201}Tl from residual blood flow activity. Eventually the intramyocardial thallium concentration in the ischemic region approaches that in the normal area, with the final concentration being dependent upon myocardial mass and intracellular cation content.

Exercise ^{201}Tl myocardial imaging is especially helpful when the exercise electrocardiogram is difficult to interpret, as in the case of an intraventricular conduction defect, old myocardial infarction, left ventricular hypertrophy, and electrolyte abnormalities. In patients with a normal resting electrocardiogram the sensitivity of ^{201}Tl for detection of ischemia is equal to or slightly greater than that of exercise electrocardiography. Exercise ^{201}Tl myocardial imaging is, however, of special value in evaluating asymptomatic patients for ischemic heart disease.[16] Since the prevalence of significant coronary artery disease and myocardial ischemia is relatively low in asymptomatic individuals, and the specificity of exercise electrocardiography less than 90 percent, a positive electrocardiographic response on exercise testing will be falsely positive in a high proportion of such patients. The availability of an independent test for myocardial ischemia such as ^{201}Tl

128

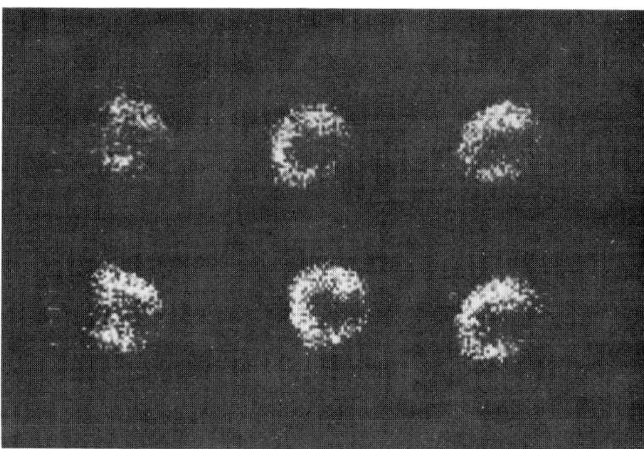

Figure 2. Thallium scan performed at peak stress (top row) and following the delay of three hours (bottom row) in the anterior position (left panel), 45 degree left anterior oblique position (middle panel), and 70 degree left anterior oblique position (right panel). The examination reveals focal decrease of thallium concentration in the apex on the anterior view and involving the posterolateral wall on the 45 degree LAO study. The tracer distribution does not change over the interval of observation. This increases the likelihood of scar. Clinically, this patient had sustained a previous myocardial infarction and now returned for evaluation of atypical chest pain. No zones of ischemia were identified on the scan and stress ECG performed simultaneously revealed no additional ischemia.

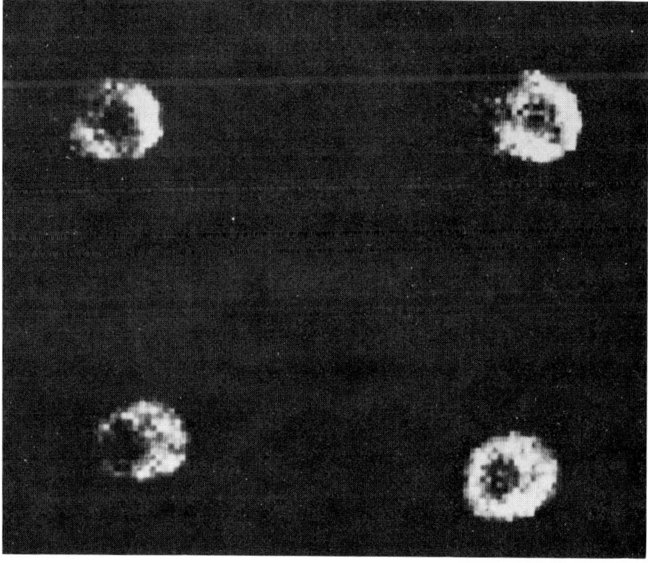

Figure 3. Thallium scan performed following the injection at peak stress (left panel) and 2½ hours later (right panel) in the anterior position (top row) and in the 45 degree left anterior oblique position (bottom row). The initial study reveals decrease of thallium concentration in the inferior and septal portions of the myocardium. On the redistribution images, these zones are normal. This study raises the likelihood of transient ischemia involving the myocardium. Stress ECG performed at the same time revealed ST segment depression in lead V_5, unaccompanied by chest pain. Coronary arteriography performed subsequently revealed 80 percent or greater narrowings of all three coronary arteries.

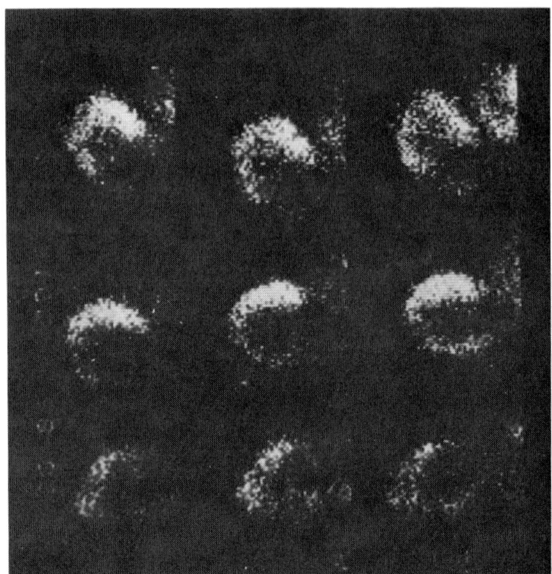

Figure 4. Thallium scan performed following the injection at peak stress (top row), 30 minutes later (middle row), and two hours later (bottom row). In the anterior position (left panel), 45 degree left anterior oblique position (center panel), and in the 70 degree left anterior oblique position (right panel). This study reveals marked decrease of thallium concentration involving the apex, inferior wall, and septal surfaces of the myocardium. Over the interval of observation there is some relative increase in thallium concentration involving the septum; however, the prominent apical inferior lesion remains. This study suggests the presence of persistent ischemia in addition to myocardial scar. Clinically, the patient had sustained a previous myocardial infarction and now returned because of increasing shortness of breath on exertion and atypical chest pain. The stress ECG had ST segment abnormalities at rest and was therefore difficult to interpret.

adds increased certainty to the diagnosis. If both the exercise electrocardiogram and exercise ^{201}Tl myocardial image are suggestive of ischemia, significant obstructive coronary artery disease will be detected in a high percentage of patients. If, however, the exercise ^{201}Tl myocardial image is negative in an asymptomatic individual with a positive exercise electrocardiogram, the certainty of significant anatomic coronary artery disease diminishes greatly. If both the exercise electrocardiogram and exercise ^{201}Tl myocardial image are negative, myocardial ischemia is unlikely. Significant coronary artery disease may be present, however, even in the absence of a positive exercise electrocardiogram or exercise ^{201}Tl myocardial image. This could reflect the presence of functional collateral vessels, or the fact that the functional significance of a given anatomic coronary artery lesion may be overestimated by conventional cineangiography.

Exercise ^{201}Tl myocardial imaging is also of value in the evaluation and followup of patients undergoing coronary artery bypass graft surgery.[17,18] As already mentioned, the location of the ^{201}Tl tracer defect can be of value in localizing the area of significant coronary artery narrowing. A number of instances have arisen in which there was uncertainty as to the functional significance of a given 40 to 60 percent coronary artery lesion following coronary arteriography. The appearance during exercise of a defect in tracer uptake within the distribution of the vessel under question is evidence of a functionally significant lesion.

The distribution of myocardial blood flow within the region of the three major coronary arteries can be assessed from the anterior, 40 and 60 degree LAO ^{201}Tl myocardial images. Cineangiographic studies by Bailey and coworkers in patients with single, double,

130

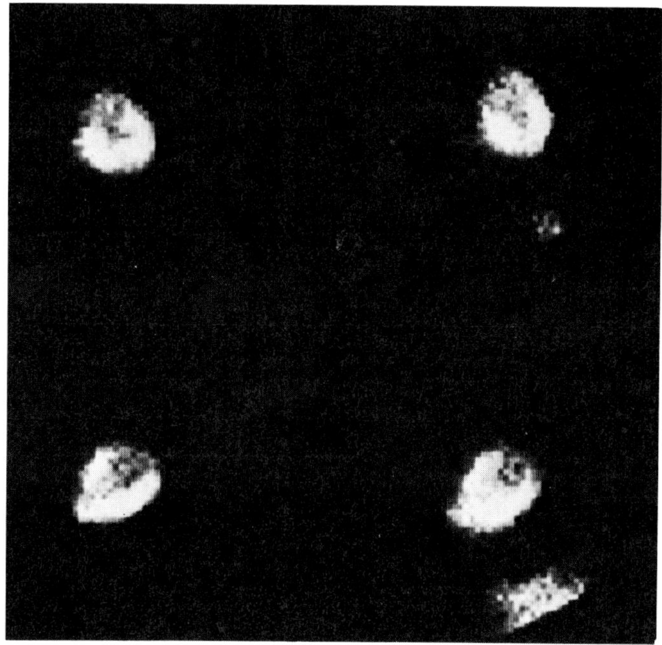

Figure 5. Thallium scan performed following injection at peak stress (left panel) and 3 hours later (right panel) in the anterior position (top row) and in the 45 degree left anterior oblique position (bottom row). There is a homogeneous distribution of thallium in the anterior views, but the LAO views reveal a focal zone of decreased thallium concentration. Due to the orientation of the heart, this was considered a variant of apical thinning. The examination was considered normal. Stress ECG revealed 4-mm ST segment depression at a peak heart rate of 170, without accompanying chest pain. Coronary arteriography revealed no significant narrowings of the coronary arteries.

and triple vessel disease, showed the distribution of the left anterior descending coronary artery to be reflected in the anterior apical portion of the anterior ^{201}Tl myocardial image and in the septal portion of the 40 and 60 degree LAO images.[19] The right and left circumflex coronary artery distribution appears to overlap in the inferior border of the anterior image and the lateral border of the 60 degree LAO image. However, the high lateral border of the 40 degree LAO image appears to be relatively specific for the left circumflex coronary artery. Defects in the apex of the 40 and 60 degree LAO images have relatively little localizing value since they may occur with lesions of any of the major coronary arteries. Significant coronary artery lesions of the left anterior descending coronary artery above the first septal perforator can be distinguished from those below the first septal perforator by the finding of a defect involving the anterior apical area of the anterior image and the septal area of the 40 and 60 degree LAO images in those with lesions above the first septal perforator. In patients with lesions distal to the first septal perforator, tracer defects are found more distally on the anterior view and involve the apex, but do not involve the septum in the 40 and 60 degree LAO images. This distinction may be of clinical importance in that lesions of the left anterior descending coronary artery above the first septal perforator generally indicate a worse prognosis than those below the first septal perforator.

Although useful in localizing the area of ischemia, ^{201}Tl has only limited utility in determining whether single or multiple vessel disease is present. The appearance of a defect in two or more vascular areas is strong evidence for the presence of double or triple vessel disease. The appearance of a defect in tracer uptake in a single vascular area cannot, however, be used to exclude multivessel disease, since one vascular area may become ischemic before another and exercise may have been terminated before the other vascular areas became ischemic. In some instances a significant coronary artery lesion may not become apparent during exercise because of the presence of functional collateral vessels.

Thallium-201 myocardial imaging may be of value in determining whether a segment of myocardium is viable or contains only scar. Absence of tracer uptake at rest with a myocardial ^{201}Tl-to-background ratio of approximately one and a failure to see increased ^{201}Tl uptake on redistribution imaging suggests scar. Persistent but diminished uptake at rest with an increase on redistribution imaging suggests viable myocardium. This differentiation is often important in deciding whether or not a bypass graft should be attempted in a particular region. Perhaps the most important use of ^{201}Tl myocardial imaging in patients undergoing coronary artery bypass graft surgery is in assessing the results of surgery. Symptomatic relief has been found to occur in more than 85 percent of patients undergoing coronary artery bypass graft surgery. Relief of symptoms following bypass graft surgery may, however, be the result of a number of causes besides relief of myocardial ischemia (e.g., placebo effect, destruction of sensory coronary artery nerves, and perioperative infarction with resultant loss of previously ischemic myocardium). In a patient who has an exercise-induced ^{201}Tl defect preoperatively, the postoperative finding of a completely normal rest and exercise ^{201}Tl image is evidence for graft patency and successful relief of ischemia. Postoperative persistence of an exercise-induced tracer defect despite relief of symptoms suggests a placebo effect—often with an occluded graft— while postoperative appearance of a new resting ^{201}Tl defect suggests perioperative infarction as the cause for relief of symptoms.

Although exercise has been of great value in provoking myocardial ischemia in patients with suspected ischemic heart disease, not every patient is capable of achieving sufficient exercise to determine whether significant coronary artery disease is present. Another approach involves the use of maximum coronary vasodilatation.[20] In a patient with a subcritical coronary artery narrowing (e.g., subcritical left anterior descending coronary artery stenosis with normal right and left circumflex coronary arteries), resting coronary blood flow—and hence tracer uptake—will be similar in the distribution of the narrowed left anterior descending and the normal right and left circumflex coronary artery vascular regions. Following administration of a maximal coronary artery vasodilator such as ethyl adenosine or dipyridamole, coronary flow will increase greater than twofold in the normal vascular regions while flow in the distribution of the subcritically narrowed left anterior descending coronary artery will increase only slightly. Administration of ^{201}Tl during this period will reveal a defect in tracer uptake since flow and tracer uptake will be relatively greater in the distribution of the normal coronary arteries. This technic has proved clinically useful for the detection of patients with suspected ischemic heart disease in whom exercise is not feasible or desirable.[21]

CORONARY ARTERY SPASM

In patients with myocardial ischemia due to coronary artery spasm ^{201}Tl myocardial imaging has also been useful. Injection of ^{201}Tl during a spontaneous episode of Prinzmetal variant angina reveals a defect in tracer uptake which fills in on redistribution imaging several hours later.[22] The finding that the myocardial ^{201}Tl-to-background ratio falls to one during the episode of spontaneous variant angina lends indirect support to the premise that transmural ischemia due to coronary spasm is an important mechanism in these patients. Thallium-201 myocardial imaging may also be of value in provocative testing for coronary artery spasm. Ergonovine has been used to provoke coronary artery spasm during coronary arteriography and has also been used with ^{201}Tl myocardial imaging. In a patient with suspected coronary artery spasm ergonovine is injected in increasing increments with the patient beneath the detector of the scintillation camera. At the onset of precordial chest pain or after the maximum dose of ergonovine has been administered, ^{201}Tl should be injected and imaging begun within five minutes. If a defect in ^{201}Tl myocardial uptake is detected, imaging should be repeated approximately three hours later to determine whether the initial tracer defect was caused by coronary artery spasm as manifested by transient myocardial ischemia or by previous myocardial infarction.

MYOCARDIAL INFARCTION

Thallium-201 myocardial imaging has found wide use in the detection and evaluation of patients with acute myocardial infarction.[23] Nearly all patients who are evaluated within the first six hours of onset of symptoms will reveal transmural and nontransmural infarction if present. Increasing the time between onset of symptoms and testing moderately reduces the sensitivity for detection of transmural infarction and markedly reduces detection of nontransmural infarction. Wackers and associates detected nontransmural infarction in less than half of their stricken patients studied 24 hours or more after onset of symptoms.[23] Despite this loss of sensitivity with time, the sensitivity of ^{201}Tl myocardial imaging is greater than that of standard electrocardiography. In a study of 101 patients with angiographically significant coronary artery disease and a history of myocardial infarction, Bailey found a resting ^{201}Tl myocardial defect in a significantly greater number of patients than had shown definite electrocardiographic evidence of infarction.[24] Although more sensitive than the electrocardiogram for the detection of old myocardial infarction, smaller infarcts (< 7 gm) may be missed by myocardial imaging.[25]

The clinical value of an independent technique for detection of myocardial infarction is evident in patients entering the coronary care unit with a history suggestive of myocardial infarction but in whom the electrocardiogram is difficult or impossible to interpret because of left bundle branch block. The location and extent of the initial defect in ^{201}Tl uptake may also be of value in predicting the subsequent occurrence of congestive heart failure, complex ventricular arrhythmias, and death. In a recent study ^{201}Tl myocardial imaging was used to assess the effect of therapeutic interventions during the early stages of infarction.[9] In a double-blind study of nitroglycerin versus placebo in patients with acute myocardial infarction, patients receiving nitroglycerin had a significantly greater reduction in the extent of their defect in ^{201}Tl uptake from the time of initial imaging (on entry into the coronary care unit) to imaging one week later than did those receiving placebo. Caution should be taken when interpreting serial ^{201}Tl myocardial imaging and the effect of therapeutic interventions in patients with acute myocardial infarction since, as mentioned previously, the extent of the initial ^{201}Tl defect reflects the extent of ischemia as well as infarction.

Thallium-201 myocardial imaging is also of value in patients with cardiogenic shock. In the patient with anterior infarction and cardiogenic shock due to massive infarction of the left ventricle, the ^{201}Tl myocardial image will reveal a large defect in tracer uptake corresponding to the location and extent of the infarction. In patients with cardiogenic shock and electrocardiographic evidence of inferior or posterior infarction, ^{201}Tl myocardial imaging may reveal a number of clinically useful patterns. Patients with predominant left ventricular damage will be found to have a defect in tracer uptake on the inferior or posterolateral portion of the myocardium. Those with predominant right ventricular infarction may have a normal ^{201}Tl myocardial image, since in most instances only the left ventricular myocardium is visualized on the ^{201}Tl image. In others with right ventricular infarction increased right ventricular activity may be detected occasionally with a defect in tracer uptake. The presence of right ventricular infarction can be confirmed by hemodynamic monitoring, ^{99m}Tc-PYP imaging, or gated blood pool imaging. In patients with cardiogenic shock or severe left ventricular dysfunction due to papillary muscle rupture or rupture of the intraventricular septum, ^{201}Tl myocardial imaging alone or in combination with gated blood pool imaging is useful in decisions concerning surgical intervention. The prognosis for surgical repair is good in those with a relatively small defect in tracer uptake, while in those with an extensive defect in tracer uptake—suggesting massive myocardial infarction—the prognosis is poor, even if the patient initially survives surgery.

Initially it was thought that ^{201}Tl myocardial imaging would not be of value in distinguishing acute and old infarction. Recent experience suggests that this distinction may be possible. Pohost and coworkers have shown that redistribution of ^{201}Tl into ischemic areas occurs over time.[15] A defect in tracer uptake within an ischemic area occurs since the initial delivery to and extraction of tracer by the ischemic area is diminished. Over several hours the ischemic myocardium continues to extract ^{201}Tl from residual blood pool activity until the intracellular cation space and transport system is saturated. On reimaging of the initial defect in tracer uptake several hours later, uniform distribution of tracer may be encountered. This technique has been used to evaluate patients entering the coronary care unit with suspected acute myocardial infarction in whom the initial electrocardiograms and serum enzymes were nondiagnostic.[26] Patients with unstable angina pectoris studied within six hours of onset of symptoms who did not show evidence of subsequent infarction by serial electrocardiograms and serum enzymes were found to have a defect on initial tracer uptake which on reimaging three hours later could not be detected. In contrast, those with unstable angina pectoris who subsequently were found to have acute infarction on serial enzyme study were found to have a changing but persistent defect on reimaging. In patients with a history of old infarction and chest pain not associated with an episode of acute infarction an initial defect in tracer uptake could be detected which on reimaging several hours later was unchanged. Patients with chest pain of noncardiac origin were found to have a normal distribution of ^{201}Tl both on their initial and redistribution images. Although experience with this approach is as yet still limited, the early results appear promising. Application of this approach to the triage of patients with suspected acute myocardial infarction could result in considerable savings by minimizing the use of high-cost intensive care unit beds and by facilitating a more rapid application of appropriate diagnostic and therapeutic interventions to those with infarctions.

CARDIOMYOPATHY

Myocardial imaging with ^{201}Tl has also been found to be of value in the evaluation of patients with ischemic cardiomyopathy.[27] The diagnosis of ischemic cardiomyopathy is often difficult on clinical grounds alone. Patients with ischemic cardiomyopathy may present with congestive heart failure without a typical anginal history or electrocardiographic changes suggestive of previous myocardial infarction. These patients are often suspected of having idiopathic congestive cardiomyopathy. The presence of significant coronary artery disease as a cause of chronic congestive heart failure in these patients is recognized in many instances only at postmortem examination. Conversely, patients with idiopathic congestive cardiomyopathy may have a history of precordial chest discomfort and loss of R wave voltage across the anterior precordium suggestive of previous myocardial infarction.

Thallium-201 myocardial imaging may be useful in detecting ischemic cardiomyopathy and distinguishing it from idiopathic congestive cardiomyopathy. In ischemic cardiomyopathy a large defect in tracer uptake of greater than 40 percent of the image circumference can be detected reflecting massive myocardial replacement by fibrosis, while in idiopathic congestive cardiomyopathy ^{201}Tl uptake is relatively uniform. Small scattered defects in tracer uptake may be found in patients with idiopathic congestive cardiomyopathy, but the extent of these defects is usually less than 20 percent of the circumference of the image and does not account for the extent of left ventricular dysfunction. In one study of 13 patients with ischemic cardiomyopathy and 8 patients with idiopathic congestive cardiomyopathy, ^{201}Tl myocardial imaging permitted the correct diagnosis in 20 of the 21 patients. However, caution should be exercised in assigning an etiologic diagnosis to a given tracer defect. In the United States the most likely explanation for a defect in myocardial ^{201}Tl uptake is ischemic heart disease; in other parts of the world other etiologies may be of equal or greater importance. For example, a defect in ^{201}Tl uptake in patients in Argentina, Venezuela, and Brazil may be caused by Chagas disease of the myocardium. In fact, any process resulting in loss of viable myocardium (e.g., tumors, infiltrates, granuloma formation) may produce a defect in ^{201}Tl uptake. Thus, ^{201}Tl myocardial imaging is valuable in the evaluation of patients with sarcoidosis.[28] Patients with pulmonary sarcoidosis have been found to have large defects in tracer uptake of the left ventricle resulting from granuloma formation. Many of these patients are asymptomatic and were not suspected of having myocardial involvement on clinical grounds alone. Myocardial involvement may subsequently become evident with the occurrence of an intraventricular conduction defect, dysrhythmia, congestive heart failure, or sudden death. In other patients with pulmonary sarcoidosis ^{201}Tl myocardial imaging has suggested right ventricular hypertrophy and dilatation compatible with the diagnosis of cor pulmonale.

Thallium-201 myocardial imaging has also been found to be of value in the evaluation of patients with suspected hypertrophic obstructive cardiomyopathy.[29] In our initial experience with ^{201}Tl myocardial imaging in patients with suspected hypertrophic obstructive cardiomyopathy, a number of patients were encountered with either a false positive or false negative diagnosis by conventional M-mode echocardiography. Asymmetrical septal hypertrophy (ASH), a diagnostic clue for hypertrophic obstructive cardiomyopathy, also may be encountered in patients with right ventricular hypertrophy. Right ventricular hypertrophy may not be appreciated on standard M-mode echocardiography but is obvious on LAO ^{201}Tl myocardial images in which both the right and left ventricles can be independently assessed. Conversely, asymmetric septal hypertrophy which also can be missed occasionally by M-mode echocardiography depending upon technic and angulation of the transducer, may be detected on ^{201}Tl myocardial imaging by the finding of an intraventricular septum and apex which is thicker than the left ventricular wall. Gated ^{201}Tl myocardial imaging should be used in assessing the relative thickness of the myocardium, since with ungated imaging myocardial wall motion will blur the outline of the myocardial image such that hyperkinetic regions will appear relatively thicker than akinetic or hypokinetic areas.

CONCLUSION

From the experiences cited it is evident ^{201}Tl myocardial imaging is finding application in a wide variety of clinical situations. The clinical utility of ^{201}Tl myocardial imaging should increase even further with the additional development and application of tomographic myocardial imaging. Several approaches are being explored; the most promising for widespread clinical application is the use of new collimators with computer reconstruction of tomographic slices through the myocardium. Initial experience with a multiple-pinhole collimator and tomographic reconstruction has shown increased sensitivity for detection of myocardial ischemia and infarction.[30] The use of ^{201}Tl myocardial imaging alone or in combination with other radioactive tracer studies such as infarct-avid imaging with ^{99m}Tc-PYP and/or gated cardiac blood pool imaging adds a new dimension to cardiovascular diagnosis which at this early stage of development and application can only be partially appreciated.

REFERENCES

1. PITT, B., AND STRAUSS, H. W.: *Myocardial imaging in the noninvasive evaluation of patients with suspected ischemic heart disease.* Am. J. Cardiol. 37:797, 1976.

2. STRAUSS, H. W., AND PITT, B.: *Thallium-201 as a myocardial imaging agent.* Semin. Nucl. Med. 7:49, 1977.

3. STRAUSS, H. W., HARRISON, K., LANGAN, J., ET AL.: *Thallium-201 for myocardial imaging: Relation of thallium-201 to regional myocardial perfusion.* Circulation 51:641, 1975.

4. WEICH, H. F., STRAUSS, H. W., AND PITT, B.: *The extraction of thallium-201 by the myocardium.* Circulation 56:188, 1977.

5. HAMILTON, G. W., NARAHARA, K. A., YEE, H., ET AL.: *Myocardial imaging with thallium-201: Effect of cardiac drugs on myocardial images and absolute tissue distribution.* J. Nucl. Med. 19:10, 1978.

6. COOK, D. J., BAILEY, I., STRAUSS, H. W., ET AL.: *Thallium-201 for myocardial imaging: Appearance of the normal heart.* J. Nucl. Med. 17:184, 1976.

7. COHEN, H. A., BAIRD, M. G., ROULEAU, J. R., ET AL.: *Thallium-201 myocardial imaging in patients with pulmonary hypertension.* Circulation 54:790, 1976.

8. ROULEAU, J., GRIFFITH, L., STRAUSS, H. W., ET AL.: *Detection of diffuse coronary artery disease by quantification of thallium-201 myocardial images.* Circulation 51 and 52:111, 1975.

9. BECKER, L. C., BULKLEY, B. H., PITT, B., ET AL.: *Enhanced reduction of thallium-201 defects in acute myocardial infarction by nitroglycerin treatment: Initial results of a prospective randomized trial.* Clin. Res. 26:219A, 1978.

10. NEILL, C., KELLY, D., BAILEY, I., ET AL.: *Thallium-201 myocardial scintigraphy in single ventricle.* Circulation 53 and 54:174, 1976.

11. COWLEY, M. J., COGHLAN, H. C., AND LOGIC, J. R.: *Visualization of atrial myocardium with thallium-201.* J. Nucl. Med. 18:984, 1977.

12. BAILEY, I. K., ROULEAU, J. R., GRIFFITH, L. S. C., ET AL.: *Myocardial perfusion imaging to detect patients with single and multivessel disease.* Herz 2:135, 1977.

13. RITCHIE, J. L., ZARET, B. L., STRAUSS, H. W., ET AL.: *Myocardial imaging with thallium-201 at rest and exercise—a multicenter study: Coronary angiographic and electrocardiographic correlations.* J. Nucl. Med. 18:642, 1977.

14. BOTVINICK, E. H., TARADASH, M. R., SHAMES, D. M., ET AL.: *Thallium-201 myocardial perfusion scintigraphy for the clinical clarification of normal, abnormal and equivocal electrocardiographic stress test.* Am. J. Cardiol. 41:43, 1978.

15. POHOST, G. M., ZIR, L. M., MOORE, R. H., ET AL.: *Differentiation of transiently ischemic from infarcted myocardium by serial imaging after a single dose of thallium-201.* Circulation 55:294, 1977.

16. CARALIS, D. G., KENNEDY, H. L., BAILEY, I. K., ET AL.: *Thallium-201 myocardial perfusion scanning in the evaluation of asymptomatic patients with ischemic ST segment depression.* Am. J. Cardiol. 39:320, 1977.

17. COHEN, H., ROULEAU, J., GRIFFITH, L., ET AL.: *Myocardial perfusion and wall motion pre- and post coronary bypass surgery.* Circulation 51 and 52:170, 1975.

18. RITCHIE, J. L., NARAHARA, K. A., TROBAUGH, G. B., ET AL.: *Thallium-201 myocardial imaging before and after coronary revascularization.* Circulation 56:830, 1977.

19. BAILEY, I., BUROW, R., GRIFFITH, L. S. C., ET AL.: *Localizing value of thallium-201 myocardial perfusion imaging in coronary artery disease.* Am. J. Cardiol. 39:320, 1977.

20. STRAUSS, H. W., AND PITT, B.: *Noninvasive detection of subcritical coronary arterial narrowings with a coronary vasodilator and myocardial perfusion imaging.* Am. J. Cardiol. 39:403, 1977.

21. ALBRO, P. C., GOULD, K. L., WESTCOTT, R. J., ET AL.: *Noninvasive assessment of coronary disease in man by myocardial imaging during pharmacologic vasodilatation.* J. Nucl. Med. 19:743, 1978.

22. MASERI, A., PARODI, O., SEVERI, S., ET AL.: *Transient transmural reduction of myocardial blood flow demonstrated by thallium-201 scintigraphy as a cause of variant angina pectoris.* Circulation 54:280, 1976.

23. WACKERS, F. J., SOKOLE, E. B., SAMSON, G., ET AL.: *Value and limitations of thallium-201 scintigraphy in the acute phase of myocardial infarction.* N. Engl. J. Med. 295:1, 1976.

24. BAILEY, I. K.: *Unpublished observations.*

25. MUELLER, T. M., MARCUS, M. L., EHRHARDT, J. C., ET AL.: *Limitations of thallium-201 myocardial perfusion scintigrams.* Circulation 54:640, 1976.

26. POND, M., REHN, T., BUROW, R., ET AL.: *Early detection of myocardial infarction by serial thallium-201 imaging.* Circulation 56:893, 1977.

27. BULKLEY, B. H., HUTCHINS, G. M., BAILEY, I., ET AL.: *Thallium-201 imaging and gated cardiac blood pool scans in patients with ischemic and idiopathic cardiomyopathy: A clinical and pathologic study.* Circulation 55:753, 1977.

28. BULKLEY, B. H., ROULEAU, J. R., WHITAKER, J. Q., ET AL.: *The use of thallium-201 for myocardial perfusion imaging in sarcoid heart disease.* Chest 72:27, 1977.

29. BULKLEY, B. H., ROULEAU, J. R., STRAUSS, H. W., ET AL.: *Idiopathic hypertrophic subaortic stenosis: Detection by thallium-201 myocardial perfusion imaging.* N. Engl. J. Med. 293:1113, 1975.

30. VOGEL, R. A., KIRCH, D. L., LEFREE, M. T., ET AL.: *Improved diagnostic results of myocardial perfusion tomography using a new rapid inexpensive technique.* J. Nucl. Med. 19:731, 1978.

Technetium-99m Stannous Pyrophosphate "Hot Spot" Imaging to Detect Acute Myocardial Infarcts*

*James T. Willerson, M.D., Robert W. Parkey, M.D.,
L. Maximilian Buja, M.D., and Frederick J. Bonte, M.D.*

The recognition of acute myocardial infarction is not always easily accomplished.[1-4] Infarct recognition is especially difficult using electrocardiography in individuals who have had previous myocardial infarcts, in those with left bundle branch block, in those that have been cardioverted and in those with nontransmural (subendocardial) myocardial infarcts. Even the most sophisticated enzymatic techniques presently available have certain limitations in identifying the presence or absence of acute myocardial infarction in patients, including the following: (1) there is a temporal dependency in the ability of enzyme markers to detect severe myocardial cellular injury, and (2) certain clinical circumstances preclude using traditional enzyme techniques (creatine kinase and MB creatine kinase) determinations for infarct recognition (and in this regard the perioperative and postoperative setting after coronary artery revascularization should be mentioned). Therefore, it does seem important to develop additional noninvasive means of identifying the presence of myocardial necrosis which may be used in conjunction with standard techniques for infarct recognition. The additional noninvasive means that are developed to identify the presence or absence of acute myocardial infarcts should, ideally, be relatively noninvasive, as inexpensive as possible, and usable in diverse clinical situations; it should also be possible to repeat the test with whatever frequency is necessary to evaluate the possibility of extension of myocardial damage or relatively delayed evolution of it.

We have recently developed and extensively evaluated a "hot spot" myocardial imaging technique, technetium-99m stannous pyrophosphate (^{99m}Tc-PYP) for these purposes. This radionuclide myocardial imaging approach has been evaluated both in experimental animals and in patients at our institution over the past four years to test its ability to document the presence or absence of acute myocardial necrosis.[5-21]

The potential for using ^{99m}Tc-PYP myocardial scintigrams to recognize myocardial necrosis was first suggested by the observations of Shen and Jennings[22] and D'Agostino[23] when they demonstrated that calcium is deposited in crystalline and subcrystalline form in irreversibly damaged myocardial cells. These earlier observations led Bonte and associates to question whether ^{99m}Tc-PYP might be able to identify irreversibly damaged myocardial cells on the basis of pyrophosphate complexing with calcium in crystalline or precrystalline form.[5] We have carried out an extensive evaluation in patients and in large numbers of experimental animals at our institution over the past several years, and have examined the ability of this imaging agent to identify irreversibly damaged myocardium and to further

---

*Supported in part by NIH Ischemic SCOR HL-17669.

elucidate pathophysiological factors involved in the uptake and concentration of ^{99m}Tc-PYP in injured myocardium.[6-21] Our studies have identified that the three most important determinants of ^{99m}Tc-PYP myocardial uptake in the heart are (1) the presence of myocardial necrosis, (2) persistent residual collateral coronary blood flow into the area of irreversible myocardial damage, and (3) time elapsed after the onset of myocardial damage prior to obtaining the ^{99m}Tc-PYP myocardial scintigram.[5-21] These observations have been generally confirmed by others.[24-26] At least 3 grams of myocardial necrosis must be present to scintigraphically identify acute myocardial infarcts in experimental animals with fixed coronary occlusion utilizing ^{99m}Tc-PYP myocardial imaging.[17,18,21,27] Collateral coronary blood flow alterations are also particularly important determinants of ^{99m}Tc-PYP myocardial uptake in irreversibly damaged myocardium.[10,11,19] ^{99m}Tc-PYP myocardial uptake is greatest in myocardial regions in which coronary blood flow is reduced to levels of 20 to 40 percent of normal with fixed coronary occlusion and experimental myocardial infarction. In experimental animals, if the occlusion is positioned very proximally around the left anterior descending coronary artery, there is initially reduced ^{99m}Tc-PYP uptake in the center of the area of gross necrosis; the scintigraphic pattern that results has been referred to by us as the "doughnut" pattern of ^{99m}Tc-PYP uptake.[7,10,11] This same phenomenon occurs in man and appears also to be associated with very proximal severe narrowing or complete occlusion of the left anterior descending coronary artery (Figs. 1 and 2).[7,28] In experimental animals, when the coronary occlusion is positioned more distally around the left anterior descending coronary artery or when collateral flow is relatively abundant, the doughnut pattern does not occur. Instead, homogeneous and intense ^{99m}Tc-PYP myocardial uptake is observed when transmural myocardial infarction follows. The doughnut pattern generally becomes at least partially filled in during the first few days after acute myocardial infarction in both experimental animals and in man (Fig. 2); this presumably is the result of collateral blood flow reaching the more central portions of the infarct and reperfusing an area that previously received almost no blood flow. These observations emphasize that ^{99m}Tc-PYP uptake is dependent upon coronary flow and that there must be some flow reaching the area of myocardial damage in order for increased ^{99m}Tc-PYP myocardial uptake to be detectable. It seems likely that coronary flow–dependent uptake will also be a feature of every other "hot spot" myocardial imaging agent that is presently available or that may be developed in the future to image irreversibly damaged myocardial cells. Fortunately, in both experimental animals and in man our empirical observations utilizing the ^{99m}Tc-PYP imaging technique suggest that infarcts associated with no coronary flow over a period of several days either do not occur or occur so uncommonly as to not be clinically significant problems.[6-9,11-13] This is particularly true if serial myocardial scintigrams are obtained so that one may detect the occasional patients in whom collateral coronary flow adjustments

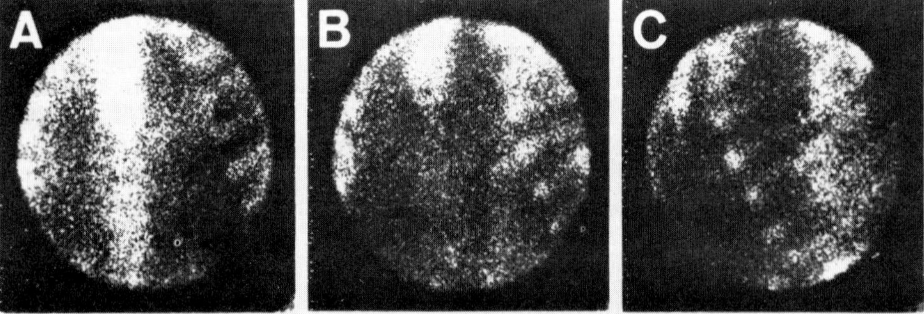

Figure 1. A negative ^{99m}Tc-PYP myocardial scintigram in the (A) anterior, (B) left anterior oblique, and (C) left lateral imaging projections. Note that there is a skeletal uptake of ^{99m}Tc-PYP but none in the region of the heart.

140

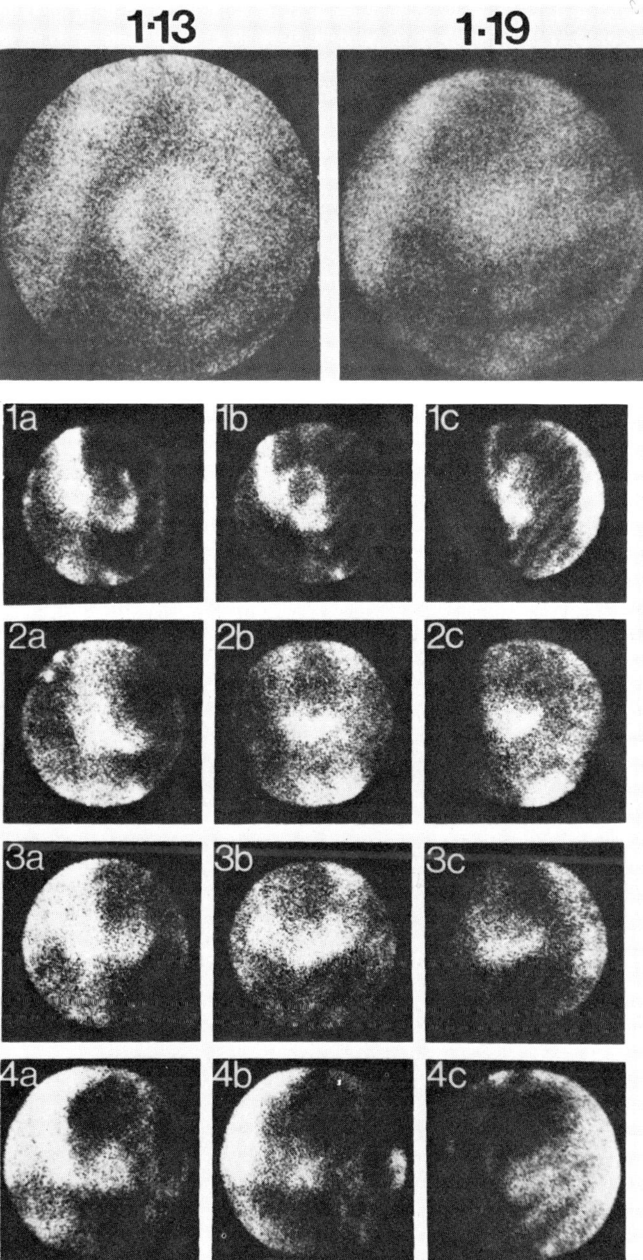

Figure 2. Above: Abnormal ^{99m}Tc-PYP myocardial scintigrams in the left anterior oblique projection. The scintigraphic "doughnut" pattern has a central area of decreased ^{99m}Tc-PYP uptake surrounded by an area of intense pyrophosphate uptake. The scintigram in the right panel demonstrates that the central portion of the doughnut lesion has become partially filled in six days after the initial scintigram (left panel) was obtained. Below: ^{99m}Tc-PYP scintigraphic display of representative images obtained with various types of acute transmural myocardial infarcts. In each set of scintigrams the figure on the left demonstrates the anterior view, the one in the middle, the left anterior oblique view and the one on the right the left lateral view. Panels 1a through 1c show a larger anterolateral ("doughnut lesion") myocardial infarct, 2a through 2c an inferolateral myocardial infarct, 3a through 3c an antero-inferolateral and posterior myocardial infarct, and 4a through 4c a true posterior myocardial infarct (From Rude et al.,[28] with permission of the American Heart Association, Inc.)

Table 1. Procedural steps in obtaining ^{99m}Tc-PYP myocardial scintigrams

1. Ensure proper labeling of pyrophosphate with ^{99m}Tc by checking either all or representative batches of material for free ^{99m}Tc.
2. Careful intravenous injection into a vein (not into plastic I.V. tubing) with good flush.
3. Obtain ^{99m}Tc-PYP myocardial images in *at least* 3 projections including the anterior, left anterior oblique and left lateral positions; images are routinely obtained at 1 hour following IV injection but if there is any question concerning whether a particular image represents a "blood pool" scintigram, then repeat images should be obtained at 2 and, if necessary, at 3 hours after IV injection. Blood pool scintigrams due to delayed renal clearance (severe renal disease, etc.) or severely deranged ventricular function would be expected to clear within that period of time. However, those due to in vivo labeling of red blood cells because of excessive free ^{99m}Tc-pertechnetate in the IV injectate would not clear within that time period and it is necessary to wait until the next day to repeat the test again.
4. In positioning the patient and interpreting the ^{99m}Tc-PYP myocardial scintigrams, it is necessary to be certain that the position of the heart in the chest is known and also that the imaging projections have allowed all of the various regions to be seen. In particular, the inferior or diaphragmatic portion of the heart must be visualized with certainty.

require longer periods of time for their development in order to allow visualization of the damaged myocardial area. We recommend that when using this technique a ^{99m}Tc-PYP myocardial scintigram be obtained within 24 to 48 hours after symptoms suggestive of infarction. However, if clinical suspicion of the presence of an infarct is strong and the initial scintigram is negative or equivocal, then the test should be repeated at least once, approximately 96 hours after the onset of symptoms suggestive of infarction. This serial imaging technique will allow detection of an occasional patient in whom the initial scintigram was negative and who develops a positive or abnormal ^{99m}Tc-PYP myocardial scintigram in a slightly more delayed fashion than is classical. Most patients that require longer periods of time for the development of an abnormal ^{99m}Tc-PYP myocardial scintigram after acute myocardial infarction will have severe intrinsic three-vessel coronary artery disease.

There are certain additional factors that are important for proper utilization of the ^{99m}Tc-PYP imaging test. Table 1 outlines those steps that we believe are important to perform the test properly. Figure 3 demonstrates our grading scheme for the interpretation of ^{99m}Tc-PYP myocardial scintigrams. We regard those ^{99m}Tc-PYP myocardial scintigrams that are "2 to 4+" as indicating increased ^{99m}Tc-PYP uptake and the presence of myocardial

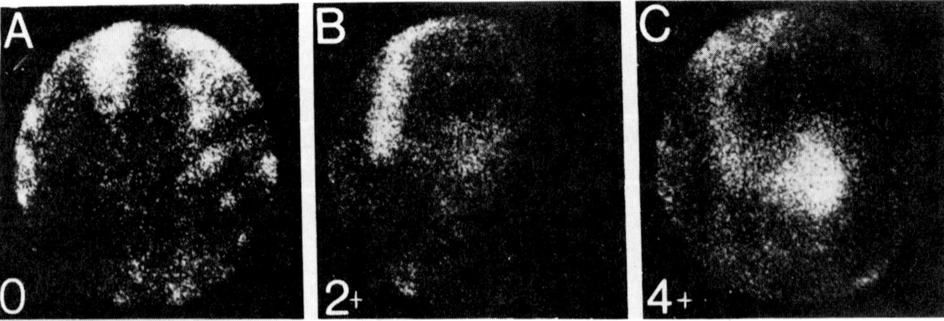

Figure 3. The grading scheme that we utilize for interpreting ^{99m}Tc-PYP myocardial scintigrams is shown. We consider ^{99m}Tc-PYP scintigrams that have 2 to 4+ uptake as being abnormal. (From Circulation[7] with permission of the American Heart Association, Inc.)

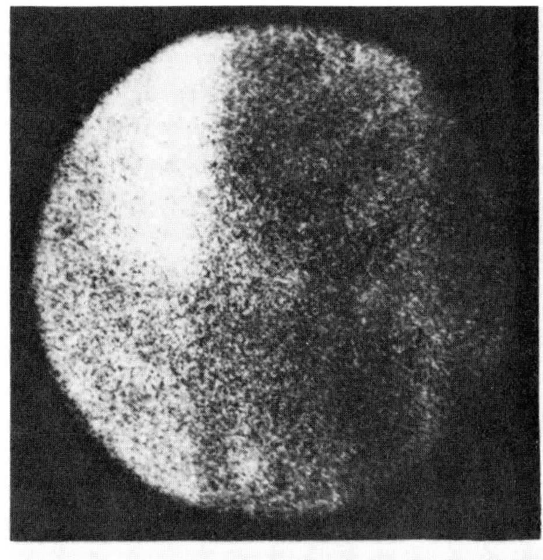

Figure 4. An abnormal ^{99m}Tc-PYP myocardial scintigram demonstrating 2+, faint and poorly localized uptake of ^{99m}Tc-PYP is evident. This scintigram was obtained from a patient with an acute subendocardial myocardial infarction.

necrosis. Two-plus ^{99m}Tc-PYP uptake is the faintest but definitely increased uptake in the region of the heart. Three-plus ^{99m}Tc-PYP uptake is uptake equivalent to that noted in bones, and 4+ uptake is greater than that in surrounding bony skeleton. In our experience most acute transmural myocardial infarcts evolve to a pattern of being 3 or 4+ but many nontransmural (subendocardial) infarcts demonstrate 2+ and poorly localized ^{99m}Tc-PYP uptake (Fig. 4).[7-9]

The proper utilization of this test requires that excessive ^{99m}Tc-pertechnetate not be present in the ^{99m}Tc-PYP injectate.[29,31] If this occurs the ^{99m}Tc-pertechnetate may be taken up in red cells and result in a persistent radionuclide blood pool scintigram; this might be erroneously interpreted as an abnormal ^{99m}Tc-PYP myocardial scintigram indicative of myocardial necrosis. There are several different procedures for testing the ^{99m}Tc-PYP injectate for excessive free ^{99m}Tc-pertechnetate; the one that we utilize is outlined in Table 2.

We have found that dead and dying cells are the primary cell types that take up increased amounts of ^{99m}Tc-PYP following experimental myocardial infarction. This implies that any circumstance in which a significant region of heart muscle is irreversibly injured could result in an abnormal ^{99m}Tc-PYP myocardial scintigram. Such clinical circumstances would include not only acute myocardial infarcts but also viral myocarditis, cardioversion,[15]

Table 2. Testing ^{99m}Tc-PYP for excessive free ^{99m}Tc-pertechnetate

---

1. Silica gel chromatographic paper cut into 5 by 20 mm strips and dried in desiccator.
2. Small drop of ^{99m}Tc-PYP placed one half inch on opposite end of paper held with forceps; allow drop to dry in N_2 atmosphere.
3. Set dried paper strip in a bottle with approximately one quarter inch of methylethyl ketone; drop remains above the level of MEK.
4. After 30 seconds remove strip and allow it to dry. Then, either image the strip, or cut paper into pieces and count it in a well counter.
5. ^{99m}Tc-PYP batch should be discarded if there is significant migration of ^{99m}Tc-pertechnetate away from the drop.

---

metastatic tumor to the heart,[16] trauma, etc. This should be kept in mind as ^{99m}Tc-PYP myocardial scintigrams are used to evaluate patients with chest pain.

As noted previously, the timing between the onset of infarction and obtaining the ^{99m}Tc-PYP myocardial scintigram is of considerable importance. In experimental animals we have found that the scintigrams become positive within 10 to 12 hours after acute myocardial infarcts, but that they become increasingly positive during the first 24 to 72 hours following the event.[5,10,11] These same observations have also been made in patients with acute myocardial infarcts in whom the time periods in which the scintigram becomes abnormal initially and during which it becomes increasingly abnormal are essentially the same as those documented in experimental animals (Fig. 5).[7,9,29] With fixed coronary occlusion and acute myocardial infarcts in both experimental animals and in man, the ^{99m}Tc-PYP myocardial scintigrams ordinarily become negative approximately 10 to 14 days following the infarct. The development of a positive ^{99m}Tc-PYP myocardial scintigram and its evolution to a negative one correlates temporally and topographically with calcium deposition and resorption in the damaged myocardial area.[10] These observations do not necessarily exclude the possibility that mechanisms other than calcium binding contribute to concentration of ^{99m}Tc-PYP in irreversibly damaged myocardial cells. However, our recent studies suggest that a major mechanism involved in the ability of this imaging technique to identify acute myocardial infarcts is related to complexing of pyrophosphate with calcium in crystalline and subcrystalline form within the damaged myocardial tissue.[19]

In many patients and in most animals the ^{99m}Tc-PYP myocardial scintigrams fade and/or become negative approximately six days after acute myocardial infarction.[6-9]

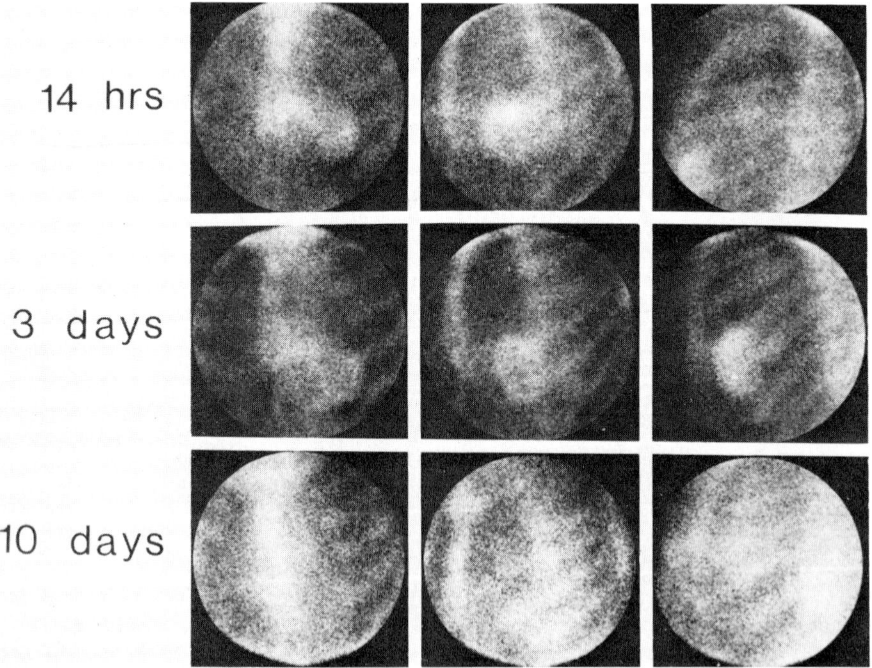

Figure 5. Temporal relationships involved in a properly performed imaging test. The ^{99m}Tc-PYP myocardial scintigram obtained approximately 14 hours after the onset of symptoms suggestive of infarction is positive (top panels), but the scintigram obtained 72 hours after the onset of infarction is even more definitely positive in all three imaging projections. The ^{99m}Tc-PYP myocardial scintigram obtained from the same patient 10 days postinfarction has now begun to fade and is no longer clearly positive.

144

However, in a subgroup of patients with acute myocardial infarcts the ^{99m}Tc-PYP myocardial scintigrams remain persistenly abnormal for weeks or months following the infarct (Fig. 6).[30] The reason(s) for persistently abnormal scintigrams in approximately 10 to 40 percent of patients with acute myocardial infarcts are presently uncertain, but dystrophic cardiac calcification and/or continued cellular evolution of myocardial injury at the margins of the infarct could be responsible. Indeed, in a small subgroup of patients that died months following their myocardial infarcts and had persistently positive scintigrams, microscopic evidence of continued cellular evolution of myocardial injury at the margins of the infarct was present at the time of their postmortem examination.[30] In our experience, persistently positive ^{99m}Tc-PYP myocardial scintigrams for a period of three months or longer following infarction appear to be a relatively poor prognostic sign in the sense that at least some of these patients continue to have frequent limiting chest pain and many are readmitted on numerous occasions during the several months' followup period to "rule out" myocardial infarction. A small percentage of these patients with persistently positive scintigrams have severe congestive heart failure rather than recurrent chest pain as their major persistent clinical problem. Since persistently positive ^{99m}Tc-PYP myocardial scintigrams occur in a certain percentage of patients following infarction, and since ^{99m}Tc-PYP myocardial scintigrams become increasingly positive during the first 72 hours following infarction and less positive (even if not absolutely negative) 6 to 10 days after infarction, it is imperative that serial myocardial images be obtained to derive maximal diagnostic information from this test. Serial ^{99m}Tc-PYP myocardial scintigrams that become increasingly positive and then fade (even without becoming absolutely negative over a period of time) should identify acute myocardial necrosis in contrast to a persistently positive scintigram which is a consequence of an earlier infarction, ongoing myocardial damage, dystrophic cardiac calcification from whatever etiology, etc.

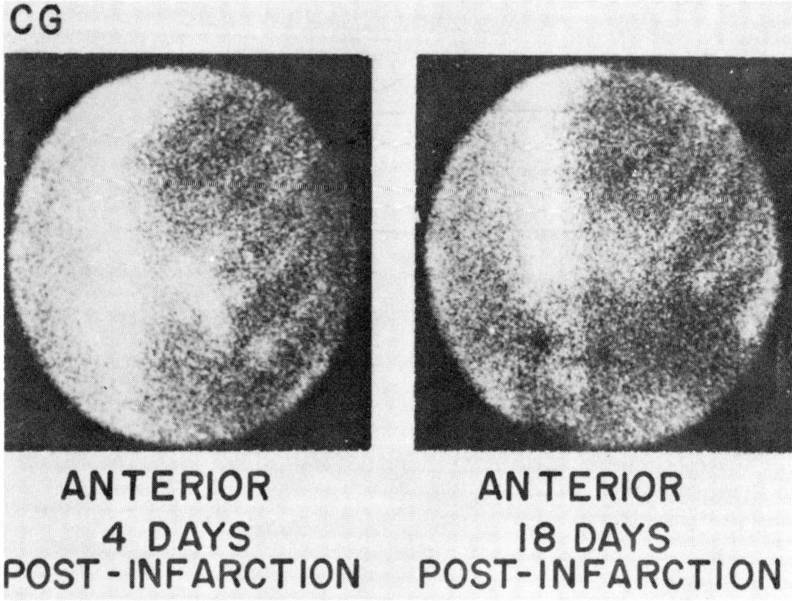

Figure 6. A persistently positive ^{99m}Tc-PYP myocardial scintigram is shown in a patient with an acute anterolateral myocardial infarction. The left hand panel demonstrates the area of increased uptake of ^{99m}Tc-PYP in the anterior region of the heart 4 days postinfarction. A subsequent ^{99m}Tc-PYP myocardial scintigram obtained 18 days later is still positive although the pattern of ^{99m}Tc-PYP uptake has changed.

It should also be emphasized that it is presently possible to accurately size acute anterior or anterolateral transmural infarcts in experimental animals and in man using the ^{99m}Tc-PYP myocardial imaging technique.[17,18,21,27] The ^{99m}Tc-PYP myocardial scintigraphic estimates of infarct size with acute anterior or anterolateral infarcts in experimental animals have a close relationship to histologically determined infarct weight (Fig. 7).[17,18,21,27] It is possible to size acute anterior or anterolateral infarcts with planimetric area (two-dimensional) measurements of infarct size; with currently available imaging equipment and the corresponding projections (anterior, left anterior oblique, and left lateral) one can see the overall extent of the infarct with transmural infarcts that are positioned in this location. However, it is not yet possible to precisely size acute nontransmural infarcts or acute inferior infarcts because the overall extent of these infarcts is not seen in a two-dimensional projection. We are optimistic that the development of tomographic cameras which allow three-dimensional estimates of infarct size with non-positron-emitting radionuclides (such as ^{99m}Tc-PYP) will allow an accurate estimation of the size of even these types of infarcts, but this is a matter that will have to be tested as equipment developments occur in the future. Alternatively, it may be possible using computer techniques to simulate three-dimensional reconstruction of infarct size and provide a better estimate of the size of the lesion.[21] Whether or not this approach will work for sizing acute inferior and nontransmural infarcts is a subject that is presently under study.

It should be noted that others have described additional "hot spot" imaging techniques which may be utilized to detect acute myocardial infarcts.[32-34] Included among these other "hot spot" imaging techniques are technetium-labeled tetracycline and glucoheptonate[32,33] and a labeled antibody to cardiac myosin.[34] Each of these approaches has been shown to be useful for infarct recognition in experimental animals, but whether any or all of them will ultimately find major clinical usefulness for this purpose remains to be determined.

Our experience indicates that ^{99m}Tc-PYP myocardial imaging is a useful addition to the noninvasive armamentarium for recognizing and generally localizing acute myocardial

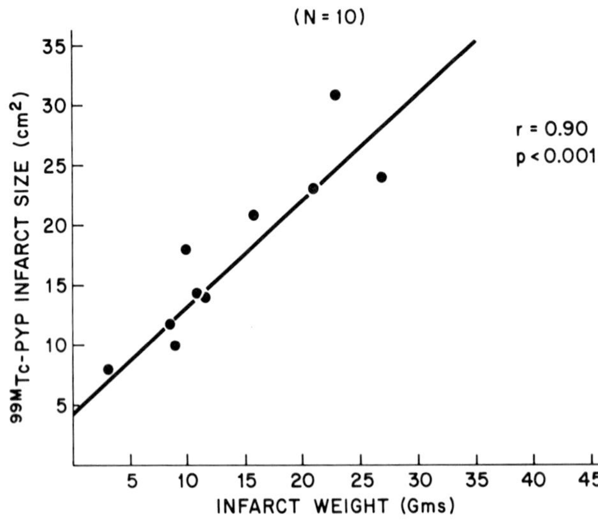

Figure 7. The significant relationship between maximal projected ^{99m}Tc-PYP infarct size in cm^2 (vertical axis) and histological infarct weight in grams (horizontal axis) is shown. This relationship was evaluated in 10 dogs with acute anterior or anterolateral infarcts produced by proximal left anterior descending coronary artery ligation. The animals had their acute myocardial infarcts for approximately 30 hours prior to the ^{99m}Tc-PYP myocardial scintigrams being obtained. (From Willerson et al.,[18] with permission of Cardiovascular Research.)

infarcts. Others have also found this myocardial imaging technique to be useful in identifying the presence of acute myocardial infarcts,[35-39] but a few have experienced difficulty.[40,41] In addition, this imaging technique is able to estimate infarct size for acute anterior or anterolateral transmural infarcts. It also promises to be helpful in infarct sizing with inferior and nontransmural infarcts as additional equipment development occurs allowing three-dimensional estimates of infarct size for these lesions. There are several important technical details that must be adhered to in the proper utilization of the test, but if these are observed, this imaging technique should be generally useful for the purposes cited.

REFERENCES

1. ROBERTS, W. C.: *The coronary arteries and left ventricle in clinically isolated angina pectoris. A necropsy analysis.* Circulation 54:388, 1976.

2. ALISON, H. W., MORASKI, R. E., MANTLE, J. A., ET AL.: *Coronary anatomy and arteriography in patients with unstable angina pectoris.* Am. J. Cardiol. 35:118, 1975.

3. ELIOT, R. S., AND EDWARDS, J. E.: *Pathology of coronary atherosclerosis and its complications.* In Hurst, J. W. (ed.): *The Heart.* McGraw-Hill, 1974, p. 1003.

4. POLINER, L. R., BUJA, L. M., PARKEY, R. W., ET AL.: *Clinicopathologic correlates of technetium-99m stannous pyrophosphate myocardial scintigraphy in patients.* Circulation 54(Suppl. II):80, 1975.

5. BONTE, F. J., PARKEY, R. W., GRAHAM, K. D., ET AL.: *A new method for radionuclide imaging of acute myocardial infarcts.* Radiology 110:473, 1974.

6. PARKEY, R. W., BONTE, F. J., MEYER, S. L., ET AL.: *A new method for radionuclide imaging of acute myocardial infarction in humans.* Circulation 50:540, 1974.

7. WILLERSON, J. T., PARKEY, R. W., BONTE, F. J., ET AL.: *Technetium stannous pyrophosphate myocardial scintigrams in patients with chest pain of varying etiology.* Circulation 51:1046, 1975.

8. WILLERSON, J. T., PARKEY, R. W., BONTE, F. J., ET AL.: *Acute subendocardial myocardial infarcts detected by technetium-99m stannous pyrophosphate myocardial scintigrams.* Circulation 51:436, 1975.

9. WILLERSON, J. T., PARKEY, R. W., BONTE, F. J., ET AL.: *Technetium-99m stannous pyrophosphate myocardial scintigrapy: A new method of proven value for the diagnosis and localization of acute myocardial infarcts and for the detection of infarct extension in patients.* Tex. Med. 72:61, 1976.

10. BUJA, L. M., PARKEY, R. W., DEES, J. H., ET AL.: *Morphologic correlates of technetium-99m stannous pyrophosphate imaging of acute myocardial infarcts in dogs.* Circulation 52:596, 1975.

11. BUJA, L. M., PARKEY, R. W., STOKELY, E. M., ET AL.: *Pathophysiology of technetium-99m stannous pyrophosphate and thallium-201 scintigraphy of acute anterior myocardial infarcts in dogs.* J. Clin. Invest. 57:1508, 1976.

12. PLATT, M. R., PARKEY, R. W., WILLERSON, J. T., ET AL.: *Technetium-99m stannous pyrophosphate myocardial scintigrams in the recognition of myocardial infarction in patients undergoing coronary artery revascularization.* Ann. Thorac. Surg. 21:311, 1976.

13. PLATT, M. R., MILLS, L. J., PARKEY, R. W., ET AL.: *Perioperative myocardial infarction diagnosed by technetium-99m stannous pyrophosphate myocardial scintigrams.* Circulation 54(Suppl. III):24, 1976.

14. DONSKY, M. S., CURRY, G. C., PARKEY, R. W., ET AL.: *Unstable angina pectoris: Clinical, angiographic, and myocardial scintigraphic observations.* Br. Heart J. 38:257, 1976.

15. PUGH, B. R., BUJA, L. M., PARKEY, R. W., ET AL.: *Cardioversion and its potential role in the production of "false positive" technetium-99m stannous pyrophosphate myocardial scintigrams.* Circulation 54:399, 1976.

16. HARFORD, W., WEINBERG, M., BUJA, L. M., ET AL.: *Positive technetium-99m stannous pyrophosphate myocardial scintigram in a patient with carcinoma of the lung.* Radiology 122:747, 1977.

17. STOKELY, E. M., BUJA, L. M., LEWIS, S. E., ET AL.: *Measurement of acute myocardial infarcts in dogs with technetium-99m stannous pyrophosphate scintigrams.* J. Nucl. Med. 17:1, 1976.

18. WILLERSON, J. T., PARKEY, R. W., STOKELY, E. M., ET AL.: *Infarct sizing with technetium-99m stannous pyrophosphate scintigraphy in dogs and man; the relationship between scintigraphic and precordial mapping estimates of infarct size in patients.* Cardiovasc. Res. 11:291, 1977.

19. BUJA, L. M., TOFE, A. J., KULKARNI, P. V., ET AL.: *Sites and mechanisms of localization of technetium-99m phosphorus radiopharmaceuticals in acute myocardial infarcts and other tissues.* J. Clin. Invest. 60:724, 1977.

20. PARKEY, R. W., BONTE, F. J., STOKELY, E. M., ET AL.: *Acute myocardial infarction imaged with technetium 99m stannous pyrophosphate and thallium 201: A clinical evaluation.* J. Nucl. Med. 17:771, 1976.

21. Lewis, M., Buja, L. M., Saffer, S., et al.: *Experimental infarct sizing utilizing computer processing and a three-dimensional model.* Science 197:167, 1977.

22. Shen, A. C., and Jennings, R. B.: *Myocardial calcium and magnesium in acute ischemic injury.* Am. J. Pathol. 67:417, 1972.

23. D'Agostino, A. N.: *An electron microscopic study of cardiac necrosis produced by 9 a-fluorocortisol and sodium phosphate.* Am. J. Pathol. 45:633, 1964.

24. Perez, L. A.: *Clinical experience: Technetium 99m labeled phosphates in myocardial imaging.* Clin. Nucl. Med. 1:2, 1976.

25. Bruno, F. P., Cobb, F. R., Rivas, F., et al.: *Evaluation of* ^{99m}*technetium stannous pyrophosphate as an imaging agent in acute myocardial infarction.* Circulation 54:71, 1976.

26. Zaret, B. L., DiCola, V. C., Donabedian, R. K., et al.: *Dual radionuclide study of myocardial infarction. Relationship between myocardial uptake of potassium-43, technetium-99m stannous pyrophosphate, regional myocardial blood flow and creatine phosphokinase depletion.* Circulation 53:422, 1976.

27. Poliner, L. R., Buja, L. M., Parkey, R. W., et al.: *Comparative evaluation of several different noninvasive methods of infarct sizing during experimental myocardial infarction.* J. Nucl. Med. 18:517, 1977.

28. Rude, R., Parkey, R. W., Bonte, F. J., et al.: *Clinical implications of the "doughnut" pattern of uptake in myocardial imaging with technetium-99m stannous pyrophosphate.* Circulation 56 (Suppl. III):561, 1977.

29. Willerson, J. T., Parkey, R. W., Buja, L. M., et al.: *Are technetium-99m stannous pyrophosphate myocardial scintigrams clinically useful?* Clin. Nucl. Med. 2:137, 1977.

30. Buja, L. M., Poliner, L., Parkey, R. W., et al.: *Clinicopathologic findings in patients with persistently positive technetium-99m stannous pyrophosphate myocardial scintigrams and myocytolytic degeneration after acute myocardial infarction.* Circulation 56:1016, 1977.

31. Stokely, E. M., Parkey, R. W., Bonte, F. J., et al.: *Gated blood pool imaging following technetium-99m phosphate scintigraphy.* Radiology 120:433, 1976.

32. Holman, B. L., Lesch, M., Zweiman, F. G., et al.: *Detection and sizing of acute myocardial infarcts with* ^{99m}*Tc(Sn) tetracycline.* N. Engl. J. Med. 291:159, 1974.

33. Rossman, D. J., Strauss, H. W., Siegel, M. E., et al.: *Accumulation of* ^{99m}*Tc-glucoheptonate in acutely infarcted myocardium.* J. Nucl. Med. 16:875, 1975.

34. Beller, G. A., Khaw, B. A., Haber, E., et al.: *Localization of radiolabeled cardiac myosin-specific antibody in myocardial infarcts: comparison with technetium-99m stannous pyrophosphate.* Circulation 55:74, 1977.

35. Weber, P. M., VanDyke, D., dos Remedios, L. V., et al.: *Radionuclide tomographic scanning in acute myocardial infarction (AMI).* J. Nucl. Med. 16:581, 1975.

36. Mason, J. W., Myers, R. W., Kriss, J. P., et al.: *Technetium-99m pyrophosphate myocardial scanning in coronary surgery patients.* Circulation 52 (Suppl. II):54, 1975.

37. Berman, D. S., Amsterdam, E. A., Salel, A. F., et al.: *Improved diagnostic assessment of acute myocardial infarction: sensitivity and specificity of Tc-99m pyrophosphate scintigraphy.* J. Nucl. Med. 17:523, 1976.

38. Campeau, R. J., Gottlieb, S., Chandarplapaty, S. K. C., et al.: *Accuracy of technetium-99m labelled phosphates for detection of acute myocardial infarction.* J. Nucl. Med. 16:518, 1975.

39. Perez, L. A.: *Clinical experience: technetium-99m labeled phosphates in myocardial imaging.* Clin. Nucl. Med. 1:2, 1976.

40. Karunaratne, H. B., Walsh, W. F., Fill, H. R., et al.: *Technetium-99m pryophosphate myocardial scintigraphy in patients with chest pain—lack of diagnostic specificity.* J. Nucl. Med. 17:524, 1976.

41. Ahmad, M., Dubiel, J. P., Verdon, T. A., Jr., et al.: *Technetium-99m stannous pyrophosphate myocardial imaging in patients with and without left ventricular aneurysm.* Circulation 53:833, 1976.

The Role of Radionuclide Techniques in Patients with Myocardial Disease

Gerald M. Pohost, M.D., John T. Fallon, M.D., and H. William Strauss, M.D.

Disorders affecting heart muscle are classified in several ways. One approach is to use a *functional classification* based on the pathophysiological findings as originally proposed by Goodwin.[1] Goodwin divided the cardiomyopathies into three classes: (1) congestive (or dilated) cardiomyopathy, with ventricular dilatation in excess of hypertrophy and with congestive heart failure due to poor systolic ejection function; (2) restrictive cardiomyopathy, with reduced ventricular compliance resulting in diminished filling ability and a syndrome resembling constrictive pericarditis; and (3) hypertrophic cardiomyopathy, characterized generally by asymmetric hypertrophy of the left ventricle.

Myocardial disease may also be classified according to whether its cause is *secondary* to a systemic disease process (e.g., sarcoidosis) or to extramyocardial cardiac disease (e.g., coronary artery disease), or *primary*, in which case direct involvement of the myocardium by a pathologic process leads to dysfunction (e.g., alcoholic cardiomyopathy) (Table 1).[2] Radionuclide methods are helpful in determining both the physiology and etiology of a variety of disorders affecting the myocardium (Table 2).

RADIONUCLIDE IMAGING TECHNIQUES USEFUL IN EVALUATION OF THE CARDIOMYOPATHIES

Radionuclide techniques of value in the diagnostic evaluation of patients with cardiomyopathy include the gated cardiac blood pool scan, the "first-pass" radionuclide angiogram, and myocardial imaging with thallium-201. The gated or multigated cardiac blood pool scan is performed following the intravenous injection of 15 to 20 mCi of ^{99m}Tc labeled human serum albumin or red blood cells to evaluate both chamber size and function. Images are usually recorded in synchrony with the patient's heartbeat either at two points in the cardiac cycle, end-systole (shortly after the peak of the T wave) and end-diastole (near the peak of the QRS complex), or encompassing the entire cardiac cycle with multiple images. From these images left ventricular ejection fraction and regional wall motion can be reliably determined.[3] Another approach to determination of left ventricular ejection fraction is the first-pass radionuclide angiogram.[4]

Thallium-201 is a monovalent cationic radiopharmaceutical with many biological properties similar to potassium. Following intravenous administration of 1.5 to 2.0 mCi, the initial distribution of ^{201}Tl in the heart is related to blood flow in the myocardium.[5] Frequently, the ^{201}Tl scan reveals a homogeneous distribution in patients with myocardial disease, suggesting diffuse rather than focal involvement. Focal regions of scar or infiltrate appear as discrete zones of decreased tracer concentration, i.e., "cold spots" on gamma

Table 1. Classification of most common disorders affecting heart muscle

Myocardial Involvement (etiology)	Congestive (dilated) CCM	Restrictive RCM Infiltrative	Restrictive RCM Obliterative	Hypertrophic HCM
PRIMARY				
Idiopathic	Idiopathic CCM Postpartum cardio-myopathy		Endocardial fibroelastosis	Obstructive HCM
	Endocardial fibroelas-tosis (dilated type)		Endomyocardial fibrosis —Davies' disease —Loeffler's disease	Nonobstructive HCM
Toxic	Alcohol Heavy metals (e.g., arsenic, cobalt) Adriamycin Diphtheria			
Inflammatory	Chagas' disease Rheumatic myocarditis			
SECONDARY (associated with systemic disease)				
Inflammatory	Infective Viral Bacterial Mycoplasma Rickettsial (e.g., Q fever) Noninfective Lupus erythematosus Sarcoidosis Polyarteritis nodosa	Scleroderma		
Metabolic	Thiamine deficiency Acromegaly	Amyloidosis Hemochroma-tosis Glycogen storage disease		
Neuromuscular	Friedreich's ataxia Muscular dystrophy			
SECONDARY (associated with cardiac disease)				
Valvular	Chronic volume and/or pressure overload		Endocardial fibroelastosis	
Congenital				
Hypertensive				
Coronary	"Ischemic cardio-myopathy"			
Pericardial		Constrictive pericarditis		

Table 2. Characteristics of cardiomyopathies and disorders simulating cardiomyopathies which can be detected by radionuclide imaging

	CCM *(Figs. 2 & 3)*	RCM *(infiltrative)*	RCM *(obliterative)*	HCM *(Figs. 5–7)*	*LV Pressure Overload*	*LV Volume Overload (Fig. 2)*	*Coronary Heart Disease (Fig. 4)*
Blood Pool Imaging							
LV size	↑	N	N,↓	N,↓	N	↑	↑
LV ejection fraction	↓	↓	↓	N,↑	N	N	↑
IVS thickness	N,↓	N,↑	N,↑	↑	↑	N,↓	N,↓
IVS DUST or flattening	0	0	0	∠ (50%)	0	0	0
Myocardial Distribution of ^{201}Tl							
Normal	∠	∠	∠	∠	∠	∠	0
Inhomogeneous	∠	∠	∠	0	∠	∠	0
Defects (≥40% of perimeter)	0	∠	0	0	0	0	∠
ASH	0	0	0	∠	0	0	0

↑ = increased; N = normal; ↓ = decreased; 0 = unusual; ∠ = present; LV = left ventricle; IVS = interventricular septum; DUST = disproportionate upper septal thickening; ASH = asymmetric septal hypertrophy; CCM = congestive cardiomyopathy; RCM = restrictive cardiomyopathy; HCM = hypertrophic cardiomyopathy.

camera images.[6] For example, myocardial involvement with sarcoid may produce multifocal regions of reduced thallium activity corresponding to regions of granulomata or scar.[7]

By depicting cardiac chamber size and function as well as configuration and thickness of the interventricular septum the gated blood pool scan or radionuclide angiogram and ^{201}Tl myocardial scan will suggest the functional class of the patient with heart muscle disease (Fig. 1, Table 2). Once the functional class is suggested, a number of etiological possibilities should be considered (Table 1).

CONGESTIVE CARDIOMYOPATHY (CCM)

Clinical findings of right and left heart failure are characteristic of congestive cardiomyopathy.[8] In general, the pathologic findings are nonspecific. Heart weight is increased, although ventricular dilatation in excess of myocardial hypertrophy is often apparent. Apical thrombi are common in both ventricles. The coronary arteries demonstrate no significant disease. Microscopic examination usually reveals diffuse interstitial fibrosis as well as foci of replacement fibrosis. Alcoholic cardiomyopathy[9] and postpartum cardiomyopathy[10] are examples of congestive cardiomyopathy (CCM). Other examples of CCM are given in Table 1.

An example of a gated cardiac blood pool scan from a patient with congestive cardiomyopathy is illustrated in Figure 2. Patients with CCM demonstrate biventricular dilatation and dysfunction. The left ventricle is usually more severely dilated than the right, and the left ventricular ejection fraction is severely reduced (typically less than 30 percent).

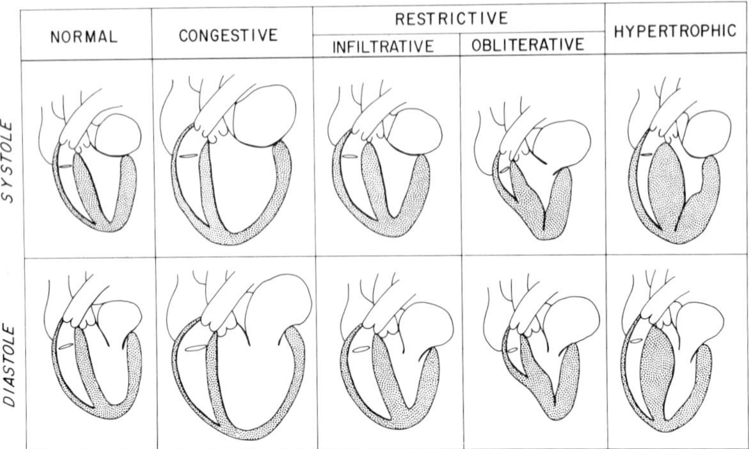

Figure 1. Schematic illustration of a functional classification of the cardiomyopathies in a view equivalent to the 40 to 50 degree left anterior oblique radionuclide images in end-systole and end-diastole. Size and contractile function of the ventricles and configuration of the chambers and interventricular septum provide characteristics which help define the functional class (Table 2).

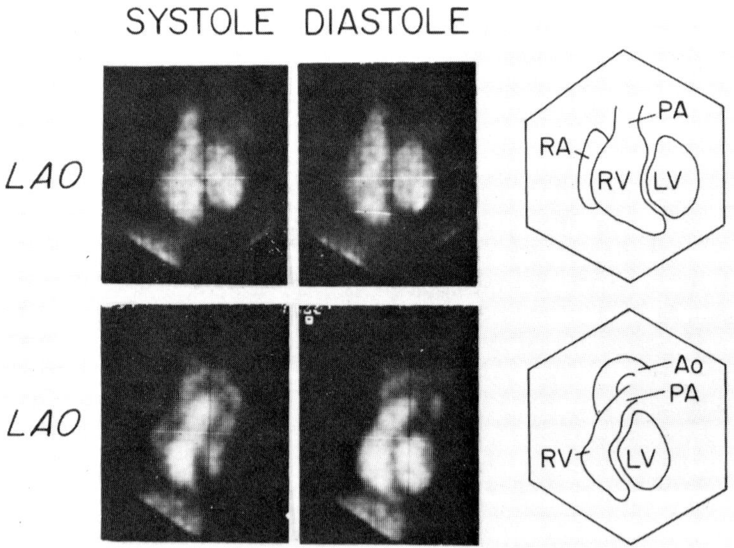

Figure 2. End-systolic and end-diastolic images with adjacent schematic diagrams of the end-diastolic images in the 50 degree left anterior oblique (LAO) projection in two patients with left heart failure and a murmur of mitral regurgitation. In the upper row, biventricular dilatation and reduced function are demonstrated consistent with congestive cardiomyopathy. The lower row shows dilatation of the left ventricle (LV) and a normal sized right ventricle (RV). The left ventricular ejection fraction is normal. This constellation of findings (i.e., dilated left ventricle and normal left ventricular ejection fraction) is consistent with LV volume overload and preserved myocardial function. The latter case suggests a surgically remediable etiology of congestive heart failure; the former case is probably not an appropriate candidate for valve replacement surgery. (RA = right atrium; PA = pulmonary artery; Ao = aorta)

Right ventricular dysfunction is related to both myocardial involvement and increased pulmonary artery pressure due to left ventricular failure. Wall motion is concentrically reduced, although the apical segment may be akinetic or dyskinetic. The anterobasal segment and the basal septum demonstrate reduced wall motion in CCM, while these segments frequently move normally in ischemic heart disease. Bulkley and associates have suggested that the percentage of the circumference involved by akinesis or dyskinesis is of diagnostic value in differentiating between CCM and ischemic heart disease.[11] Circumferential involvement of more than 40 percent in any one oblique view was identified in 84 percent of patients with "ischemic cardiomyopathy" but in only 25 percent of patients with CCM. Occasionally, akinesis and flattening of the left ventricular apical segment can be demonstrated and are suggestive of mural thrombus.

The left atrium is usually difficult to define on a normal gated blood pool study. However, biatrial dilatation in CCM frequently permits ready visualization of the left atrium in the 40 to 50 degree left anterior oblique projection and of the right atrium in the anterior projection.

Myocardial imaging with ^{201}Tl usually demonstrates left ventricular dilatation and either homogeneous or diffusely inhomogeneous uptake (Fig. 3). Severe left ventricular dysfunction resulting from coronary artery disease and multiple myocardial infarctions but without obvious ventricular aneurysm has been called "ischemic cardiomyopathy," and may have a clinical picture indistinguishable from CCM. Bulkley and coworkers have reported the utility of ^{201}Tl myocardial imaging in distinguishing between "ischemic" and congestive cardiomyopathy.[11] All patients with angiographically and/or autopsy documented "ischemic cardiomyopathy" had a defect on ^{201}Tl scan occupying over 40 percent of the circumference of the left ventricular image in any one projection. On the other hand, all but one patient with idiopathic congestive cardiomyopathy demonstrated a ^{201}Tl defect of under 20 percent of the circumference in any one projection. [201]Tl images from a normal subject

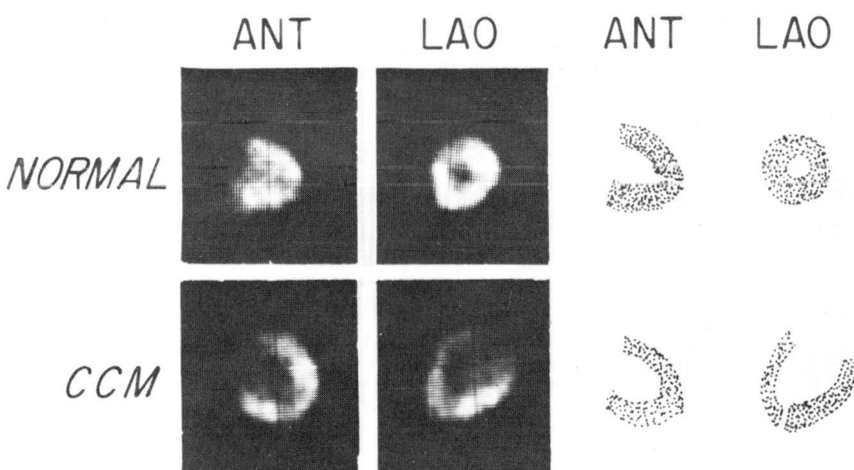

Figure 3. Anterior (ANT) and 50 degree left anterior oblique (LAO) thallium-201 myocardial images and corresponding schematic diagrams in a normal patient (upper panels) and a patient with congestive cardiomyopathy. Thallium is distributed homogeneously in the normal images, with a small region of reduced activity at the apex noted in the ANT projection due to normal apical thinning.

By comparison, the left ventricle is dilated in the patient with congestive cardiomyopathy and thallium distribution is inhomogeneous. There are no large discrete defects such as those observed in patients with left ventricular failure resulting from coronary artery disease (see Fig. 4).

and a patient with congestive cardiomyopathy are illustrated in Figure 3; images from a patient with "ischemic cardiomyopathy" are shown in Figure 4.

Serial determinations of global ventricular volumes and ejection fraction using radionuclide angiography or gated blood pool scanning provide objective methods of evaluating changes in ventricular function brought about by therapy. For example, initial therapy in a patient with alcoholic cardiomyopathy would typically consist of alcohol withdrawal, digitalis, and diuretics. If there were no significant change in the clinical status, radionuclide ejection fraction, and ventricular size, additional therapy such as afterload reduction and/or long-term bed rest would be considered. We have observed decrease in size and also improvement in ejection fraction of both ventricles in several patients with alcoholic CCM after two or more months of abstinence.

RESTRICTIVE CARDIOMYOPATHY (RCM)

Restrictive cardiomyopathy (RCM), the least common class of cardiomyopathy outside of the tropics, presents with findings simulating constrictive pericardial disease and with normal-sized or small ventricles. This form of myocardial disease results in diminished ventricular compliance due to either an infiltrative process or fibrosis. Physical findings include pulsus paradoxicus, elevated jugular venous pressure with deep X and Y descents, and increase in jugular venous pressure with inspiration (Kussmaul's sign). Congestive or dilated cardiomyopathy may have stigmata resembling constrictive disease, but to avoid confusion we also include in our definition of RCM that ventricular end-diastolic cavity

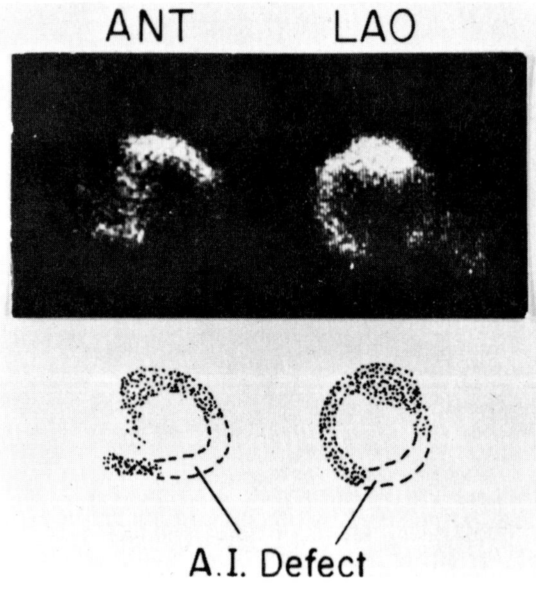

Figure 4. Thallium-201 myocardial images obtained after injection in the resting state in the anterior (ANT) and 45 degree left anterior oblique (LAO) projections in a patient with congestive heart failure. A schematic diagram beneath these images defines the large apical inferior (A.I.) defect consistent with a large myocardial scar. This large defect suggests a coronary artery etiology for the congestive failure. The gated scan showed a low ejection fraction of 33 percent but no ventricular aneurysm. Coronary and left ventricular cineangiography confirmed the presence of severe coronary artery disease with extensive akinesis. Comparison of these images with those of the normal and the patient with CCM, illustrated in Figure 3, underscore the potential utility of ^{201}Tl imaging in differentiating between coronary artery disease and CCM.

154

dimension or size (as defined by gated blood pool scan or echocardiogram) is normal or reduced. Restrictive cardiomyopathy, so defined, may result from a variety of pathologic disorders which can be subdivided into two basic types: (1) infiltrative RCM in which the left or both ventricular cavities are normal to decreased in size and the walls are normal to increased in thickness, and (2) obliterative RCM in which the involved ventricle(s) is(are) reduced in size and there is partial obliteration of one or both of the ventricular cavities. Fibrosis of the myocardium and endocardium usually leads to increased wall thickness in these disorders as well. It is important to note that in some functional classifications obliterative cardiomyopathy has been employed as a separate category.[12] Other schemes have included both infiltrative and obliterative diseases under the single heading of restrictive cardiomyopathy.[13] The pathologic findings in restrictive cardiomyopathy are determined by the underlying disease process. In infiltrative RCM the ventricular cavities are usually normal to reduced in size, although heart weight may be moderately or occasionally markedly increased. Amyloid infiltration, the most common cause of infiltrative RCM in nontropical regions, leads to a waxy, stiff, rubbery-appearing myocardium.[14] Other causes of myocardial infiltration which may lead to RCM include metastatic carcinoma,[15, 16] lymphoma,[17] or invasion with leukemic cells.[18] Usually, infiltration involves both right and left ventricles to varying degrees, so that patients with infiltrative RCM usually have clinical evidence of biventricular failure, although typically signs of right ventricular failure predominate.

Progressive fibrosis involving the endocardium and myocardium, leading to compromise of the size of one or both ventricles, is characteristic of the obliterative form of restrictive cardiomyopathy. In tropical zones, including East and West Africa, Indonesia, Ceylon, and Brazil, this process is known as endomyocardial fibrosis (EMF).[19] Obliterative cardiomyopathy occurring in temperate regions is rare, and is known as Loeffler's fibroplastic eosinophilic endocarditis.[20] The clinical picture in obliterative RCM is similar to that of infiltrative RCM.

Differentiation between constrictive pericarditis and restrictive cardiomyopathy is frequently difficult on clinical grounds. Radionuclide imaging may be of some value. In constrictive pericarditis both ventricles usually demonstrate normal systolic function, whereas in the restrictive cardiomyopathies biventricular function is usually reduced although typically not as severely as in CCM (i.e., the ejection fraction usually exceeds 30 percent). In obliterative disease the unusual configuration of the ventricles suggests the underlying pathologic process. Since mural thickness is generally increased with infiltration, myocardial imaging may be helpful in depicting increased mural thickness, as with amyloid disease, or focal myocardial involvement, as with metastatic disease.

Conversely, constrictive pericardial disease can involve the myocardium and lead to depressed function, whereas early RCM with stigmata of constrictive disease by physical examination may display normal or only mildly reduced ventricular function. If the clinical, noninvasive, and catheterization data cannot clearly distinguish between constrictive pericardial disease, a surgically curable disease, and RCM, it may be necessary to perform myocardial biopsy or pericardiectomy to ultimately resolve the issue.

The gated blood pool scan is useful in distinguishing between congestive or dilated cardiomyopathy and RCM by depicting the size of the cardiac chambers. As noted earlier, usually in CCM all four chambers are dilated whereas in RCM the ventricles are normal to reduced in size while the atria are usually considerably dilated. In addition, the gated cardiac blood pool scan may be helpful in differentiating between the infiltrative and obliterative forms of restrictive cardiomyopathy. In the infiltrative type, e.g., amyloidosis, both ventricles are usually normal or reduced in size. In contrast, in the obliterative type either or both ventricles are usually smaller than normal. The right ventricle is frequently diminutive in EMF. [201]Tl imaging is of value when neoplastic involvement of the myocardium with solid tumors is a possibility since the thallium image may reveal multiple large defects present at rest or with exercise. In contrast, leukemic infiltration is usually homogeneous and myocardial images display uniform ^{201}Tl activity. The treatment of malignant disease with chemotherapeutic agents, such as adriamycin, may result in heart failure due to myocardial toxicity. The gated blood pool scan will show this toxic cardiomyopathy to be of the congestive or dilated type, and thallium distribution is homogeneous, helping to differentiate it from the restrictive cardiomyopathy due to direct neoplastic involvement with solid tumors. Although we have not studied patients with obliterative RCM by radionuclide techniques, it is anticipated from a knowledge of the pathology[19,20] that the gated blood pool scan would demonstrate small, poorly contractile ventricles (Fig. 1), and the atria, especially the right, would be greatly dilated secondary to increased ventricular filling pressures and atrioventricular valve involvement. In addition, it is anticipated that large defects in myocardial thallium distribution would not be present.

HYPERTROPHIC CARDIOMYOPATHY (HCM)

Hypertrophic cardiomyopathy (HCM) is characterized by left ventricular hypertrophy of unknown etiology. Hypertrophy is generally asymmetric, with the interventricular septum disproportionately thickened in comparison to the remainder of the ventricle.[22-25] HCM may be obstructive or nonobstructive. Left ventricular outflow obstruction is caused by hypertrophy of the anterior superior aspect of the interventricular septum and abnormal motion of the anterior leaflet of the mitral valve.[26-28] Before echocardiography the detection of HCM was largely dependent on the presence of obstruction with its resulting physical findings (e.g., bisferiens pulse, systolic ejection murmur which increased with Valsalva's maneuver and decreased with squatting).[26,29] Early names for the disease included mention of the obstructive component. Braunwald and coworkers coined the term idiopathic hypertrophic subaortic stenosis (IHSS) while Brent and associates used the designation muscular subaortic stenosis, both emphasizing obstruction to left ventricular outflow.[30,31] On the other hand, Goodwin and coworkers labeled the disorder hypertrophic obstructive cardiomyopathy (HOCM), emphasizing the primary myocardial disease.[32]

Noninvasive imaging techniques now permit the diagnosis of HCM with or without obstruction. Asymmetric septal hypertrophy (ASH) is easily documented by echocardiography and has been employed as a marker of HCM;[24] however, it is not specific. For example, right ventricular hypertrophy may also lead to ASH.[33] Abnormal systolic anterior motion (SAM) of the anterior leaflet of the mitral valve which can be documented echocardiographically and angiographically has been thought to be highly specific for HCM with obstruction to left ventricular outflow,[34] although recent reports suggest that SAM is not as specific as it was once considered.[35,36]

Both the gated cardiac blood pool scan (Fig. 5) and myocardial imaging with [201]Tl (Fig. 6) are useful alternatives to the echocardiogram in the detection of HCM.[37,38] The end-diastolic image of the gated blood pool scan in the left anterior oblique view allows evaluation of the interventricular septal configuration. The appearance of the interventricular septum by gated scan is not only dependent upon septal configuration but is also affected by angle of view. The normal interventricular septum is curvilinear with its concavity facing the left ventricle. In a shallow left anterior oblique view, the apical septum appears thickest, while in a high oblique view, the upper septum appears thickest.[36] This phenomenon occurs because of the spiral geometry of the interventricular septum. The view in which the septum appears most uniform from top to bottom (usually 40 to 50 degree left anterior oblique) is selected for evaluation of configuration.

In HCM, the interventricular septum has an abnormal configuration which can be demonstrated by gated scan when evaluated as described above. In our experience, 50 percent of patients with hypertrophic cardiomyopathy demonstrate disproportionate upper septal thickening (DUST) by gated blood pool scan.[37] The middle or upper septum appears

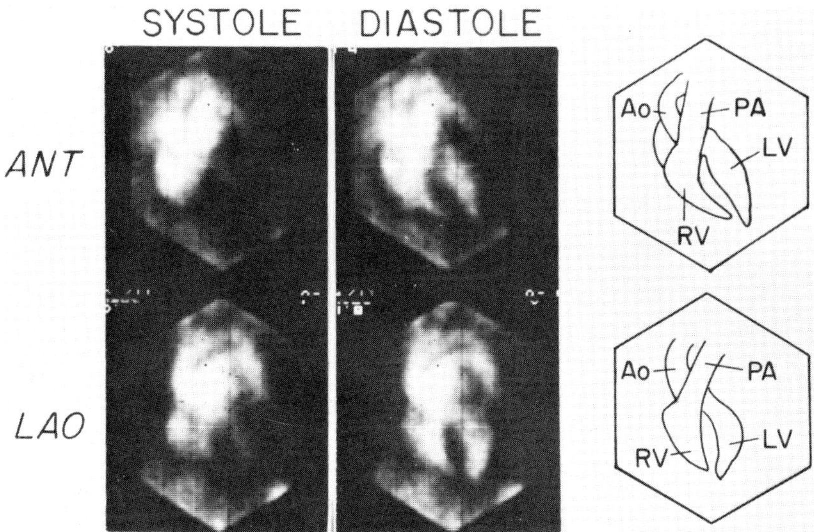

Figure 5. Anterior (ANT) and 50 degree left anterior oblique (LAO) gated blood pool images in end-systole and end-diastole in a representative patient with hypertrophic obstructive cardiomyopathy. In the right hand column, the end-diastolic images are depicted schematically illustrating the left ventricle (LV), right ventricle (RV), aorta (Ao) and pulmonary artery (PA). The end-diastolic LAO image demonstrates disproportionate upper septal thickening (DUST). Comparing end-diastolic to end-systolic images in both projections, the cavity obliteration which characterizes hypertrophic cardiomyopathy can be seen.

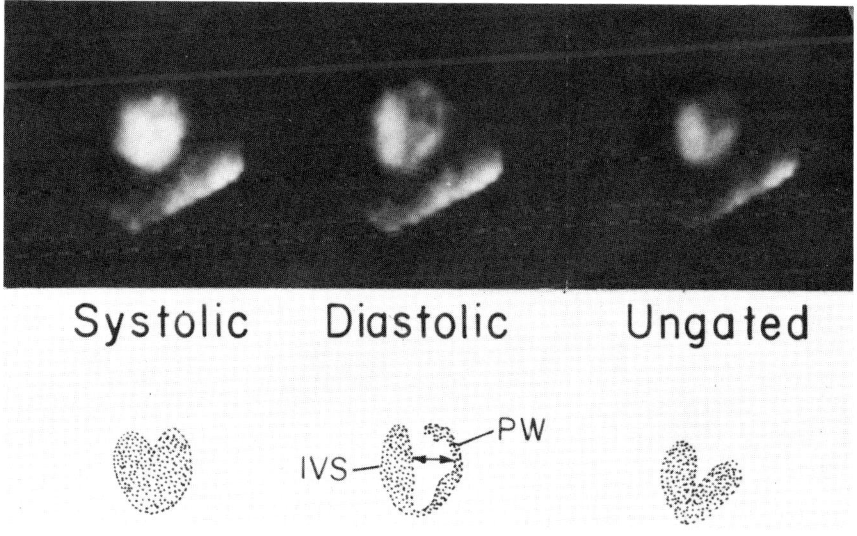

Figure 6. Thallium-201 myocardial images showing end-systolic, end-diastolic and ungated configuration in the 50 degree left anterior oblique projection in a patient with hypertrophic obstructive cardiomyopathy. The images are illustrated schematically in the lower row. The ungated image suggests hypertrophy, although asymmetric septal hypertrophy (ASH) is not as clearly defined. The end-diastolic image clearly documents ASH and demonstrates the septum to be homogeneously thickened from top to bottom. The end-systolic image demonstrates cavity obliteration and, as a consequence, septal configuration cannot be determined. (Illustration courtesy of Dr. Geoff Mews.)

thicker than the lower septum in the left anterior oblique projection. In addition, the majority of patients with HCM (73 percent) demonstrate loss of the normal concavity on the left ventricular aspect of the interventricular septum. While this flattening of the septum is a frequent finding in HCM, it is less specific than DUST and can be observed in some patients with aortic stenosis as well as in some normal patients. The gated scan may also demonstrate left ventricular systolic cavity or apex obliteration and a circular defect in the left ventricular outflow most likely related to the displacement of blood pool activity by the thickened upper septum. These abnormalities are best seen in the anterior or 30 degree right anterior oblique projection.

Bulkley and coworkers employed ^{201}Tl myocardial imaging to evaluate the ratio of septal to posterior wall thicknesses in patients with HCM.[38] As anticipated from echocardiographic studies, asymmetric septal hypertrophy was noted on thallium scan in all 10 patients with HCM and confirmed in 9 by cardiac catheterization. In addition, it was noted that the basal and midposterior walls were of equal thickness in patients with obstructive HCM, but that the basal wall was thinner than the midposterior wall in patients with nonobstructive HCM. Three patients with chronic pulmonary hypertension had ASH by ^{201}Tl scan, but in these the right ventricular free wall thickness was equal to the septal thickness consistent with right ventricular hypertrophy. The configuration of the interventricular septum on ^{201}Tl scan in patients with HCM was noted to be triangular with the maximal thickness inferoapically in contrast to DUST noted in end-diastole on gated blood pool studies. Since electrocardiographic gating was not employed with ^{201}Tl imaging, the images obtained represented a composite view of the beating heart over multiple cardiac cycles. As a result, these images included end-systole when the muscle is thicker and the count density higher. In end-systole, obliteration of the left ventricular apex may give the erroneous impression that the apical septum is thickest (Figs. 6 and 7). Gated ^{201}Tl scanning demonstrates that the end-diastolic septal configuration in HCM is similar to that noted on gated blood pool studies, i.e., the septum is flattened and frequently shows DUST (Fig. 6). The inferoapical septal thickening, present in end-systole, disappears in end-diastole. ASH is clearly defined on the end-diastolic image.

MYOCARDIAL DISEASE SECONDARY TO CHRONIC VOLUME OR PRESSURE OVERLOAD

Myocardial dysfunction may result from valvular or congenital heart disease as a consequence of chronic volume or pressure overload.[39] Although the clinical presentation in patients with heart muscle dysfunction secondary to volume or pressure overload is usually dominated by the primary process, the myocardial pathology in the late stage may closely resemble that of congestive cardiomyopathy.[40]

Volume Overload

In the majority of patients with chronic left ventricular volume overload, such as that due to aortic or mitral regurgitation, end-diastolic volume is increased in proportion to stroke volume with maintenance of a normal ejection fraction[41-44] (Fig. 2). However, chronic increase in stroke volume and concomitant dilatation may lead to myofibrillar damage.[39] Similarly, right ventricular dysfunction may result from prolonged right ventricular volume overload, as most commonly observed in the presence of atrial septal defect. Left ventricular volume overload states may lead to dilatation and reduction in ejection fraction[45-47] which may be difficult to distinguish from congestive cardiomyopathy. Preoperative ventricular function is an important determinant of potential surgical benefit.[46-49] Radionuclide techniques provide a means of determining ventricular function, and thus prognosis, by quantification of ejection fraction and ejection velocity.[50,51] In addition, these

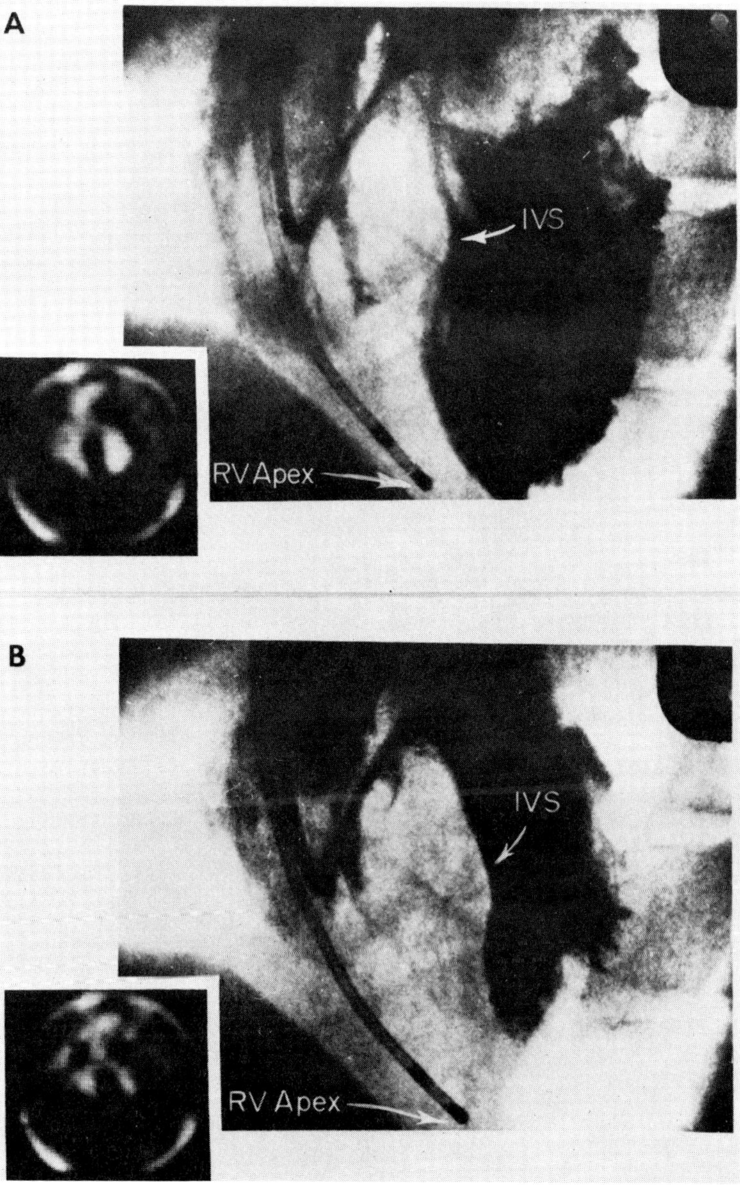

Figure 7. End-diastolic (top, "A") and end-systolic (bottom, "B") gated blood pool images and biventricular angiographic frames depict the configuration of the interventricular septum (IVS). The right ventricular (RV) apex is defined by the tip of a bipolar pacing electrode. Note that both the radionuclide and the cineroentgenographic studies depict disproportionate upper septal thickening in end-diastole while the lower septum appears more prominent by gated scan in end-systole. Obliteration of both right and left ventricular apices lead to this apparent change. Furthermore, the right ventricular apex is never opacified by the contrast injected into its outflow tract. Thus, biventricular angiography frequently does not outline the true configuration of the IVS. (From Pohost et. al.[37] by permission of the American Heart Association, Inc.)

methods allow repeated measurements of ventricular size and function and thus a means of following patients with volume overload, which might be helpful in deciding when to operate, e.g., when ejection fraction begins to fall. Hammermeister and coworkers recently found that survival in chronic mitral valve disease is related to end-diastolic volume and ejection fraction.[52] Furthermore, patients with a moderately reduced ejection fraction demonstrated the most significant benefits from surgery compared to their medically treated cohorts. These data suggest that serial noninvasive evaluation of ejection fraction and end-diastolic volume by radionuclide techniques may be useful in determining when surgery is indicated in patients with mitral valve disease. Multiple gated acquisition scanning affords the opportunity to determine the ejection velocity,[51] which has been suggested to be a sensitive marker of myocardial dysfunction in volume overload states.[44]

Borer and associates have recently reported their experience with multigated blood pool imaging at rest and with supine exercise in a series of patients with aortic regurgitation.[53] These investigators found that many asymptomatic patients with aortic regurgitation develop a fall in ejection fraction with exercise, while normal subjects uniformly increase their ejection fraction. Although exercise induced a fall in ejection fraction postoperatively, this fall was significantly less than that documented preoperatively. The magnitude of the fall in ejection fraction occurring with exercise may provide an index of impending left ventricular failure and thus allow a more sensitive means of determining when surgery should be employed than studies performed only in the resting state.

Pressure Overload

Pressure overload due to chronic increase in systolic pressure usually leads to secondary hypertrophy with maintenance of a normal ejection fraction and end-diastolic volume.[54] Myocardial failure is associated with decreasing stroke volume and ejection fraction, and some dilatation. Left ventricular failure may complicate the clinical picture in advanced aortic stenosis.[38] Pressure overload in the left ventricle due to systemic hypertension and in the right ventricle due to pulmonary hypertension (primary or secondary) or pulmonic stenosis may result similarly in failure of the respective ventricle.

Concentric hypertrophy and normal ventricular ejection function are characteristic of uncomplicated pressure overload states. In an appropriate left anterior oblique projection (40 to 60 degrees), the interventricular septum can be evaluated and if hypertrophy is moderate to marked, septal hypertrophy may be suggested by the end-diastolic gated blood pool or the thallium image. As mentioned earlier, interventricular septal configuration can be evaluated by gated blood pool scan, and in concentric hypertrophy due to chronically increased afterload the septum is concave with its concavity facing the left ventricle. Radionuclide scanning will also demonstrate ventricular cavity size and function. In chronic severe aortic stenosis, ventricular dilatation and dysfunction may occur[54] and is associated with a poor prognosis. Late in the course of aortic stenosis, this dilatation and reduced ejection fraction may resemble congestive cardiomyopathy on gated scan. At rest, patients with compensated aortic stenosis usually have a supernormal ejection fraction.[55] In a majority of patients with normal resting ventricular function, Borer found a fall in ejection fraction with exercise gated scanning.[55] It was speculated that the magnitude of the fall in ejection fraction with exercise was related to surgical prognosis. Although this concept is provocative, more data is needed before the utility of stress gated blood pool imaging for prognostication in valvular heart disease can be defined.

ISCHEMIC HEART DISEASE

The clinical presentation of patients with ventricular failure related to coronary artery disease generally includes a history of angina pectoris and/or myocardial infarction.

160

Myocardial damage occurs as a consequence of reduced coronary blood flow leading to ischemia and infarction. As a result, large regions of myocardium are replaced by scar. Chronic myocardial failure resulting from coronary artery disease is at times clinically indistinguishable from congestive cardiomyopathy. As has been discussed, both gated blood pool and myocardial imaging (Fig. 4) are useful in differentiating between ischemic and congestive cardiomyopathy.[11]

CONCLUSIONS

Radionuclide techniques provide information which is helpful in determining diagnosis, prognosis and therapeutic response in patients with myocardial disease. Gated blood pool scanning or radionuclide angiography permit evaluation of ejection fraction, ventricular size and wall motion, and interventricular septal thickness and configuration. With this information, the functional class of the cardiomyopathy can be established (congestive, restrictive [infiltrative or obliterative] or hypertrophic) (Table 2), and a number of etiologic possibilities are suggested (Table 1). Nonmyocardial forms of heart disease which secondarily involve the myocardium can simulate a primary cardiomyopathy. The diagnosis can frequently be clarified by radionuclide techniques. In valvular heart disease, exercise gated blood pool imaging is a provocative new modality which may be of value in detecting subclinical myocardial dysfunction and in determining prognosis. Myocardial imaging with thallium-201 is useful in detecting pathology which replaces (e.g., fibrosis) or displaces (e.g., metastatic neoplasm) myocardium. It is of particular value in differentiating myocardial dysfunction resulting from coronary artery disease from congestive cardiomyopathy and in depicting septal abnormalities in hypertrophic cardiomyopathy.

Finally, although these radionuclide techniques can be quite helpful in the evaluation of patients with myocardial disease, application of other diagnostic modalities should be considered initially. These include electrocardiography, chest roentgenography, fluoroscopy, phonocardiography and echocardiography. In the final analysis, the information provided by the appropriate noninvasive technique might establish the diagnosis and obviate the need for cardiac catheterization, angiography, and myocardial biopsy. If invasive studies are required to establish the diagnosis of a potentially treatable disease, serial imaging studies may be useful in following therapeutic effect and thereby in helping to delineate the appropriate therapeutic program.

ACKNOWLEDGMENTS

The authors wish to express their appreciation to Drs. Robert E. Dinsmore, Allen B. Nichols, Andrew Hoffman, Charles Homcy and Henry Gewirtz for their supply of clinical material and their critical review and constructive recommendations, and to Ms. Eileen Fitzgerald for valuable help in the preparation of the manuscript.

REFERENCES

1. GOODWIN, J. F., GORDON, H., HOLLMAN, A., ET AL.: *Clinical aspects of cardiomyopathy.* Br. Med. J. 1:69, 1961.

2. HUDSON, R. E. B.: *The cardiomyopathies: Order from chaos.* Am. J. Cardiol. 25:70, 1970.

3. STRAUSS, H. W., AND PITT, B.: *Gated cardiac blood pool scan: Use in patients with coronary artery disease.* Prog. Cardiovasc. Dis. 20:207, 1977.

4. SCHELBERT, H. R., VERBA, J. W., JOHNSON, A. D., ET AL.: *Non-traumatic determination of left ventricular ejection fraction by radionuclide angiocardiography.* Circulation 51:902, 1975.

5. STRAUSS, H. W., HARRISON, K., LANGAN, J. K., ET AL.: *Thallium-201 for myocardial imaging. Relation of thallium-201 to regional myocardial perfusion.* Circulation 51:641, 1975.

6. WACKERS, F. J., BECKER, A. E., SAMSON, G., ET AL.: *Location and size of acute transmural myocardial infarction estimated from thallium-201 scintiscans: A clinicopathologic study.* Am. J. Cardiol. 56:72, 1977.

7. BULKLEY, B. H., ROULEAU, J., STRAUSS, H. W., ET AL.: *Sarcoid heart disease: Diagnosis by thallium-201 myocardial perfusion imaging.* Am. J. Cardiol. 37:125, 1976.

8. ROBERTS, W. C., AND FERRANS, V. J.: *Pathologic anatomy of the cardiomyopathies. Idiopathic dilated and hypertrophic types, infiltrative types, and endomyocardial disease with and without cosinophilia.* Human Pathology 6:287, 1975.

9. TOBIN, J. R., JR., DRISCOLL, J. F., LIM, M. T., ET AL.: *Primary myocardial disease and alcoholism: Clinical manifestations and course of the disease in a selected population of patients observed for three or more years.* Circulation 35:754, 1967.

10. DEMAKIS, J. G., AND RAHIMTOOLA, S. H.: *Peripartum cardiomyopathy.* Circulation 55:753, 1977.

11. BULKLEY, B. H., HUTCHINGS, G. M., BAILEY, I., ET AL.: *Thallium-201 imaging and gated cardiac blood pool scans in patients with ischemic and congestive cardiomyopathy: A clinical and pathologic study.* Circulation 55:753, 1977.

12. GOODWIN, J. F.: *Prospects and predications for the cardiomyopathies.* Circulation 50:210, 1974.

13. OAKLEY, C. M.: *Clinical recognition of the cardiomyopathies.* Circ. Res. 35(Suppl. II):152, 1974.

14. BUJA, L. M., KHOI, N. B., AND ROBERTS, W. C.: *Clinically significant cardiac amyloidosis. Clinicopathologic findings in 15 patients.* Am. J. Cardiol. 26:394, 1970.

15. GLANCY, D. L., AND ROBERTS, W. C.: *The heart in malignant melanoma. A study of 70 autopsy cases.* Am. J. Cardiol. 21:555, 1968.

16. HANFLING, S. M.: *Metastatic cancer to the heart: Review of the literature and report of 12 cases.* Circulation 22:474, 1960.

17. ROBERTS, W. C., GLANCY, D. L., AND DEVITA, V. T., JR.: *Heart in malignant lymphoma (Hodgkin's disease, lymphoma, reticulum cell sarcoma and mycosis fungoides). A study of 196 autopsy cases.* Am. J. Cardiol. 22:85, 1968.

18. ROBERTS, W. C., BODEY, G. P., AND WERTLAKE, P. I.: *The heart in acute leukemia. A study of 420 autopsy cases.* Am. J. Cardiol. 21:388, 1968.

19. PARRY, E. H. O.: *Endomycological fibrosis.* In Wolstenholme, G. E. W., O'Connor, M., London, J., and Churchill, A. (eds.): *Cardiomyopathies: A Ciba Foundation Symposium.* 1964, pp. 322–346.

20. LOEFFLER, W.: *Endocarditis parietalis fibroplastica mit Bluteosinophile.* Schweiz. Med. Wschr. 66:817, 1936.

21. MEANEY, E., SHABETAI, R., BHARGAVA, V., ET AL.: *Cardiac amyloidosis, constrictive pericarditis and restrictive cardiomyopathy.* Am. J. Cardiol. 38:547, 1976.

22. TEARE, D.: *Asymmetrical hypertrophy of the heart in young patients.* Br. Heart J. 20:1, 1958.

23. ABBASI, A. S., MACALPIN, R. N., EBER, L. M., ET AL.: *Echocardiographic diagnosis of idiopathic hypertrophic cardiomyopathy without obstruction.* Circulation 46:897, 1972.

24. HENRY, W. L., CLARK, C. E., AND EPSTEIN, S. E.: *Asymmetric septal hypertrophy (ASH): Echocardiographic identification of the pathognomonic anatomic abnormality of IHSS.* Circulation 47:225, 1973.

25. EPSTEIN, S. E., HENRY, W. L., CLARK, C. E., ET AL.: *Asymmetric septal hypertrophy.* Ann. Intern. Med. 81:650, 1974.

26. BRAUNWALD, E., LAMBREW, C. T., ROCKOFF, S. D., ET AL.: *Idiopathic hypertrophic subaortic stenosis. I. A description of the disease based upon an analysis of 64 patients.* Circulation 30(Suppl. IV):3, 1964.

27. HENRY, W. L., CLARK, C. E., GRIFFITH, J. M., ET AL.: *Mechanism of left ventricular outflow obstruction in patients with obstructive asymmetric septal hypertrophy.* Am. J. Cardiol. 35:337, 1975.

28. WIGLE, E. D., ADELMAN, A. G., AND SILVER, M. D.: *Pathophysiological consideration in muscular subaortic stenosis.* In Wolstenholme, G. E. W., O'Connor, M., London, J., and Churchill, A. (eds.): *Hypertrophic Obstructive Cardiomyopathy.* CIBA Foundation Study Group No. 37, 1971, p. 63.

29. NELLEN, M., BECK, W., VOGELPOEL, L., ET AL.: *Auscultatory phenomena in hypertrophic obstructive cardiomyopathy.* In Wolstenholme, G. E. W., O'Connor, M., London, J., and Churchill, A. (eds.): *Hypertrophic Obstructive Cardiomyopathy.* CIBA Foundation Study Group No. 37, 1971, p. 77.

30. BRAUNWALD, E., MORROW, A. G., CORNELL, W. P., ET AL.: *Idiopathic hypertrophic subaortic stenosis: Clinical, hemodynamic, and angiographic manifestations.* Am. J. Med. 29:924, 1960.

31. BRENT, L. B., ABURANO, A., FISHER, D. L., ET AL.: *Familial muscular subaortic stenosis: An unrecognized form of "idiopathic heart disease" with clinical and autopsy observations.* Circulation 21:167, 1960.

32. GOODWIN, J. F., HOLLMAN, A., CLELAND, W. P., ET AL.: *Obstructive cardiomyopathy simulating aortic stenosis.* Br. Heart J. 22:169, 1960.

33. TARTER, W. E., ALLEN, H. D., SAHN, D. J., ET AL.: *The asymmetrically hypertrophied septum: Further differentiation of its causes.* Circulation 53:19, 1976.

162

34. SHAH, P. M., GRAMIAK, R., AND KRAMER, D. H.: *Ultrasound localization of left ventricular obstruction in hypertrophic obstructive cardiomyopathy.* Circulation 40:3, 1969.

35. MINTZ, G. S., KOTLER, M. N., SEGAL, B. L., ET AL.: *Systolic anterior motion in the absence of asymmetric septal hypertrophy.* Circulation 57:256, 1978.

36. MARON, B. J., GOTDIENER, J. S., ROBERTS, W. C., ET AL.: *Left ventricular outflow tract obstruction due to systolic anterior motion of the anterior mitral valve leaflet in patients with concentric left ventricular hypertrophy.* Circulation 57:527, 1978.

37. POHOST, G. M., VIGNOLA, P. A., McKUSICK, K. A., ET AL.: *Hypertrophic cardiomyopathy: Evaluation by gated cardiac blood pool scanning.* Circulation 55:92, 1977.

38. BULKLEY, B. H., ROULEAU, J., STRAUSS, H. W., ET AL.: *Idiopathic hypertrophic subaortic stenosis: Detection by thallium-201 myocardial perfusion imaging.* N. Engl. J. Med. 50:421, 1971.

39. SCHLANT, R. C., AND NUTTER, D. O.: *Heart failure in valvular heart disease.* Medicine 50:421, 1971.

40. MARON, B. J., FERRANS, V. J., AND ROBERTS, W. C.: *Myocardial ultrastructure in patients with chronic aortic valve disease.* Am. J. Cardiol. 35:725, 1975.

41. DODGE, H. T., AND BAXLEY, W. A.: *Hemodynamic aspects of heart failure.* Am. J. Cardiol. 22:24, 1968.

42. DODGE, H. T., AND BAXLEY, W. A.: *Left ventricular volume and mass and their significance in heart disease.* Am. J. Cardiol. 23:528, 1969.

43. RACKLEY, C. E., AND HOOD, W. P., JR.: *Quantitative angiographic evaluation and pathophysiologic mechanisms in valvular heart disease.* Prog. Cardiovasc. Dis. 15:427, 1973.

44. ECKBERG, D. L., GAULT, J. H., BOUCHARD, R. L., ET AL.: *Mechanics of left ventricular contraction in chronic severe mitral regurgitation.* Circulation 47:1252, 1973.

45. GAULT, J. H., COVELL, J. W., BRAUNWALD, E., ET AL.: *Left ventricular performance following correction of free aortic regurgitation.* Circulation 42:773, 1970.

46. MILLER, G. A. H., KIRKLIN, J. W., AND SWAN, H. J. C.: *Myocardial function and left ventricular volumes in acquired valvular insufficiency.* Circulation 31:374, 1965.

47. SHEA, W. H., BOUCHER, C. A., CURFMAN, G. D., ET AL.: *Early reversibility of left ventricular dilatation following relief of chronic valvular regurgitation.* Circulation 53, 54(Suppl. II):215, 1976.

48. BONOW, R. O., HENRY, W. L., KENT, K. M., ET AL.: *Predictors of late deaths due to congestive heart failure following operation for aortic regurgitation.* Am. J. Cardiol. 41:382, 1978.

49. SCHULER, G., ROSS, J., JR., JOHNSON, A., ET AL.: *Factors affecting recovery of left ventricular function after mitral valve surgery.* Am. J. Cardiol. 41:383, 1978.

50. STRAUSS, H. W., ZARET, B. L., HURLEY, P. J., ET AL.: *A scintiphotographic method for measuring left ventricular ejection fraction in man without cardiac catheterization.* Am. J. Cardiol. 28:275, 1971.

51. BUROW, R. D., STRAUSS, H. W., SINGLETON, R., ET AL.: *Analysis of left ventricular function from multiple gated (MUGA) cardiac blood pool imaging: Comparison to contrast angiography.* Circulation 56:1024, 1977.

52. HAMMERMEISTER, K. E., FISHER, L., KENNEDY, J. W., ET AL.: *Prediction of late survival in patients with mitral valve disease from clinical, hemodynamic, and quantitative angiographic variables.* Circulation 57:341, 1978.

53. BORER, J. S., BACHARACH, S. L., GREEN, M. V., ET AL.: *Exercise induced left ventricular dysfunction in symptomatic and asymptomatic patients with severe aortic regurgitation assessed by radionuclide cineangiography.* Am. J. Cardiol. 39:284, 1977.

54. KENNEDY, J. W., TWISS, R. D., BLACKMAN, J. R., ET AL.: *Quantitative angiocardiography: III. Relationship of left ventricular pressure, volume, and mass in aortic valve disease.* Circulation 38:838, 1968.

55. BORER, J. S., BACHARACH, S. L., GREEN, M. V., ET AL.: *Left ventricular function in aortic stenosis: response to exercise and effects of operation.* Am. J. Cardiol. 41:382, 1978.

Fatty Acid Uptake and "Metabolic Imaging" of the Heart*

Milton S. Klein, M.D., and Burton E. Sobel, M.D.

The consequences of acute myocardial infarction appear to depend in part on the extent of tissue damage sustained.[1] Accordingly, interventions have been designed to improve prognosis and protect ischemic myocardium pharmacologically by favorably altering metabolism and hemodynamics.[2] To determine whether such interventions are indeed beneficial, quantification of the amount of jeopardized and irreversibly injured myocardium is necessary. Despite their utility for some investigative purposes, methods presently available exhibit some limitations. Thus, electrocardiographic indexes although useful in gauging directional changes when applied serially, lack some specificity and may be influenced by pericarditis and the actions of drugs such as digitalis, verapamil, or propranolol.[3-6] Estimates based on analysis of serial plasma creatine kinase values require acquisition of data over relatively prolonged intervals and provide no insight regarding the locus or distribution of infarction.[7] Conventional scintigraphic techniques employing single photon detection suffer from superimposition of overlapping regions of myocardium in a two-dimensional display, difficulties due to variable attenuation as a function of distance of the tracer from the detector, dependence of accumulation of tracers on both altered perfusion and metabolism, and limited resolution.[8] However, because imaging techniques, particularly when coupled with computer reconstruction tomography, offer promise of elucidating locus, distribution, and extent of infarction promptly and noninvasively, they have been explored intensively.[9,10] To provide a framework for considering the use of physiological substrates of the heart such as palmitate, labeled with positron-emitting isotopes, for imaging the heart, it may be useful to briefly review some of the experience gained with gamma-emitting tracers.

TRACERS USED TO ASSESS PERFUSION

Radioactive potassium (^{43}K) and its analogs such as thallium (^{201}Tl) and rubidium (^{81}Rb) have been employed to assess myocardial perfusion in vivo.[11-13] However, extraction of these tracers by the heart is dependent not only on flow but also on cation transport mechanisms requiring energy. Thus, factors that alter potassium flux across the cell membrane such as ischemia, acidosis or alkalosis, digitalis, or insulin availability may influence accumulation of the tracer independent of perfusion.[14] Furthermore, limited diffusion of the tracer may result in its retention in the interstitium rather than in intra-

---

*Work supported in part by National Institutes of Health Grant HL 19746, SCOR in Ischemic Heart Disease.

cellular fluid.[15] Redistribution into relatively ischemic zones may occur also, limiting estimates of relative and absolute regional perfusion with these agents.

Washout of inert gases such as krypton (^{85}Kr) and xenon (^{133}Xe) has been used as an index of myocardial perfusion since efflux of the tracer from tissue is directly related to blood flow rather than to myocardial metabolism.[16,17] Unfortunately, however, intracoronary injection has generally been required with this approach and analysis depends on the assumptions that perfusion is homogeneous and that mixing of the radionuclide in blood is complete. Since these assumptions may not be satisfied in all cases and since most analyses utilize nonregional detection of washout in coronary sinus blood, results have been relatively insensitive in detecting perfusion abnormalities.[18] Heterogeneity of flow related to hypoxia, or regional release of potassium, catecholamines, and possibly prostaglandins probably contributes to both limitation of sensitivity and some inaccuracy.[19] Multicrystal detector systems permit analysis of washout curves of tracers in each of several precordial zones. However, overlap of myocardium underlying each zone, overlap of the fields of view of each crystal, and difficulties in correcting for attenuation as a function of depth of the tracer within the tissue lead to limitations in estimation of regional blood flow.[8] Thus, monoexponential fits to clearance curves from each precordial zone do not adequately describe heterogeneity resulting from differences in coronary flow in superimposed regions from epicardium to endocardium. In addition, the lipophilic nature of the inert gases and heterogeneous contributions of adipose tissue within or surrounding the heart make analysis of the later portion of washout curves difficult.[20]

TRACERS WITH AVIDITY FOR ZONES OF INFARCTION

Radionuclides that accumulate in zones of recent myocardial infarction permit localization of infarction with imaging techniques.[21] However, agents such as technetium pyrophosphate (^{99m}Tc(Sn) PYP) consistently delineate infarction only after intervals of 24 to 48 hours.[22] Because ^{99m}Tc has a relatively long half-life (4.7 hours), serial studies for evaluation of rapid changes are impractical. Factors which hamper accurate estimation of the extent of infarction with such agents include: superimposition of normal myocardium, abnormal tissue, and bone (which also accumulates tracer); limited delivery of tracer to areas of markedly decreased flow with consequent decreased accumulation;[8,23] and variable attenuation of the gamma emission dependent on the distance of the tracer from the scintillation camera. Since detection of counts is inversely proportional to the distance of the tracer from the detector (with 50 percent loss of efficiency with each 5 cm distance) accumulation of large amounts of tracer in an infarct of the posterior left ventricular wall may appear comparable to accumulation of much less tracer in anterior regions of infarction. Normal tissues overlying abnormal myocardium may limit both resolution and image contrast, and in some cases, attenuation of radiation by surrounding structures may obscure abnormalities entirely.

For these reasons, efforts have been made to delineate zones of infarction with imaging techniques based on computer reconstruction tomography providing displays of a series of cross (or longitudinal) sections and reconstruction of the organ of interest in three dimensions.[24,25] Because positron-emitting isotopes offer major advantages for this purpose,[26] some of their properties merit special consideration.

POSITRON EMISSION AND TOMOGRAPHY

Positrons are positively charged, subatomic particles with mass comparable to that of electrons. In the process of positron decay, a positron-neutrino pair is released and collision with an electron occurs. Annihilation of the positron and electron results in the production of two 511-keV gamma photons directed 180 degrees apart. An event is recorded only when both photons remain within the cylindrical field of view between two NaI(Tl) crystals aligned 180 degrees apart, and when each of the two photons impinges on the corresponding crystal. Coincidence detection thus provides "electronic collimation" of the emission, defining its location accurately. The combined attenuation of the two photons (sensed by both crystals) depends on the total amount of absorbing material encountered. Thus, when the source is moved closer to one detector, the increased attenuation is offset by a proportional decrease with respect to radiation detected by the other. Accordingly, variable attenuation due to the depth of the tracer within the tissue does not distort the reconstructed image[10,27] as it does with gamma emission detected with a scintillation camera.[28]

The positron emission transaxial tomograph developed by the Radiation Sciences group at Washington University employs a hexagonal array of 48 detectors (24 pairs) and programmed rotation around the patient to obtain radiation profiles from a series of cylindrical fields of view. Computer reconstruction is employed to depict the distribution of tracer throughout 1.5-cm-thick cross sections, permitting quantitative delineation of accumulation of tracer in myocardium without interference due to superimposition or accumulation in adjacent tissue. Unlike single photon gamma ray emission tomograms,[28] the positron emission computer reconstructed tomographic images of cross sections of the heart do not exhibit distortion related to heterogeneity of attenuation of radiation.

MYOCARDIAL IMAGING WITH POSITRON-EMITTING RADIONUCLIDES

Several positron-emitting radiopharmaceuticals have been synthesized for use in assessing myocardial perfusion and irreversible ischemic injury.[29] Most require a cyclotron on-site because of their short half-lives. ^{82}Rb, with a half-life of 75 seconds, has been employed in serial studies of regional myocardial perfusion,[30,31] using intravenous or intracoronary injection. However, inaccuracies may result because of recirculation and variable extraction of the tracer by ischemic myocardial cells and effects of altered residence time of the tracer.[14] Since extraction increases when residence time is prolonged, diminished uptake by ischemic myocardium may be offset by increased extraction resulting from prolonged exposure of cell surfaces to the tracer during conditions of low flow.

Gallium (^{68}Ga) linked to microspheres can be used to assess regional blood flow, but the need for intracoronary injections is a disadvantage. On the other hand, this agent is attractive since it can be produced from a germanium (^{68}Ge) generator. Because the half-life of ^{68}Ge is so long (267 days), a single generator can provide ^{68}Ga for several years.[32]

Another agent used to assess perfusion is ^{13}N-labeled ammonia (^{13}NH$_3$).[33-35] Its short half-life (10 minutes) permits serial studies. Unfortunately, inhibition of glutamine synthetase leads to a marked reduction of the uptake of ^{13}NH$_3$ by myocardial cells, suggesting that accumulation is dependent on metabolism as well as perfusion.[33] After ^{13}NH$_3$ is incorporated in the large tissue ammonia pool, egress appears to reflect the distribution of myocardial blood flow.[36] However, apparent perfusion defects may be influenced by relative rates of synthesis of glutamine from ammonia. Furthermore, after peripheral intravenous injection, significant incorporation of ammonia into the lungs of heavy smokers may impair delineation of the cardiac uptake and obscure border recognition.[37]

^{68}Ga-labeled ethylene diamine tetramethylene phosphonate or diethylene triaminepentamethylene phosphonate is avidly accumulated in areas of myocardial infarction analogously to ^{99m}Tc-PYP.[38] Although it can therefore be used to assess regions of infarction (as opposed to ischemia), comparable limitations apply related to the delay after the onset of irreversible injury before images become abnormal and related to the flow-dependence of accumulation.

MYOCARDIAL STUDIES WITH ^{11}C-LABELED FATTY ACIDS

Accumulation of many tracers in normal or abnormal myocardium depends on a complex combination of factors affecting delivery, extraction, and metabolism. In view of the extensive information available regarding myocardial metabolism of fatty acids, the preferred physiological substrate of the heart,[39] it seemed likely to us that interpretation of results would be facilitated if quantitative imaging techniques could be applied to assess the distribution of positron-emitting labeled fatty acids in the heart in vivo, particularly if labeling could be accomplished with a short-lived tracer, suitable for sequential studies and one which did not alter the chemical structure of the fatty acid. Furthermore, since oxidation of free fatty acids diminishes promptly in ischemic myocardium, early detection of jeopardized tissue would be feasible.

^{11}C (half-life approximately 20 minutes) can be produced easily in a cyclotron by bombardment of boron with deuterons of energy greater than 6 MeV. With boric oxide as the target material, ^{11}C was synthesized by the ^{10}B(d,n)^{11}C reaction, swept from the field as a mixture of carbon monoxide and carbon dioxide with zinc dust on to asbestos heated to 700 °C, and used to synthesize ^{11}C-labeled palmitate by carbonation of pentadecyl magnesium bromide from ^{11}CO$_2$ followed by hydrolysis to produce an albumin-fatty acid complex.[40]

In previous studies, Poe utilized fatty acids labeled with iodine (^{123}I) injected into the coronary arteries of dogs[41] and found that the initial extraction fraction of the tracer was 35 to 39 percent, falling to 20 to 25 percent after two minutes. Initial extraction fractions of ^{11}C-oleic acid or ^{18}F-tetradecanoate were similar, but significantly lower after two minutes suggesting more rapid metabolism. All of these tracers were extracted less avidly than ^{43}K, suggesting that an alteration in the molecular structure of iodinated fatty acids might be responsible for a blunted utilization. ^{123}I or ^{131}I-labeled oleic or linoleic acid is extracted twice as avidly if the radioisotope is located on the terminal portion of the molecule, suggesting that terminal labeling results in less stearic hindrance, permitting more physiological utilization of the tracer by myocardium.[29,42]

MYOCARDIAL METABOLISM OF FATTY ACIDS

Fatty acid is a preferred substrate of myocardium[39] and fatty acid oxidation normally accounts for 60 to 80 percent of energy production by the heart. Even when moderate hypoxia or ischemia supervenes, exogenous fatty acid and endogenous fatty acid liberated from triglycerides are metabolized in preference to glucose. Anaerobic metabolism provides a substantial proportion of energy via glycolytic mechanisms,[43] only with marked hypoxia (oxygen delivery less than 20 percent of normal) or early after the onset of severe ischemia. Extraction and intracellular transport of fatty acids by the heart is related to chain length.[44]

The metabolic pathway of fatty acid has been well clarified. In intact organisms, long chain fatty acids are synthesized in the liver and adipose tissue, transported in blood bound primarily to albumin,[45] and extracted by myocardium as a function of several factors including: chain length, molarity of both albumin and the fatty acid,[45,46] metabolic integrity of the cell, perfusion (since regional coronary flow determines residence time), and myocardial oxygen requirements. Both ischemia and hypoxia lead to decreased extraction.[47]

Circulating triglycerides and other lipoproteins may provide fatty acid after hydrolysis mediated by lipoprotein lipases present in myocardial cells and in capillary endothelium.[48]

Fatty acids in interstitial or intracellular fluid are bound to soluble proteins, and uptake of fatty acids into the cell appears to depend on competition between cellular binding sites and binding sites on albumin. Intracellular fatty acids are activated and converted to thioester derivatives in the cytosol in reactions requiring both coenzyme A

168

(CoA) and ATP. Esterified fatty acids may undergo oxidation or incorporation into triglycerides. Oxidative metabolism of fatty acids derived from endogenous triglycerides may occur during conditions in which exogenous fatty acid supply is insufficient to meet the needs of aerobic metabolism. On the other hand, accumulation of triglycerides is a hallmark of repetitive ischemia,[49] in part because the glycerol skeleton derived from anaerobic glycolysis accumulates and fatty acid oxidation is impaired. Activated fatty acids in the cytosol cannot be oxidized directly. They are first transported across the mitochondrial membrane by acyl CoA carnitine transferases specific for chain length and intimately associated with mitochondrial membranes.[50,51] Carnitine-dependent translocation facilitates ingress of acyl CoA into the mitochondrial matrix where beta oxidation occurs. Acyl CoA is oxidized to produce acetyl CoA which is oxidized via the Krebs cycle with liberation of CO_2, formation of reduced pyridine nucleotides and flavoproteins, and synthesis of intracellular ATP.[52]

ASSESSMENT OF MYOCARDIAL FATTY ACID METABOLISM WITH RADIOACTIVELY LABELED TRACERS

In recent studies in our laboratory,[53] isovolumically beating rabbit hearts were perfused with equimolar albumin and ^{14}C-labeled palmitate (0.4 mM) with heart rate maintained constant at 180 beats/min. The metabolic fate of ^{14}C-labeled palmitate was evaluated in hearts perfused at flow rates of 15 ml/min (control) or 5 ml/min (ischemia) for five minutes. Although coronary flow in rabbit hearts in vivo under physiological conditions is approximately 5 ml/min, flows of this magnitude do not provide sufficient oxygenation to hearts perfused with nonhemoglobin containing solutions. After perfusion of the hearts, lipid extracts were prepared from samples of myocardium frozen rapidly in liquid nitrogen. Rabbit hearts perfused at high flow (control) incorporated approximately 40 percent of the ^{14}C-palmitate radioactivity as neutral lipids, 5 percent as polar lipids, and 39 percent as acyl CoA, fatty acid, and short chain CoA derivatives. Hearts perfused at low flow exhibited a decrease in total ^{14}C-palmitate accumulated, and a nearly twofold increase in the percentage of radioactivity in neutral lipids, due to an increase of tracer present in both unesterified fatty acid and triglyceride. These data confirm the impression that in ischemic tissue, incorporation of fatty acid is decreased with relatively increased incorporation into triglyceride due to decreased fatty acid oxidation. Thus, it seemed likely to us that imaging of zones of ischemia in vivo should be possible by using a positron emission transaxial tomograph and injecting ^{11}C-palmitate intravenously.

ANALYSIS OF ^{11}C-PALMITATE CLEARANCE CURVES

Bolus injection of ^{11}C-palmitate in isolated perfused rabbit hearts resulted in characteristic time-activity curves of radioactivity in the heart (Fig. 1).

Analysis of such curves provides estimates of two parameters: (1) the extraction fraction of the tracer which relates to flow and metabolic activity of the myocardium, the two determinants of substrate utilization, and (2) the slope of disappearance of counts representing an index of utilization of intracellular fatty acid, primarily attributable to oxidation.

The first portion of a curve of this type is related to the vascular washout of tracer from the coronary bed. When the injectate contains a tracer confined to the vascular compartment (such as ^{11}CO-hemoglobin), washout occurs rapidly.[54] Rapid washout of this type is seen with injections into the left atrium of anesthetized open chest rabbits. Much longer "vascular" washout curves are seen in isolated hearts,[25] probably because of capillary endothelial damage associated with high nonphysiological flow rates, result-

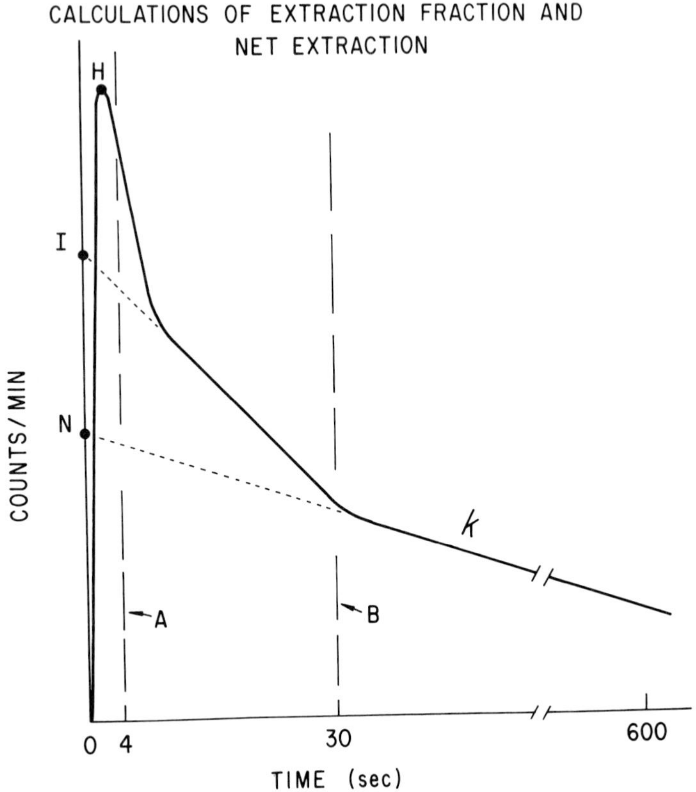

Figure 1. A diagrammatic representation of the time-activity curve of counts per minute recorded within the field of view of one pair of detectors, 180 degrees apart, surrounding an isolated perfused rabbit heart. The ordinate represents counts per minute on a logarithmic scale and the abscissa, time in seconds. The peak height of the curve (H) represents the total radioactivity associated with positron emission reaching the heart after a bolus injection of ^{11}C-palmitate into the perfusate at zero time. Prompt initial washout during the interval bracketed by line A represents primarily washout through the vascular space. Distribution of isotope in the interstitial fluid and intracellular spaces is the main factor accounting for the mid-portion of the curve contained between lines A and B. The back extrapolation of this portion of the curve to the ordinate provides an intercept (I). I/H × 100 therefore represents the unidirectional extraction fraction of isotope by the heart. The third portion of the curve, succeeding line B, reflects intracellular metabolism of tracer trapped within the heart. Back extrapolation of this portion provides an intercept (N). N/H × 100 represents the net extraction of tracer into a metabolizable pool.

ant edema, and mild regional hypoxia characteristic of such preparations, permitting diffusion of some tracer into interstitium. The second portion of typical washout curves reflects the fate of the tracer residing in the interstitial and intracellular spaces. Extrapolation of this part of the curve to the Y-intercept provides an estimate of the unidirectional, initial extraction fraction of tracer, which may substantially exceed net extraction of carrier. Extraction fraction is calculated from: I/H × 100 where H is the height of the curve (peak counts) resulting from injection of the tracer and I is the Y-intercept.[25] The third portion of the curve, generally a straight line on a semilog plot, begins about 30 seconds after injection and appears to be an index of the rate of metabolism of intracellular ^{11}C-palmitate by the left ventricle. Estimates of extraction fraction based on back extrapolation of this straight line to the Y-intercept may correspond to net extraction since substrate is activated in the cytosol and trapped within the cell primarily as triglyceride or fatty acyl CoA. Although the fate of incorporated labeled fatty acid may

170

entail oxidation resulting in $^{11}CO_2$ production, retention in triglycerides, or retention as fatty acid or acyl CoA, decline of radioactivity in the heart, normalized for radioactive decay, is attributable to oxidation and liberation of $^{11}CO_2$.[55]

EXTERNAL ASSESSMENT OF CARDIAC METABOLISM

At constant rates of delivery of exogenous fatty acid, increased utilization of fatty acid accompanies increased cardiac work.[39] Thus, uptake of ^{14}C-palmitate increases when left ventricular pressure is elevated. Furthermore, the fractional rate of oxidation of incorporated labeled fatty acids would be expected to increase in proportion to metabolic demands, in turn dependent on cardiac work. Accordingly, it seemed likely to us that labeled fatty acid utilization assessed externally might permit noninvasive evaluation of global or regional metabolism. In preliminary experiments,[55] paced isovolumic isolated rabbit hearts were perfused with equimolar albumin and palmitate. ^{11}C-palmitate was injected into the perfusate, and coincidence detection of the clearance of isotope from the heart was accomplished with two NaI(TI) crystals interfaced with an on-line digital computer. After initial washout of the tracer from the vascular space, determinants of cardiac work were altered and the rate of clearance of radioactivity from the heart was measured. The rate of clearance of counts was shown to be an index of the rate of metabolism of palmitate. Interventions designed to increase cardiac work such as acceleration of heart rate from 100 (in preparations paced after the SA node was crushed) to 210 beats/min, elevation of left ventricular end-diastolic pressure to 20 mm Hg, or paired pacing to augment contractility led to accelerated clearance of radioactivity. Furthermore, the slope appeared to vary directly with the calculated extent of increased oxygen demands. In contrast, interventions designed to diminish the oxygen requirements of the heart such as decreasing left ventricular end-diastolic pressure, reducing heart rate from control to 70 to 90 beats/min by crushing the sinus node, or inducing cardioplegia with potassium chloride resulted in markedly decreased rates of clearance of radioactivity. Thus, in isolated heart preparations, the alteration of substrate utilization reflecting metabolic demands and oxygen requirements was readily detectable externally.

EXTERNAL DETECTION AND IMAGING OF METABOLIC CONSEQUENCES OF ISCHEMIA WITH ^{11}C-LABELED FATTY ACIDS

Since oxidation and extraction of fatty acids by severely hypoxic or ischemic myocardium are markedly depressed,[47] external detection of analogous changes was undertaken with ^{11}C-labeled fatty acids. ^{11}C-labeled palmitate was injected into the perfusate of isolated hearts perfused at low (5 ml/min) and high flow (20 ml/min, control) and radioactivity was determined by coincidence detection. Extraction of substrate, detectable externally, decreased by 75 percent during low flow lasting for 30 minutes. The decrease in flow was not solely responsible for altered tracer kinetics since low flow maintained for only one minute did not result in observed changes in extraction even though residence time was prolonged comparably.[25]

To extend these preliminary observations to intact animals, closed chest dogs, instrumented chronically, were injected intravenously with ^{11}C-palmitate before and after transitory coronary occlusion.[25] Gated, computer reconstructed images obtained by positron emission transaxial tomography exhibited decreased uptake of tracer in regions corresponding to zones of ischemia verified at autopsy by enzymatic and histologic analysis. Since radioactivity in the blood pool cleared within three minutes after injection of the tracer, and since there was no substantial accumulation of the radionuclide in the lungs, cardiac endocardial and epicardial borders were well delineated. Reperfusion prior to the development of irreversible ischemic injury followed by a subsequent intravenous injection of ^{11}C-palmitate led to accumulation of tracer in the zone previously subjected to ischemia.

In another group of dogs given [14]C- and ^{11}C-palmitate intravenously 48 hours after infarction produced by occlusion of the left anterior descending coronary artery, infarct size estimated from positron emission transaxial tomograms correlated closely with estimates based on morphological and enzymatic analysis of the same hearts at necropsy. The correlation between myocardial creatine kinase depletion (an index of cell death) and decreased ^{14}C-palmitate content was close (r = .92) and morphological estimation of the extent of infarction in contiguous slices of left ventricular tissue correlated closely with creatine kinase depletion (r = .92). Estimates of the extent of infarction based on tomograms obtained after intravenous injection of ^{11}C-palmitate in vivo correlated closely with both morphologically (r = .97) and enzymatically estimated (r = .93) infarct size. Thus, detection, localization, and quantification of the extent of acute myocardial infarction could be accomplished accurately by positron emission transaxial tomography with ^{11}C-labeled physiological substrates injected intravenously in conscious dogs (Fig. 2).[24]

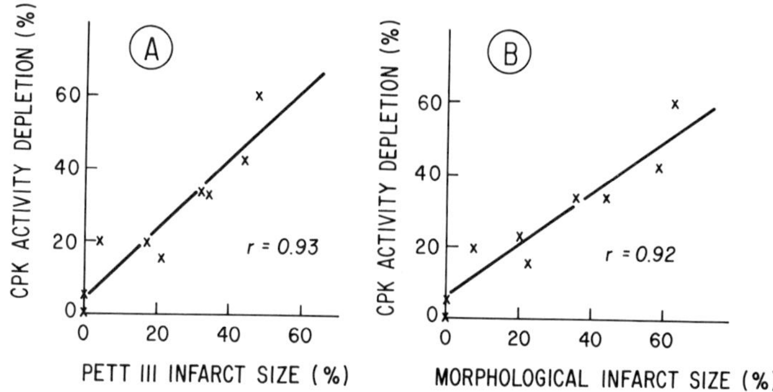

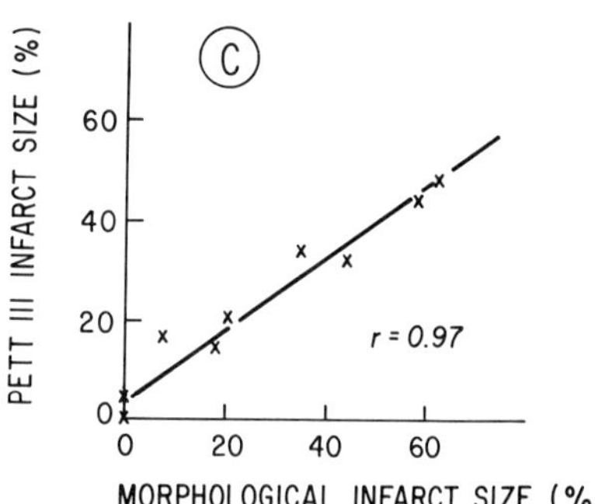

Figure 2. Correlations between myocardial creatine phosphokinase (CPK) depletion and tomographic estimates of infarct size obtained antemortem in the same cross sections of canine hearts subjected to coronary occlusion (A), between CPK depletion and morphological estimates of infarction (B), and between tomographic estimates and morphological estimates (C). For these studies, dogs were subject to coronary occlusion by constriction of an exteriorized snare. Tomograms were obtained after intravenous injection of ^{11}C-palmitate to lightly sedated animals 48 hours later, and when computer reconstructions of the heart were complete, the animals were sacrificed and the hearts subjected to enzymatic and morphological analysis. As can be seen, tomographic estimates of infarction obtained antemortem correlated closely with enzymatic and histological estimates of infarction obtained at necropsy.

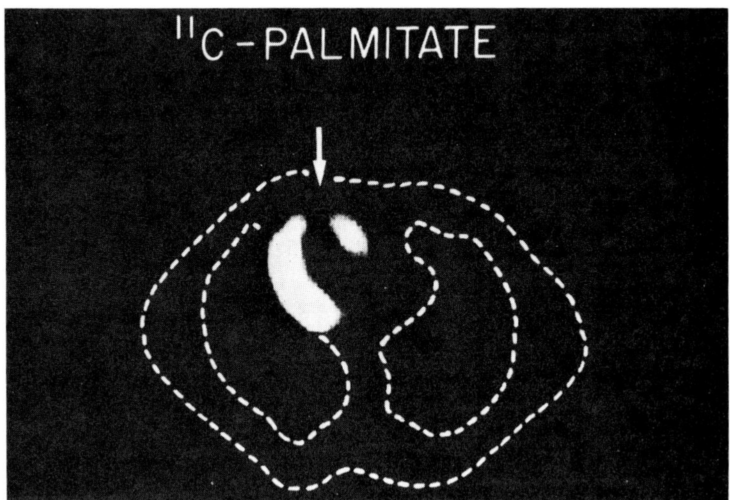

Figure 3. Positron emission transaxial tomogram from a patient with acute anterior myocardial infarction after injection of ^{11}C-palmitate intravenously. The infarct was localized electrocardiographically. As can be seen in this cross section of the heart at the level of the A-V valves, decreased uptake of ^{11}C-palmitate is associated with a region of anterior infarction. The gap posteriorly is normal and results from contributions to the posterior border of the heart at this level by the mitral valve apparatus and the left atrium.

Analogous results have been obtained recently in patients with remote myocardial infarction.[56] Ten normal volunteers and twelve patients with remote myocardial infarction documented by enzyme and electrocardiographic criteria at the time of the acute episode were given 5 to 10 mCi of ^{11}C-palmitate intravenously 3 to 6 months after infarction. Images were obtained with positron emission transaxial tomography three minutes after injection. Diminished accumulation of tracer indicative of infarction was observed in tomograms from all patients despite the normalization of the electrocardiogram (regression of Q waves) in some cases. The locus of the infarct detected tomographically corresponded to the electrocardiographic locus in each patient (Fig. 3). Uptake of tracer was homogeneous in cross sections of the heart from all normal subjects.

Positron emission transaxial tomographic images obtained after intravenous injection of ^{11}C-palmitate do not distinguish fresh from old infarction or zones of jeopardized ischemic myocardium from tissue irreversibly injured. However, persistent defects in accumulation are indicative of infarction and serial tomograms can be obtained readily. In addition, positron emission transaxial tomography should permit precise assessment of flow with radiopharmaceuticals under development that remain confined to the vascular space.

The positron emission transaxial tomograph now available at our institution reconstructs seven 1.5-cm-thick cross-sectional images simultaneously with total imaging time of less than 10 minutes. At the level of the atrioventricular valve, the cross section of the left ventricle exhibits a horseshoe pattern, since the posterior border is comprised of mitral valve apparatus and atrium. The right ventricle is visualized also, but its accumulation of ^{11}C-palmitate is markedly less because its metabolic demands are less than those of the left ventricle and because its wall is so much thinner. At the mid-ventricular level, the left ventricular image appears to be circular.[57]

FUTURE DIRECTIONS

Fatty acids labeled with short-lived, positron-emitting radionuclides appear to be useful in delineating regional myocardial intermediary metabolism in quantitative terms

173

and hence in evaluating not only patients with coronary artery disease but also those with cardiomyopathic processes. Assessment of metabolism of fatty acids, amino acids, or carbohydrates, all of which can be labeled with positron-emitting isotopes, should permit recognition and characterization of cardiac metabolic derangements associated with chronic ethanol abuse, steroid administration, viral infections, or diabetes mellitus. Assessment of the balance between myocardial oxygen supply and demand may be simplified by quantification of regional substrate extraction and utilization, and hence objective assessment of effects of drugs used to alter cardiac work in the presence of acute myocardial infarction should be facilitated by serial studies in patients with positron emission transaxial tomography. In view of the importance of early quantification of the mass of myocardium in jeopardy soon after the onset of ischemia for objective evaluation of potentially therapeutic interventions, positron emission transaxial tomography should be particularly useful in studies designed to evaluate the feasibility and efficacy of protecting ischemic myocardium in man.

ACKNOWLEDGMENT

We gratefully appreciate Ms. Carolyn Lohman's preparation of the manuscript.

REFERENCES

1. SOBEL, B. E., BRESNAHAN, G. F., SHELL, W. E., ET AL.: *Estimation of infarct size in man and its relation to prognosis.* Circulation 46:640, 1972.

2. BRAUNWALD, E. (ed.): *Protection of the ischemic myocardium.* Circulation 53(Suppl. I), 1976.

3. ROSS, J., JR.: *Electrocardiographic ST-segment analysis in the characterization of myocardial ischemia and infarction.* Circulation 53(Suppl. I):73, 1976.

4. HOLLAND, R. P., AND BROOKS, H.: *TQ-ST segment mapping: Critical review and analysis of current concepts.* Am. J. Cardiol. 40:110, 1977.

5. DAVIS, L. D., AND TEMTE, J. V.: *Effects of propranolol on the transmembrane potentials of ventricular muscle and Purkinje fibers of the dog.* Circ. Res. 22:661, 1968.

6. ROSEN, M. R., AND ILVENTO, J. P.: *The electrophysiologic basis for the suppression of cardiac arrhythmias by verapamil.* Am. J. Cardiol. 33:166, 1974.

7. ROBERTS, R., HENRY, P. D., AND SOBEL, B. E.: *An improved basis for enzymatic estimation of infarct size.* Circulation 52:743, 1975.

8. TER-POGOSSIAN, M. M.: *Limitations of present radionuclide methods in the evaluation of myocardial ischemia and infarction.* Circulation 53(Suppl. I):119, 1976.

9. KUHL, D. E., EDWARDS, R. Q., RICCI, A. R., ET AL.: *Quantitative section scanning.* In *Medical Radioisotope Scintigraphy*, Vol. I. IAEA, Vienna, 1973, p. 347.

10. TER-POGOSSIAN, M. M., PHELPS, M. E., HOFFMAN, E. J., ET AL.: *A positron-emission transaxial tomograph for nuclear imaging (PETT).* Radiology 114:89, 1975.

11. STRAUSS, H. W., HARRISON, K., LANGAN, J. K., ET AL.: *Thallium-201 for myocardial imaging. Relation of thallium-201 to regional myocardial perfusion.* Circulation 51:641, 1975.

12. ZARET, B. L., STRAUSS, H. W., MARTIN, N. D., ET AL.: *Noninvasive regional myocardial perfusion with radioactive potassium. Study of patients at rest, with exercise and during angina pectoris.* N. Engl. J. Med. 288:809, 1973.

13. MARTIN, N. D., ZARET, B. L., McGOWAN, R. L., ET AL.: *Rubidium-81: A new myocardial scanning agent.* Radiology 111:651, 1974.

14. ZARET, B. L.: *Myocardial imaging with radioactive potassium and its analogs.* Prog. Cardiovasc. Dis. 20:81, 1977.

15. ZIEGLER, W. H., AND GORESKY, C. A.: *Kinetics of rubidium uptake in the working dog heart.* Circ. Res. 29:208, 1971.

16. CANNON, P. J., DELL, R. B., AND DWYER, E. M., JR.: *Measurement of regional myocardial perfusion in man with ^{133}xenon and a scintillation camera.* J. Clin. Invest. 51:964, 1972.

17. COHEN, L. S., ELLIOTT, W. C., AND GORLIN, R.: *Measurement of myocardial blood flow using krypton 85.* Am. J. Physiol. 206:997, 1964.

18. CANNON, P. J., WEISS, M. B., AND SCIACCA, R. R.: *Myocardial blood flow in coronary artery disease: Studies at rest and during stress with inert gas washout techniques.* Prog. Cardiovasc. Dis. 20:95, 1977.

19. NEEDLEMAN, P., MARSHALL, G. R., AND SOBEL, B. E.; *Hormone interactions in the isolated rabbit heart. Synthesis and coronary vasomotor effects of prostaglandins, angiotensin, and bradykinin.* Circ. Res. 37:802, 1975.

20. BASSINGTHWAIGHTE, J. B., STRANDELL, T., AND DONALD, D. E.: *Estimation of coronary blood flow by washout of diffusible indicators.* Circ. Res. 23:259, 1968.

21. COLEMAN, R. E., KLEIN, M. S., ROBERTS, R., ET AL.: *Improved detection of myocardial infarction with technetium-99m stannous pyrophosphate and serum MB creatine phosphokinase.* Am. J. Cardiol. 37:732, 1976.

22. PARKEY, R. W., BONTE, F. J., MEYER, S. L., ET AL.: *A new method for radionuclide imaging of acute myocardial infarction in humans.* Circulation 50:540, 1974.

23. TOMANEK, R. J., MARCUS, M. L., EHRHARDT, J. C., ET AL.: *Technetium pyrophosphate uptake in infarcted cardiac muscle: Relative importance of extent of necrosis and myocardial perfusion.* Am. J. Cardiol. 37:177, 1976.

24. WEISS, E. S., AHMED, S. A., WELCH, M. J., ET AL.: *Quantification of infarction in cross sections of canine myocardium in vivo with positron emission transaxial tomography and ^{11}C-palmitate.* Circulation 55:66, 1977.

25. WEISS, E. S., HOFFMAN, E. J., PHELPS, M. E., ET AL.: *External detection and visualization of myocardial ischemia with ^{11}C-substrates in vitro and in vivo.* Circ. Res. 39:24, 1976.

26. TER-POGOSSIAN, M. M., HOFFMAN, E. J., WEISS, E. S., ET AL.: *Positron emission reconstruction tomography for the assessment of regional myocardial metabolism by the administration of substrates labeled with cyclotron produced radionuclides.* In Harrison, D. C., Sandler, H., and Miller, H. A. (eds.): *Proceedings from the Conference on Cardiovascular Imaging and Image Processing Theory and Practice*, Vol. 72. Society of Photo-Optical Instrumentation Engineers, Palos Verdes Estates, Cal., 1975, p. 277.

27. WEISS, E. S., SIEGEL, B. A., SOBEL, B. E., ET AL.: *Evaluation of myocardial metabolism and perfusion with positron-emitting radionuclides.* Prog. Cardiovasc. Dis. 20:191, 1977.

28. CORMACK, A. M.: *Reconstruction of densities from their projections, with applications in radiological physics.* Phys. Med. Biol. 18:195, 1973.

29. POE, N. D.: *A critical evaluation of myocardial imaging.* In Subramanian, G., Rhodes, B. A., Cooper, J. F., et al. (eds.): *Radiopharmaceuticals.* Society of Nuclear Medicine, New York, 1975, p. 359.

30. BELLER, G. A., HOOP, B., PARKER, J. A., ET AL.: *Sequential myocardial imaging with rubidium-82, an ultra-short lived radionuclide.* Circulation 52(Suppl. II):110, 1975.

31. BELLER, G. A., ALTON, W. J., MOORE, R. H., ET AL.: *Detection of nitroglycerin-induced changes in regional myocardial perfusion during acute ischemia by serial imaging with $^{82}Rb^{+}$.* Circulation 54(Suppl. II):216, 1976.

32. HNATOWICH, D. J.: *Labeling of tin-soaked albumin microspheres with ^{68}Ga.* J. Nucl. Med. 17:57, 1976.

33. HARPER, P. V., LATHROP, K. A., KRIZEK, H., ET AL.: *^{13}N radiopharmaceuticals* In Subramanian, G., Rhodes, B. A., Cooper, J. F., et al. (eds.): *Radiopharmaceuticals.* Society of Nuclear Medicine, New York, 1975, p. 180.

34. HARPER, P. V., LATHROP, K. A., KRIZEK, H., ET AL.: *Clinical feasibility of myocardial imaging with $^{13}NH_3$.* J. Nucl. Med. 13:278, 1972.

35. HARPER, P. V., SCHWARTZ, J., BECK, R. N., ET AL.: *Clinical myocardial imaging with nitrogen-13 ammonia.* Radiology 108:613, 1973.

36. WALSH, W. F., HARPER, P. V., RESNEKOV, L., ET AL.: *Noninvasive evaluation of regional myocardial perfusion in 112 patients using a mobile scintillation camera and intravenous nitrogen-13 labeled ammonia.* Circulation 54:266, 1976.

37. PHELPS, M. E., HOFFMAN, E. J., COLEMAN, R. E., ET AL.: *Tomographic images of blood pool and perfusion in brain and heart.* J. Nucl. Med. 17:603, 1976.

38. DEWANJEE, M. K., BEH, R., AND HNATOWICH, D. J.: *New Ga-68-labeled skeletal imaging agents for positron scintigraphy.* J. Nucl. Med. 17:566, 1976.

39. NEELY, J. R., ROVETTO, M. J., AND ORAM, J. F.: *Myocardial utilization of carbohydrate and lipids.* Prog. Cardiovasc. Dis. 15:289, 1972.

40. WIELAND, B. W.: *Development and evaluation of facilities for the efficient production of compounds labeled with carbon-11 and oxygen-15 at the Washington University Medical cyclotron,* Ph.D. thesis, Ohio State University, 1974.

41. POE, N. D., ROBINSON, G. D., AND MacDONALD, N. S.: *Myocardial extraction of variously labeled fatty acids and carboxylates.* J. Nucl. Med. 14:440, 1973.

42. ROBINSON, G. D., JR., AND LEE, A. W.: *Radioiodinated fatty acids for heart imaging: Iodine monochloride addition compared with iodide replacement labeling.* J. Nucl. Med. 16:17, 1975.

43. MORGAN, H. E., HENDERSON, M. J., REGEN, D. M., ET AL.: *Regulation of glucose uptake in heart muscle from normal and alloxan-diabetic rats: The effects of insulin, growth hormone, cortisone, and anoxia.* Ann. N.Y. Acad. Sci. 82:387, 1959.

44. FRITZ, I. B., AND YUE, K. T. N.: *Long-chain carnitine acyltransferase and the role of acylcarnitine derivatives in the catalytic increase of fatty acid oxidation induced by carnitine.* J. Lipid Res. 4:279, 1963.

45. GOODMAN, D. S.: *The interaction of human serum albumin with long-chain fatty acid anions.* J. Am. Chem. Soc. 80:3892, 1958.

46. SPECTOR, A. A., FLETCHER, J. E., AND ASHBROOK, J. D.: *Analysis of long-chain free fatty acid binding to bovine serum albumin by determination of stepwise equilibrium constants.* Biochemistry 10:3229, 1971.

47. OPIE, L. H.: *Metabolism of the heart in health and disease.* Part I. Am. Heart J. 76:685, 1968.

48. GOUSIOS, A., FELTS, J. M., AND HAVEL, R. J.: *The metabolism of serum triglycerides and free fatty acids by the myocardium.* Metabolism 12:75, 1963.

49. WARTMAN, W. B., JENNINGS, R. B., YOKOYAMA, H. O., ET AL.: *Fatty change of the myocardium in early experimental infarction.* Arch. Pathol. 62:318, 1956.

50. KOPEC, B., AND FRITZ, I. B.: *Properties of a purified carnitine palmitoyltransferase, and evidence for the existence of other carnitine acyltransferases.* Can. J. Biochem. 49:941, 1971.

51. BROSNAN, J. T., AND FRITZ, I. B.: *The permeability of mitochondria to carnitine and acetylcarnitine.* Biochem. J. 125:94P, 1971.

52. WAKIL, S. J.: *Fatty acid metabolism.* In Wakil, S. J. (ed.): *Lipid Metabolism.* Academic Press, New York, 1970, p. 1.

53. GOLDSTEIN, R. A., KLEIN, M. S., AND SOBEL, B. E.: *Altered myocardial lipid metabolism early after the onset of ischemia.* J. Nucl. Med., in press.

54. RAICHLE, M. E., LARSON, K. B., PHELPS, M. E., ET AL.: *In vivo measurement of brain glucose transport and metabolism employing glucose-[11]C.* Am. J. Physiol. 228:1936, 1975.

55. KLEIN, M. S., GOLDSTEIN, R. A., WELCH, M. J., ET AL.: *External quantification of myocardial metabolism with ^{11}C-labeled fatty acids.* Am. J. Cardiol. 41:378, 1978.

56. SOBEL, B. E., WEISS, E. S., WELCH, M. J., ET AL.: *Detection of remote myocardial infarction in patients with positron emission transaxial tomography and intravenous ^{11}C-palmitate.* Circulation 55:853, 1977.

57. SOBEL, B. E.: *External quantification of myocardial ischemia and infarction with positron-emitting radionuclides.* In Vogel, J. H. K. (ed.): *Advances in Cardiology,* Vol. 22. S. Karger, Basel, in press.

Positron Radionuclide Assessment
of Myocardial and Pulmonary Blood Flow

George A. Beller, M.D.

In recent years, considerable advances have been made in the development of methods to visualize the heart with radiolabeled agents for the assessment of regional myocardial perfusion.[1-5] However, because of the relatively long half-lives of the gamma-emitting radionuclides employed in the majority of these scanning techniques, rapid sequential imaging of the myocardium has not readily been accomplished. Additionally, myocardial scintigraphy with the Anger scintillation camera only permits two-dimensional visualization of the heart. Activity present in structures anterior or posterior to the heart is superimposed on myocardial activity, thus markedly altering image contrast. Absolute quantitation of radioactivity in heart muscle is difficult to obtain since the resolution and sensitivity of the gamma scintillation camera depend upon the depth of distribution of activity within the patient as well as upon absorption of radiation within the body. Another present drawback to gamma camera imaging is the limited useful range of gamma energy over which good resolution is obtainable unless special shielding is added to the collimator. Many of the short-lived radionuclides available for myocardial perfusion imaging have too high an energy for imaging with a gamma camera.

A new approach to the scintigraphic assessment of regional myocardial perfusion has been developed which utilizes certain short-lived positron-emitting radiopharmaceuticals. A positron is a particle which has the same mass as an electron but is positively charged. As it decays, a positron first loses its kinetic energy through multiple interactions similar to those undergone by electrons. It then undergoes annihilation where the positron interacts with an electron and their mass is converted to two gamma photons that travel in diametrically opposite directions. Each annihilation photon has an energy of 511 keV. A positron camera is a multicrystal device which is designed specifically for coincidence detection of these two annihilation quanta. The camera uses coincidence techniques to collimate positron annihilation radiation. Positron-emitting radionuclides are cyclotron or generator produced and their use permit rapid sequential assessment of regional myocardial blood flow by external imaging with a positron camera.

Positron scintigraphy also provides the means to obtain multiple computer-reconstructed tomographic images of the heart in addition to the conventional two-dimensional projection images. The tomographic reconstruction approach avoids difficulties of low contrast and resolution encountered in two-dimensional display of superimposed regions. Additionally, it allows for accurate quantitation of the amount of radioactivity per volume of interest in regions of the heart, since correction for absorption within the body can be accomplished.

The intent of this chapter is to review the various applications of positron scintigraphy, currently employed clinically or under experimental investigation, for the assessment of regional myocardial and pulmonary blood flow.

POSITRON CAMERA INSTRUMENTATION

The Massachusetts General Hospital (MGH) positron camera is a multicrystal device designed for the detection of the gamma photons emitted in opposite directions upon the annihilation of a positron.[6] This camera uses coincidence detection techniques to achieve collimation of positron annihilation radiation and is similar in principle to other positron camera devices in use or under development.[7-9] Because (for a given patient thickness) the sum of the distance traversed by the two annihilation quanta is constant, their attenuation is independent of the location of a source positioned between the two detectors. The camera has two parallel detector arrays, each of which contains 127 NaI (Tl) crystals viewed by an array of 72 photomultiplier tubes (Fig. 1). Each photomultiplier tube detects scintillations in four crystals and each crystal is viewed by two photomultiplier tubes. Each crystal can thus be uniquely identified by scintillations in two photomultiplier tubes. An event is recorded when simultaneous scintillations are detected in a crystal in each array within the twenty nanosecond resolving time of the camera.

A total of 2548 possible coincidence pairs of crystals are available for imaging, permitting a high resolution uniform field response over an area of 27 by 30 cm. Spatial resolution is increased by moving the detector arrays one intercrystal distance in ten steps in a direction at 45 degrees with respect to the crystal array. The data rate capacity of the camera (10,000 coincidence counts/sec with approximately 10 percent random coincidence counts) is sufficient for sequential imaging for dynamic studies with collection times of 0.15 second per image.[10]

An image is constructed from the coincident data using a Modular Computer Systems, Inc. Model II/10, which has 64 K bytes of core memory, 5.2 M bytes of disc storage, and a

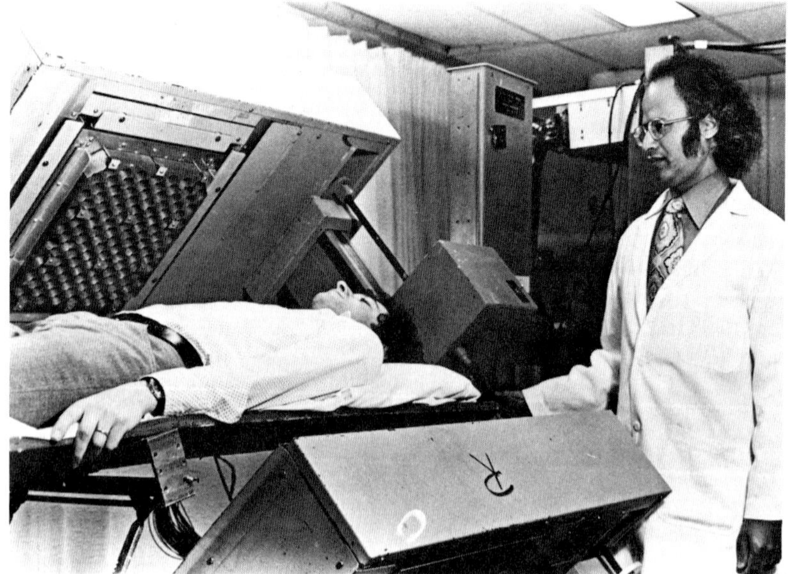

Figure 1. The MGH positron camera with two parallel detector arrays, each containing 127 NaI crystals. For transverse section emission tomography, 208 scintigrams are obtained as the camera rotates 180 degrees around the subject.

178

teletype terminal. The software system employed is the OS/S system which makes available a higher level programming language PL/S.[11] Images are viewed on an oscilloscope with a 128 by 128 raster with 64 gray levels.

TRANSVERSE TOMOGRAPHIC IMAGING OF POSITRON-EMITTING RADIONUCLIDES

Computer transverse section emission tomography with the positron camera permits the three-dimensional visualization of myocardial distribution of positron-emitting radionuclides.[12] The three-dimensional reconstruction algorithm employed is essentially identical to that of the x-ray CT body scanner. The MGH transverse section tomographic technique uses as input 208 scintigrams taken at uniformly spaced angles as the positron camera rotates 180 degrees around the subject. Computer processing of this data is performed and multiple reconstructed transverse section emission images of heart are obtained. If a correction for absorption is desired, a set of transmission images from a positron-emitting planar source is obtained corresponding to each of the emission images.

MYOCARDIAL PERFUSION IMAGING WITH POSITRON-EMITTING MONOVALENT CATIONS

The existence of active transport mechanisms for concentrating certain monovalent cations in normal myocardial tissue has led to the use of radionuclides of ionic potassium, rubidium, cesium, ammonia, and thallium for myocardial imaging for the assessment of regional myocardial blood flow. Of the many radioisotopes of potassium analogs, relatively few are positron emitters. Among these are ^{38}K, ^{81}Rb, ^{82}Rb, ^{82m}Rb, ^{84}Rb, and ^{13}NH$_4^+$. Table 1 summarizes the physical properties of positron-emitting radioisotopes of monovalent cations currently employed for clinical and experimental studies of regional myocardial perfusion. The clinical application of perfusion imaging with monovalent cations suitable for imaging with a gamma camera or rectilinear scanner has been reported.[1,13-16] The application of radionuclide monovalent cations employed for positron imaging of regional myocardial perfusion will subsequently be described in greater detail. These agents concentrate in myocardium in proportion to blood flow, cell membrane integrity, and the size of the regional cationic pool. These radiotracers are administered intravenously and their uptake by heart muscle is proportional to the fraction of the cardiac output perfusing the myocardium. Areas characterized by diminished regional myocardial flow, ischemia, or infarction appear as "cold spots" of diminished radioactivity on myocardial scintiscans.

^{13}N-ammonia

^{13}N-ammonia is a radionuclide labeled monovalent cation with interesting applications in myocardial scintigraphy. Ammonium ion labeled with ^{13}N ($T_{1/2} = 10$ min) is cyclotron produced by deuteron irradiation of methane[17] via the ^{12}C (d,n) ^{13}N nuclear reaction. With the present production method, approximately $0.20 \pm .04$ mg stable ammonia is obtained per milliliter of solution (normal blood levels of ammonia are between 0.8 and 1.1 mg/ml of blood). A typical adult dose is 5 ml of solution, or approximately 1.0 μg is NH$_3$. In a closed-chest canine experimental model, 90 percent uptake of ^{13}NH$_4^+$ was observed in the first capillary transit after intracoronary injection.[18]

^{13}N-ammonia, administered via the intravenous route, has been employed to assess regional myocardial perfusion with both an Anger[19] and positron[18] scintillation camera. Images obtained in normal dogs and normal patients with the MGH positron camera after intravenous ^{13}NH$_4^+$ demonstrated uniform uptake in the left ventricular myocardium. Figure

179

Table 1. Radionuclide monovalent cations used in positron imaging of the myocardium

Radionuclides	Half-life	Principal Useful Decay	Principal Gamma Energies (MeV)
^{38}K	7.7 min	B$^+$ 100% to 2.2 state	a.r., 2.2
^{81}Rb	4.7 hr	e.c. 67% to 0.89 state	0.450, 0.253, 0.190, (a.r.) decays to 13^{-sec} ^{81m}Kr
^{82}Rb	75 sec	B$^+$ 92% to ground state	a.r. daughter of 25-day ^{82}Sr
^{84}Rb	33 day	B$^+$ 73% to 0.880 state	a.r., 0.880
^{13}NH$_4$	10 min	B$^+$ 100% to ground state	a.r.

e.c. = electron capture; a.r. = annihilation radiation

2 shows typical images of the myocardium after intravenous administration of 2 mCi of ^{13}NH$_4^+$ in four normal dogs. Images obtained after experimental coronary occlusion demonstrated reduced ^{13}NH$_4^+$ uptake in the distribution of the occluded coronary vessels. Figure 3 shows control and postocclusion images in one such experimental preparation. Transverse section tomographic imaging with ^{13}NH$_4^+$ and positron scintigraphy has also been successfully undertaken in experimental animals[21] and in man.[22] Figures 4 and 5 show tomographic images of the heart in a normal and experimentally infarcted dog, respectively. Regional myocardial uptake of ^{13}NH$_4^+$ correlates well with distribution of radioactively labeled microspheres in pigs undergoing experimental myocardial infarction, suggesting that its distribution is indeed proportional to regional perfusion.[23]

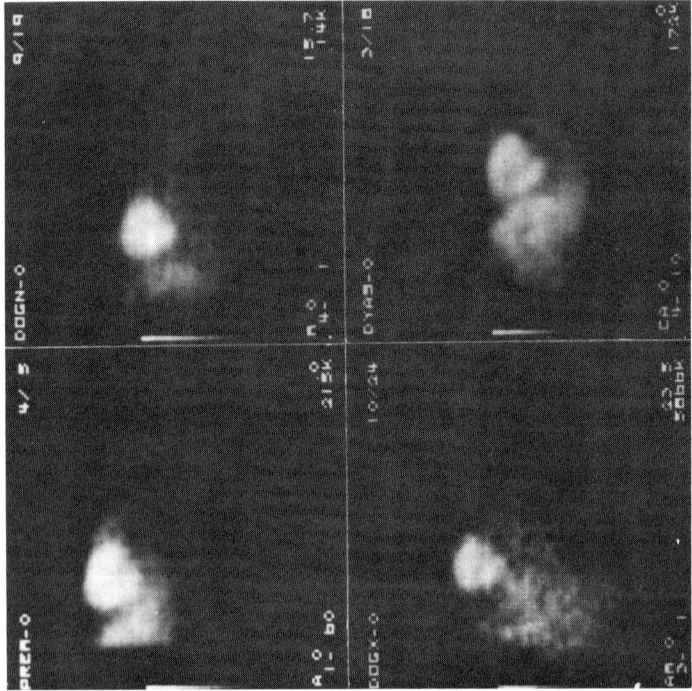

Figure 2. Myocardial images obtained in the left lateral projection in four dogs after intravenous administration of ^{13}NH$_4^+$. Homogeneous distribution of the radionuclide in the left ventricular myocardium is evident. Activity below the heart represents uptake of ^{13}NH$_4^+$ in the splanchnic organs.

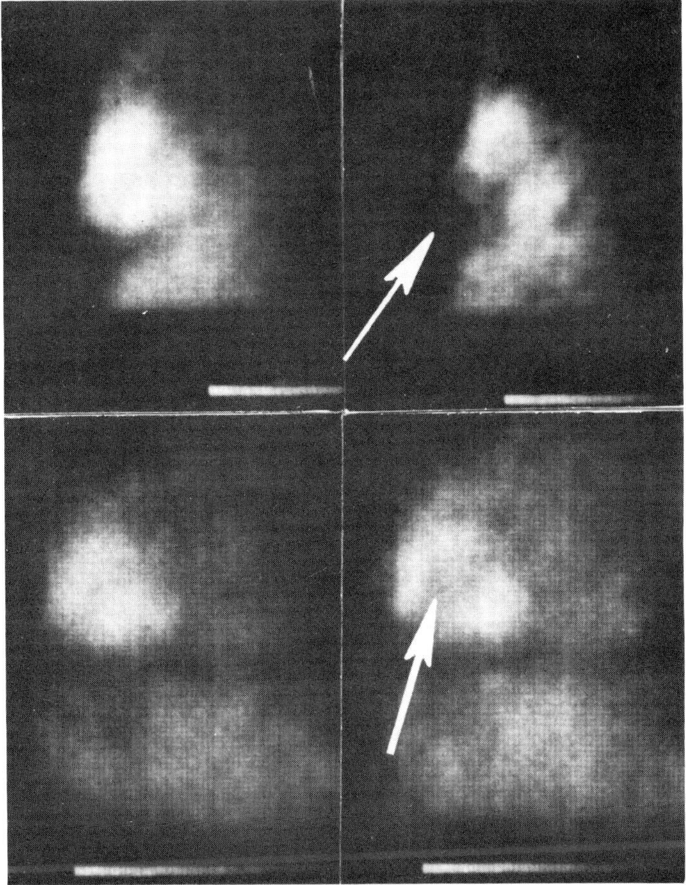

Figure 3. Lateral projection $^{13}NH_4^+$ images in two dogs prior to (left) and after one hour of coronary occlusion. Perfusion defects are designated by arrows in post occlusion images (right). In the first dog (upper row), there is diminished apical activity. In the second dog (lower row), the defect overlies the left ventricular cavity and is seen *en face.*

A potential major limitation of the clinical application of $^{13}NH_4^+$ imaging is the observation of marked pulmonary uptake of $^{13}NH_4^+$, particularly in smokers. The high lung activity may obscure adequate visualization of the heart. Despite this drawback, Walsh and coworkers demonstrated perfusion defects in 82 percent of patients with significant coronary artery stenoses and 96 percent of patients with acute myocardial infarction.[19]

Rubidium-81

Rubidium-81 is another potassium analog that has successfully been employed as an imaging agent for the assessment of regional myocardial perfusion.[13] Carrier-free $^{81}Rb^+$ is produced from Krypton gas via ^{82}K (P, 2n) ^{81}Rb reaction and contains less than 25 percent of ^{82m}Rb at calibration time.[24] Results of prior clinical studies employing an Anger scintillation camera with a pinhole collimater and heavy lead shielding have demonstrated that areas of stress-induced ischemia or infarction appear as regions of diminished rubidium uptake on gamma scintigrams.[13] Rubidium-81 ($T_{1/2} = 4.6$ hrs) decays 67 percent by elec-

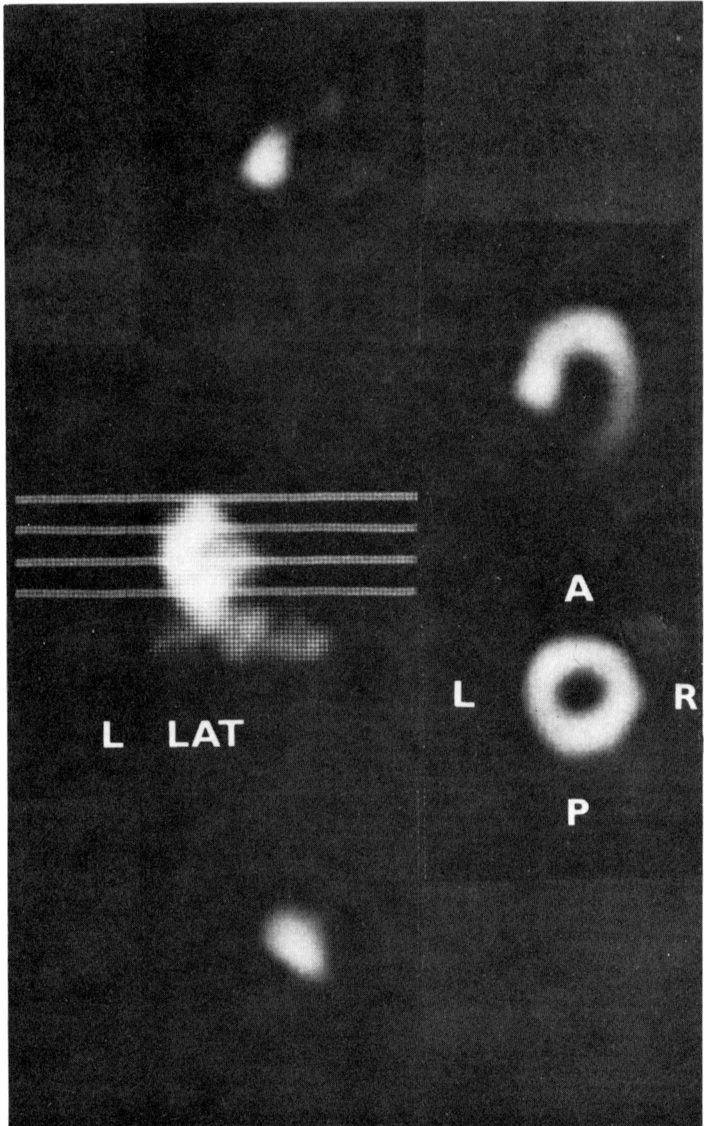

Figure 4. Four tomographic section images of normal left ventricle after intravenous administration of 2 mCi of $^{13}NH_4^+$ to a closed-chest dog. On the left is a lateral two-dimensional image with planes of tomographic sections superimposed. The lower panel shows the cut through the apex, below the left ventricular cavity. The next panel, going counterclockwise, is a cut through the center of the left ventricle (A, anterior; P, posterior; L, lateral wall; R, septum) showing homogeneous uptake of the tracer in an annular pattern around the left ventricular cavity. The next panel shows a cut through the base of the left ventricle. The horseshoe appearance of this tomographic cut is related to relatively little activity in the posteriorly positioned atrial wall. The uppermost panel is a cut through the high anteroseptal area, above the left ventricular cavity.

tron capture and 33 percent by positron emission which makes it suitable for myocardial perfusion imaging with a positron camera. Figure 6 shows two-dimensional positron images of the heart in the anterior, left lateral, and left anterior oblique projections obtained 5 minutes after the intravenous administration of 2mCi of $^{81}RbCl$ in a normal dog. Six

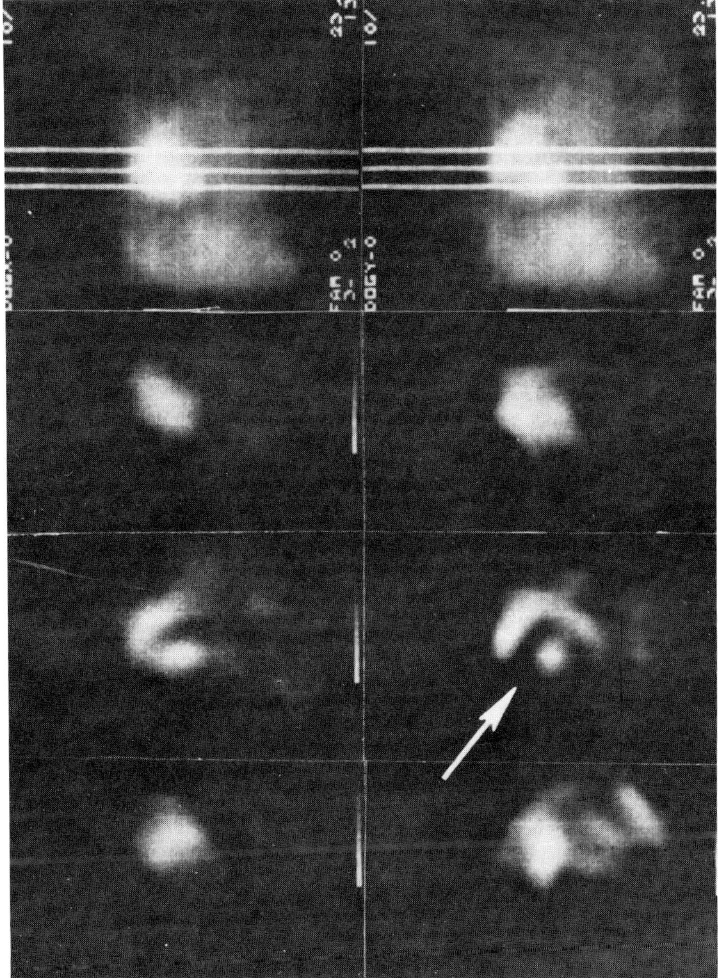

Figure 5. Tomographic images obtained after intravenous administration of 2 mCi of ^{13}NH$_4{}^+$ prior to (left), and after coronary occlusion (right) in an intact, anesthetized dog. The heart was imaged in situ, with the chest closed. The upper panels show lateral two-dimensional images with planes of tomographic sections superimposed. The upper right image demonstrates a perfusion defect in the anterolateral region. The next three pairs of images correspond to the tomographic cuts in the upper panels going from the base to the apex of the heart. In the center section, the preocclusion image shows the typical horseshoe distribution of activity with little uptake in the posterior left atrial wall. The corresponding image after occlusion (right) shows the perfusion defect in the anterolateral wall as shown by the arrow.

transverse section emission images comprising the entire left ventricle from apex to base were computer reconstructed and are shown in Figure 7. Tomographic images through the midventricular level demonstrate homogenous myocardial ^{18}Rb$^+$ distribution in an annular pattern around the left ventricular cavity. Tomographic cuts through the base of the heart demonstrate a horseshoe appearance of activity anteriorly and diminished tracer uptake posteriorly in the left atrial wall. Computerized tomographic emission imaging of the heart has not yet been reported in patients.

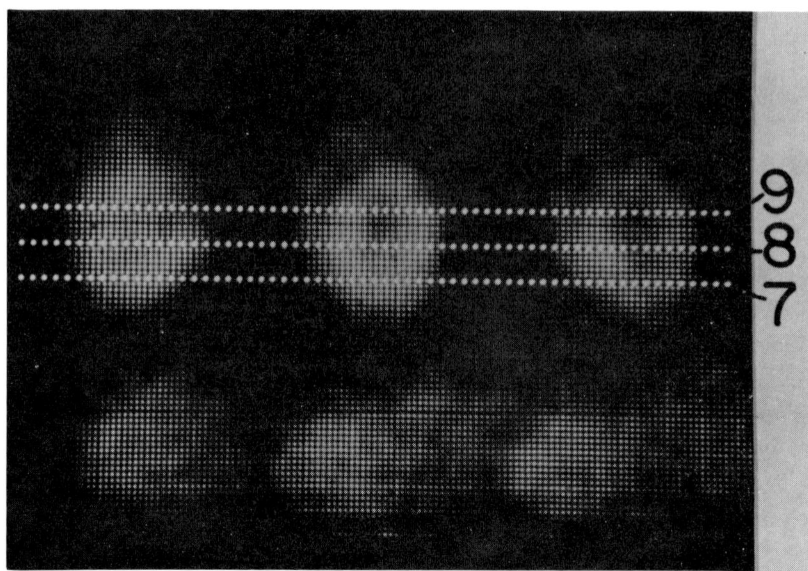

Figure 6. From left to right are shown the left lateral, 40 degree LAO, and anterior projection images obtained in a normal dog after the intravenous administration of ^{81}RbCl. Dotted lines 7, 8, and 9 delineate the levels of the computer-reconstructed tomographic cuts shown in Figure 7.

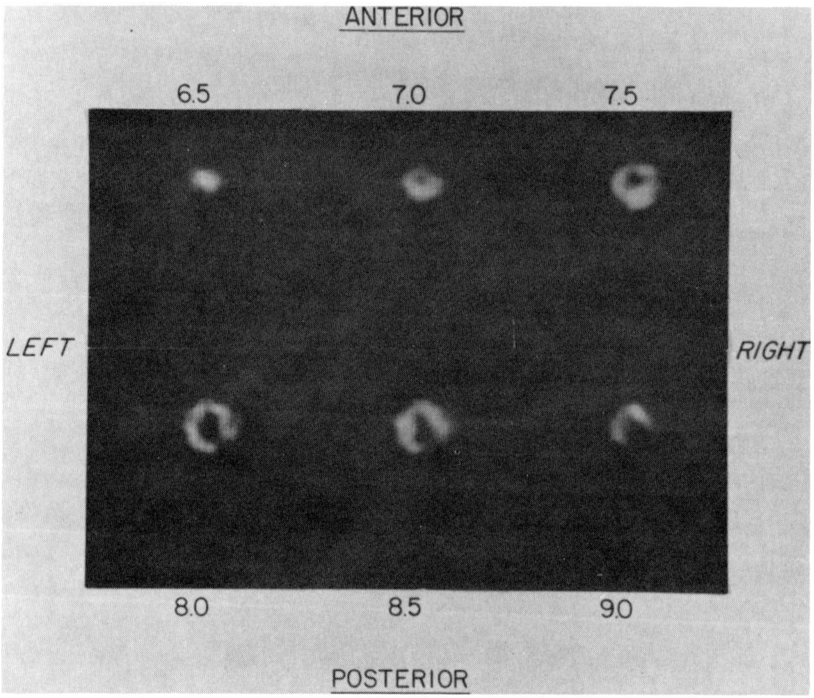

Figure 7. Multiple transverse section emission images of the left ventricle in multiple planes from apex (cut 6.5) to base (cut 9.0). The transverse section image in cut 8 is taken from the midventricular level and corresponds to line 8 shown in Figure 6. On this image, ^{81}Rb$^+$ activity is distributed in an annular pattern around the left ventricular cavity. In tomographic cut 9, the thin posterior left atrial wall is not visualized.

Rubidium-82

^{82}Rb$^+$ is an ultrashort-lived ($T_{1/2} = 75$ sec) positron-emitting monovalent cation which is obtained directly from a portable generator that contains the parent isotope ^{82}Sr deposited on an ion exchange resin (BioRex-70 or Chelex-100; BioRad Laboratories, Richmond, CA) in a stainless steel column.[25] Strontium-82 ($T_{1/2} = 25$ days) decays entirely by electron capture to ^{82}Rb$^+$. Of 100 percent ^{82}Rb$^+$ disintegration, 96.2 percent yields positrons, 29 percent yields 0.77 MeV gamma rays and 1.9 percent yields 1.47 or 0.77 and 0.70 MeV gamma rays. Since the half-life of ^{82}Rb$^+$ is much shorter than its parent ^{82}Sr^{++}, ^{82}Rb$^+$ activity is in secular equilibrium with its parent. When the ^{82}Rb$^+$ is separated from the ^{82}Sr^{++}, it again builds up with a simple exponential growth curve determined by a half-life of 75 seconds. Hence, a source of ^{82}Sr^{++} activity constitutes an equivalent source of ^{82}Rb$^+$ activity, which can be eluted from the generator every 5 minutes, providing a source of sequential doses for intravenous administration.

Rubidium-82 has certain distinct advantages over ^{13}NH$_4{}^+$ and ^{81}Rb$^+$ as a tracer for myocardial perfusion imaging. ^{13}NH$_4{}^+$ and ^{81}Rb$^+$ production require a cyclotron, whereas ^{82}Rb$^+$ can be obtained from a portable generator which can be transported to a patient's bedside. Because of its short half-life, sequential ^{82}Rb$^+$ images can be obtained at intervals as short as 5 minutes, allowing assessment of interventions that alter regional myocardial flow patterns.[26] To illustrate this unique property of ^{82}Rb$^+$, Figure 8 shows serial ^{82}Rb$^+$ images obtained with the positron camera in the left lateral projection in a dog undergoing coronary occlusion followed 5 minutes later by reperfusion. The image taken during the early reflow period demonstrated enhanced ^{82}Rb$^+$ uptake, reflecting reactive hyperemia in

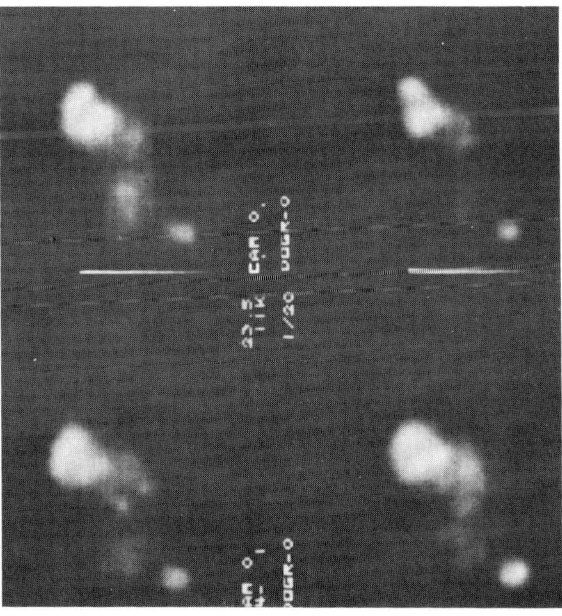

Figure 8. Serial ^{82}Rb$^+$ images obtained in the left lateral projection with the positron camera in an intact dog during control (upper left), 60 minutes after left anterior descending coronary artery occlusion (upper right), 5 minutes after reperfusion (lower left) and 45 minutes after reflow. In the postocclusion image (upper right), an anteroapical perfusion defect is observed. After reperfusion (lower left) an area of increased ^{82}Rb$^+$ uptake, perhaps reflecting reactive hyperemia, is demonstrated in the previously ischemic zone. The final image in the sequence shows only a small residual apical defect.

the region of myocardium supplied by the previously occluded vessels. Rapid sequential myocardial imaging with [82]Rb$^+$ has also recently been successfully employed in the detection of nitroglycerin-induced alterations in regional myocardial perfusion in dogs undergoing occlusion of the left anterior descending coronary artery.[27] Preliminary clinical studies with this radionuclide are presently being undertaken in patients, and its clinical utility appears promising.

Because of the shorter physical half-life, the radiation exposure with [82]Rb$^+$ is less than that of [13]NH$_4^+$ or [81]Rb$^+$, although radiation doses for all three positron-emitting monovalent cations are quite small and significantly less than doses of the radionuclides with longer half-lives currently employed clinically for perfusion imaging with an Anger scintillation camera, such as thallium-201. A potential disadvantage of [82]Rb$^+$ is that the half-life may actually be too short to be utilized for computer assisted transverse section emission imaging with the positron camera. Nevertheless, the ability to image the myocardium every 5 minutes, even in only two dimensions, is an important advance in our capability to measure serially the dynamic changes in perfusion that occur during myocardial ischemia and infarction.

MYOCARDIAL PERFUSION IMAGING WITH GALLIUM-68 LABELED ALBUMIN MICROSPHERES

Another approach to the assessment of regional myocardial blood flow involves the intracoronary administration of human serum albumin microspheres labeled with gallium-68, another positron emitter and a generator product of decay of its parent germanium-68 ($T_{1/2}$ = 68 min). Preliminary studies suggest a possible role for myocardial perfusion imaging with ^{68}Ga-labeled albumin microspheres administered at the time of coronary angiography.[28] It has previously been shown that myocardial perfusion imaging can be performed after selective intracoronary injection of either macroaggregated serum albumin or albumin microspheres labelled with technetium-99m, iodine-131, or indium-113m. These labeled aggregates or microspheres are injected directly into the coronary arteries at the time of catheterization, the rationale being that microembolization of a small percentage of capillaries occurs in proportion to nutrient blood flow. The distribution of these biodegradable particles is imaged with an appropriate scintillation camera and delineates the area of myocardial tissue supplied by the coronary artery into which the radioactive microspheres were injected. When blood flow to an area of the left ventricle is reduced, the number of particles delivered to the zone is reduced and a perfusion defect appears on the myocardial scintigram.

Perfusion scintigraphy employing this intracoronary imaging technique can also be undertaken after various interventions which assess coronary reserve, such as atrial pacing or injection of contrast media which induces hyperemia. Under these conditions, flow to an area of myocardium supplied by obstructed coronary arteries is impaired relative to that of flow to areas supplied by normal arteries. A major advantage of this invasive technique over intravenous methods of tracer administration is that improved resolution results from a substantially higher heart-to-background ratio with little or no background radioactivity in other organs. Utilizing positron scintigraphic techniques and ^{68}Ga-labeled albumin microspheres, computer reconstructed emission tomographic images of microsphere distribution in the heart can be produced. Sterile albumin microspheres coated with tin were labeled with ^{68}Ga according to the method of Hnatowich.[29] Figure 9 shows multiple transverse section images of the heart of an intact anesthetized dog after the injection of bolus doses of 200 μCi of ^{68}Ga serum albumin microspheres into the left anterior descending (LAD) and left circumflex coronary arteries, one hour after LAD ligation. A wedge-shaped perfusion defect was observed in tomographic images of the ischemic area.

Preliminary studies in patients have demonstrated the clinical feasibility of myocardial

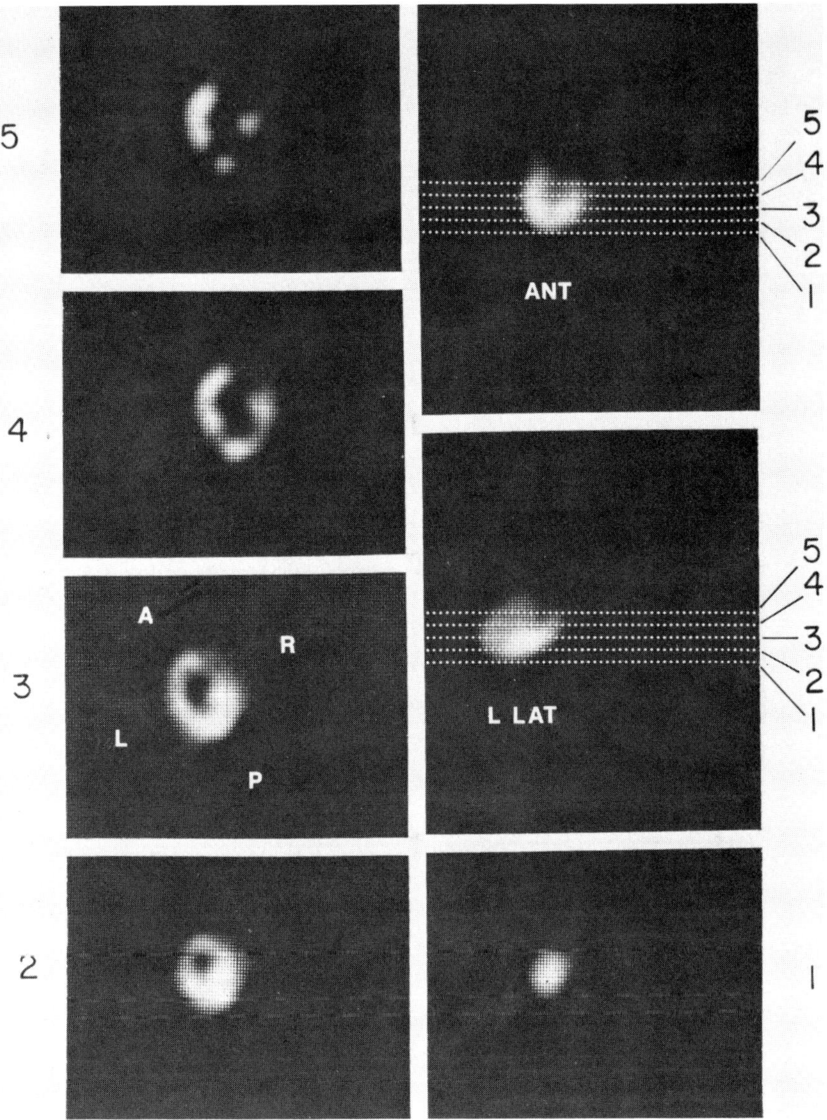

Figure 9. Anterior (ANT), left lateral (L LAT), and multiple transverse section tomographic images (identified by the numbered dotted lines depicting planes through the heart which were computer reconstructed) in a dog after the intracoronary administration of ^{68}Ga-labeled serum albumin microspheres. Tomographic cut 1 is through the left ventricular apex. Cuts 2 to 4 show relatively less activity in the anterior than posterior left ventricular myocardium. In cut 4, a wedge-shaped perfusion defect appears in the anteroseptal area. A = anterior; P = posterior; R = right; L = left.

imaging with this technique. Figure 10 shows both the anterior projection image and computer reconstructed transverse section of the mid-left ventricle image in a patient receiving 200 μCi of ^{68}Ga-serum albumin microspheres into the left main coronary artery and 100 μCi into the right coronary artery. A posterior wall defect is clearly visualized. Further studies are warranted to better evaluate the clinical utility of this invasive technique of myocardial perfusion imaging.

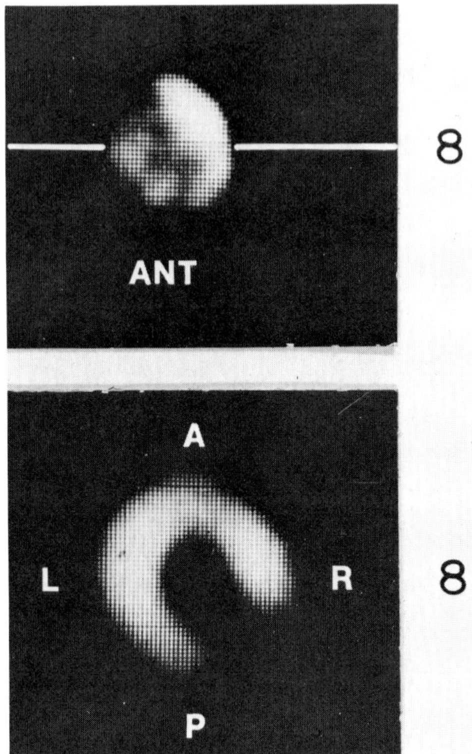

Figure 10. Upper panel: Anterior positron image of the left ventricle obtained in a patient after the intracoronary administration of ^{68}Ga-labeled serum albumin microspheres. Lower panel: The transverse section tomographic image of the midventricular plane (cut 8) shows a perfusion defect posteriorly.

MEASUREMENT OF REGIONAL MYOCARDIAL BLOOD FLOW USING ^{15}O-LABELED WATER AND ^{15}O-LABELED HEMOGLOBIN

By positron scintigraphy and computer processing, quantitative estimates of myocardial blood flow and functional images of regional perfusion can simultaneously be obtained after the intracoronary administration of ^{15}O-hemoglobin or ^{15}O-water. ^{15}O ($T_{1/2} = 2$ min) is produced in a medical cyclotron by deuteron irradiation of nitrogen gas via the ^{14}N (d,n) ^{15}O reaction.[30] For preparation of ^{15}O-hemoglobin, hemoglobin from heparinized blood is oxygenated in a tonometer using 90 percent oxygen and 10 percent CO_2. ^{15}O is then bubbled through the blood for several minutes in the cyclotron. After the intracoronary injection of either ^{15}O-hemoglobin or ^{15}O-water sequential images of regional myocardial tracer clearances are obtained over 5 minutes with a positron camera. Myocardial blood flow can be calculated from the monoexponential washout of $H_2^{15}O$ after background correction.[31] In a group of normal intact anesthetized dogs, mean ($\pm$ SE) myocardial blood flow was 78 $\pm$ cc/100 g/min.[31] Functional images of regional blood flow are derived by computer processing in which the image of peak activity is divided by the integrated image of $H_2^{15}O$ washout. Figure 11 shows the semilog plot of the time-activity curve obtained after an intracoronary injection of ^{15}O-hemoglobin in a normal dog. The fraction of the injected ^{15}O, namely, that fraction not utilized in metabolism, is rapidly washed out of the myocardium with the coronary venous blood. The fraction of oxygen-15 utilized in metabolism is rapidly converted to water and washes out at a slower rate as $H_2^{15}O$. After washout from the myocardium,

188

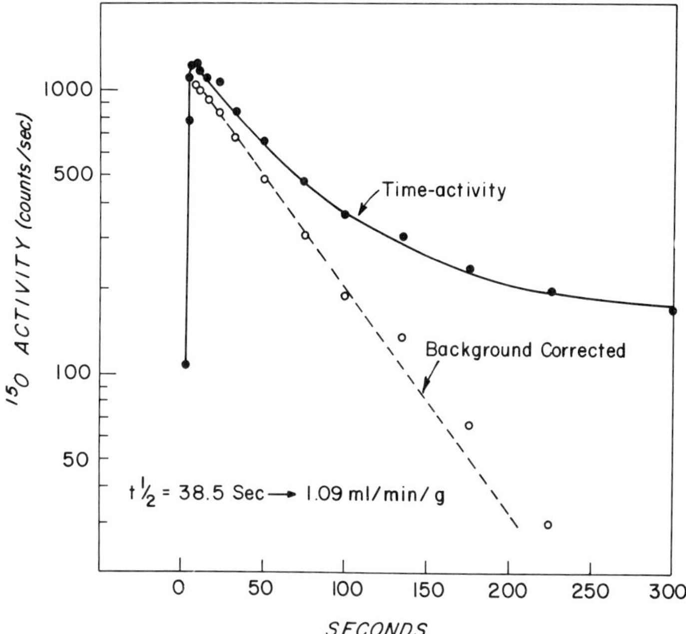

Figure 11. Time-activity curve obtained with a positron camera from the region of the myocardium after the intracoronary injection of ^{15}O-hemoglobin in a normal intact dog. The actual and background corrected curves are plotted on semilog paper. Blood flow in ml/min/g was calculated from the half-time of $H_2^{15}O$ washout.

$H_2^{15}O$ recirculates through the cardiovascular system. In order to determine the true time-activity curve for calculation of myocardial blood flow, the effect of this recirculating background activity is eliminated. Figure 11 shows both the uncorrected and recirculation corrected $H_2^{15}O$ monoexponential washout curve. In this example myocardial blood was 1.09 ml/min/g.

A functional image of regional myocardial blood flow can be derived by the stochastic analysis method (height/area). Figure 12 shows a functional image in a normal dog after intracoronary administration of ^{15}O-hemoglobin. In this example, blood flow was uniform in the myocardium as reflected by the uniform intensity of ^{15}O activity in the image.

Because of the short half-life of ^{15}O, measurement of myocardial blood flow during coronary angiography utilizing intracoronary injection of ^{15}O-hemoglobin or ^{15}O-water will require a positron camera in the catheterization laboratory.

IMAGING WITH ^{15}O-CARBON DIOXIDE FOR DETECTION AND QUANTITATION OF LEFT-TO-RIGHT SHUNTS

Noninvasive radionuclide techniques have successfully been employed for detection and quantitation of left-to-right shunts. Blood flow through a shunt is assessed by the intravenous administration of a radiotracer and external monitoring of its circulation through the lungs with a scintillation camera or probe. After an initial passage of the radiotracer through the pulmonary circulation, any early reappearance of tracer activity in the lung represents flow through the shunt. The magnitude of this abnormal recirculation curve is proportional to the magnitude of the shunt. Boucher and coworkers[10] have demonstrated that quantitation of left-to-right shunts could be determined noninvasively from the pulmonary clearance pattern of inhaled ^{15}O-carbon dioxide. ^{15}O is incorporated

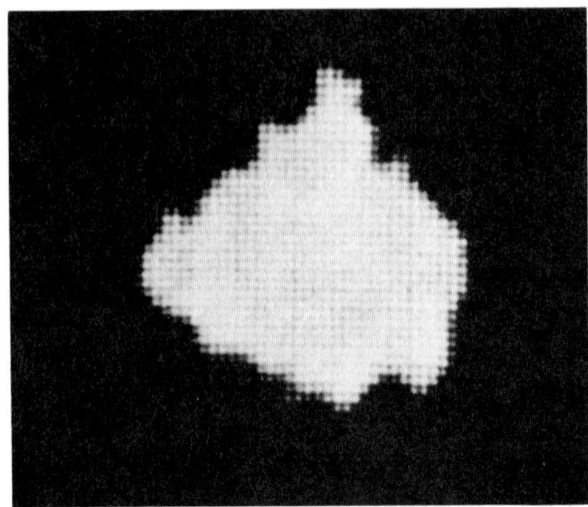

Figure 12. Posterior-anterior functional image of regional myocardial blood flow. This image was computer derived by stochastic analysis in which the image of the initial distribution of $H_2^{15}O$ is divided by the sum of the images obtained during myocardial $H_2^{15}O$ washout.

as an oxide of carbon by heating it with activated charcoal at 600 °C. Any carbon monoxide produced is converted to carbon dioxide by passing the gas over cupric oxide heated to 500 °C. The result is an air mixture with 3 percent carbon dioxide having a trace amount of sterile $C^{15}O_2$ which can be piped from a cyclotron to the clinical imaging area.[10]

After a single breath inhalation, $C^{15}O_2$ rapidly diffuses into the pulmonary blood and is almost instantaneously converted to $H_2^{15}O$ and HCO_3^- by carbonic anhydrase in red blood cells. The $H_2^{15}O$ is cleared rapidly from the lungs with a half-life of approximately 5 seconds, and the pulmonary clearance rate of $H_2^{15}O$ has been shown to be directly proportional to pulmonary blood flow.[32]

For the shunt detection technique, twenty consecutive 0.5-second positron camera images of the lung are obtained after the inhalation of 1 to 2 mCi of $C^{15}O_2$ given as bolus. Regional ^{15}O activity in the right and left upper lobes are plotted against time after inhalation. Since normal pulmonary clearance of $C^{15}O_2$ is monoexponential, time-activity curves are plotted on semilog paper. In 21 patients without left-to-right shunts studied by Boucher and coworkers,[10] the pulmonary washout of $C^{15}O_2$ showed no deviation from linearity during the 10-second imaging period. In 22 patients with left-to-right shunts, this monoexponential pulmonary clearance pattern was interrupted by an abnormal upward deviation, indicating tracer recirculation through the shunt and back to the lungs. Shunt size was derived from the ratio of the height of the recirculation curve to the height of the inhalation peak.[33] Figure 13 shows the typical $C^{15}O_2$ clearance patterns in a normal patient and in a patient with an atrial septal defect. Left-to-right shunt flow calculated from this ratio significantly correlates with shunt size as determined by oximetry.[10,33] Because of the short half-life and small radiation dose, $C^{15}O_2$ inhalation imaging for serial determinations of shunt size can be performed easily. By this technique, shunt size can also be measured in patients with heart failure and valvular regurgitation.

DETECTION OF PULMONARY EMBOLI BY POSITRON IMAGING OF INHALED ^{15}O-CARBON DIOXIDE

Inhalation imaging with $C^{15}O_2$ has been employed for the detection and localization of pulmonary emboli in dogs and more recently in patients.[34,35] $C^{15}O_2$ is an optimum radionuclide for positron scintigraphy of pulmonary emboli because of its unique physical properties. As cited in the previous section, after a single breath inhalation, $C^{15}O_2$ is rap-

190

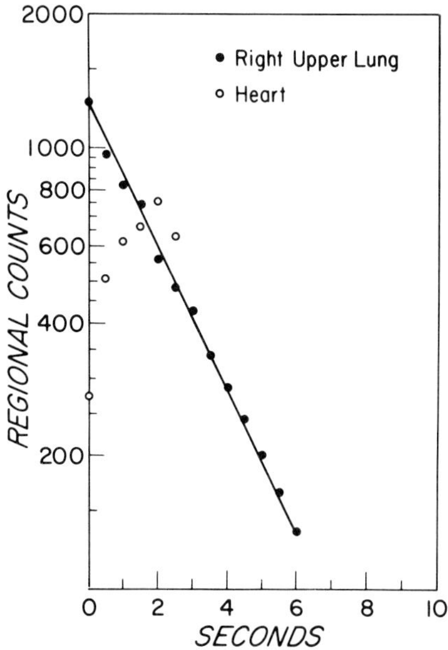

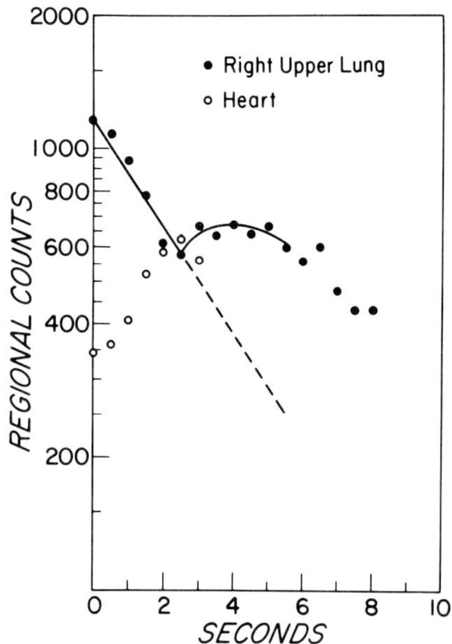

Figure 13. Left panel: Semilog plot of $C^{15}O_2$ counts from the right upper lobe versus time after peak inhalation in a normal patient. Pulmonary clearance (solid circles) of inhaled $C^{15}O_2$ is linear over time period delineated. Peak regional activity in the heart region (open circles) is noted by 3 seconds. Right panel: Semilog plot of regional $C^{15}O_2$ counts versus time after peak inhalation in a patient with an atrial septal defect. The initial pulmonary clearance of $C^{15}O_2$ from the right lung is linear, but it is interrupted by an abnormal upward deviation indicating early recirculation of tracer through the shunt to the lungs.

idly cleared from the lungs in proportion to pulmonary blood flow. After the radiotracer is cleared from the lungs by pulmonary washout, it is distributed throughout the water space of the body as $H_2^{15}O$, resulting in a low homogeneous level of background activity. In patients with pulmonary emboli obstructing regional pulmonary perfusion, ^{15}O activity is retained by stasis of blood distal to the embolus. This retained ^{15}O can be detected by a positron camera and displayed as a "hot spot" of activity against the low level of background activity.

In the study by Nichols and associates[34] of dogs receiving autologous blood clots, zones of impaired ^{15}O clearance were consistently imaged with a positron camera, and single emboli as small as 2 mm in diameter produced regions of retained ^{15}O activity. In a more recent study, inhalation imaging with $C^{15}O_2$ was undertaken in patients with suspected pulmonary emboli and compared to gamma camera ventilation/perfusion (V/Q) scans and pulmonary angiograms.[35] In comparison with angiographic presence or absence of emboli, the inhalation technique was 86 percent sensitive and 91 percent specific for the diagnosis of emboli. The V/Q scanning technique with ^{99m}Tc-labeled albumin microspheres and ^{133}Xe was 71 percent sensitive and only 64 percent specific for emboli in the same patients. Thus, the experimental and clinical studies demonstrate that the inhalation imaging of regional pulmonary perfusion with $C^{15}O_2$ and a multicrystal positron camera is a highly sensitive method for detecting pulmonary emboli and may be more specific than the conventional V/Q scan.

SUMMARY

New myocardial and pulmonary scintigraphic techniques utilizing positron-emitting radionuclides and high resolution multicrystal positron cameras are being developed and appear to hold great promise for the future. Positron scintigraphy permits the use of the short-lived radionuclide monovalent cations, ^{13}N-ammonia and ^{82}Rb$^+$, for myocardial perfusion imaging. For ^{82}Rb$^+$, which is obtained directly from a portable generator, sequential images of the heart can be obtained at intervals as short as 5 minutes, allowing assessment of pharmacological or mechanical interventions (e.g., intra-aortic balloon pumping) that alter regional flow patterns. With the advent of surgical techniques for revascularizing ischemic myocardium, and the investigation of pharmacologic interventions to increase myocardial perfusion, new clinically applicable and quantitative radionuclide methods are necessary to externally measure and document changes in regional myocardial blood flow.

The capability of computer-assisted positron emission transverse section imaging of the heart allows for the quantitation of radionuclide uptake in all regions of the myocardium because of the elimination of superimposition of structures. This three-dimensional visualization of radionuclide distribution results in greater sensitivity in detecting and localizing zones of myocardial underperfusion in patients with coronary artery disease. Clearly defined transmural areas of diminished perfusion have been demonstrated in transverse section images after the intravenous administration of ^{13}NH$_4$$^+$ and ^{81}Rb$^+$ and after the intracoronary injection of ^{68}Ga-labeled human serum albumin microspheres.

Quantitative estimates of total and regional myocardial blood flow can also be obtained by sequential positron imaging after the intracoronary administration of ^{15}O-hemoglobin and ^{15}O-water. Functional images of regional perfusion are also derived by the positron camera methodology. Myocardial metabolism can also be simultaneously assessed by sequential imaging of the heart after serial intracoronary injections of ^{15}O-hemoglobin and ^{15}O-water. Images of regional fractional oxygen extraction can be derived by dividing the functional myocardial image obtained after ^{15}O-hemoglobin injection by the functional image of myocardial H$_2$^{15}O distribution. Since the half-life of ^{15}O is only 2 minutes, measurements of myocardial blood flow and myocardial oxygen metabolism can be repeated at intervals as short as 8 minutes. Thus, for the first time, potential discrepancies between myocardial blood flow and myocardial metabolism can be assessed directly. Ultimately, this technique, although invasive, might be used to evaluate myocardial oxygen utilization and regional blood flow patterns in focally ischemic myocardium in patients undergoing cardiac catheterization.

Presently, positron imaging of pulmonary blood flow after inhalation of C^{15}O$_2$ can be employed in patients for the detection and quantitation of left-to-right shunts. Utilizing this noninvasive technique, patients with left-to-right shunts can be screened without the necessity of catheterization to ascertain the need for surgical shunt closure. This technique can be used in the postoperative patient to determine whether any residual shunt flow is present.

Finally, dynamic imaging of regional pulmonary blood flow offers a unique new approach to the scintigraphic detection and localization of pulmonary emboli. Preliminary clinical studies have demonstrated a significant improvement in both the sensitivity and specificity with which pulmonary emboli can be diagnosed noninvasively. Serial positron imaging with the short lived ^{15}O-carbon dioxide will permit the determination of resolution rates of emboli and render information relevant to the degree of residual blood flow to lung segments perfused by obstructed pulmonary arteries.

REFERENCES

1. ZARET, B. L., STRAUSS, H. W., MARTIN, N. D., ET AL.: *Noninvasive regional myocardial perfusion with radioactive potassium study of patients at rest, with exercise and during angina pectoris.* N. Engl. J. Med. 288:809, 1973.

2. ZARET, B. L., AND COHEN, L. S.: *Radionuclides and the patient with coronary artery disease.* Am. J. Cardiol. 35:112, 1975.

3. CANNON, P. J.: *Radioisotopic studies of the regional myocardial circulation.* Circulation 51:955, 1975.

4. HOLMAN, B. L.: *Radionuclide methods in the evaluation of myocardial ischemia and infarction.* Circulation 53:112, 1976.

5. PITT, B., AND STRAUSS, H. W.: *Myocardial imaging in the noninvasive evaluation of patients with suspected ischemic heart disease.* Am. J. Cardiol. 37:797, 1976.

6. BROWNELL, G. L., AND BURHAM, C. A.: *A multicrystal positron camera.* In Hine, G. J., and Sorenson, J. A. (eds.): *Instrumentation in Nuclear Medicine, Vol. 2.* Academic Press, New York, 1971, pp. 136-159.

7. TER-POGOSSIAN, M. D., PHELPS, M. E., HOFFMAN, E. J. ET AL.: *A positron-emission transaxial tomograph for nuclear imaging (PETT).* Radiology 114:89, 1975.

8. CHO, Z. H., CHAN, J. K., AND ERIKSSON, L.: *Circular ring transverse axial positron camera for 3-dimensional-reconstruction of radionuclide distribution.* IEEE Trans. Nucl. Sci. NS23/1, 613, 1976.

9. DERENZO, S. E., ZAKLAD, H., AND BUDINGER, T. F.: *Analytical study of a high-resolution positron ring detector system for trans-axial reconstruction tomography.* J. Nucl. Med. 16:1166, 1975.

10. BOUCHER, C. A., AHLUWALIA, B., BLOCK, P. D., ET AL.: *Inhalation imaging with oxygen-15 labeled carbon dioxide for detection and quantitation of left-to-right shunts.* Circulation 56:632, 1977.

11. KREVY, R., DEVEAU, L., ALPERT, N., ET AL.: *PL/S a higher level language for image processing.* In *Proceedings of the Seventh Symposium on Sharing of Computer Programs and Technology in Nuclear Medicine.* U.S. Energy Research and Development Administration Conference, 1977, in press.

12. BUDINGER, T. F., DERENZO, S. E., GULLBERG, G. T., ET AL.: *Emission computer assisted tomography with single-photon and positron annihilation photon emitters.* J. Comput. Assist. Tomog. 1:131, 1977.

13. BERMAN, D. S., SALEL, A. F., DeNARDO, G. L., ET AL.: *Noninvasive detection of regional myocardial ischemia using rubidium-81 and the scintillation camera.* Circulation 56:619, 1975.

14. ROMHILT, D. W., ADOLPH, R. J., SODD, V. C., ET AL.: *Cesium-129 myocardial scintigraphy to detect myocardial infarction.* Circulation 48:1242, 1973.

15. HARPER, P. V., SCHWARTZ, J., BECK, R. N., ET AL.: *Clinical myocardial imaging with nitrogen-13 ammonia.* Radiology 108:613, 1973.

16. STRAUSS, H. W., HARRISON, K., LANGAN, J. D., ET AL.: *Thallium-201 for myocardial imaging. Relation of thallium-201 to regional perfusion.* Circulation 51:641, 1975.

17. TILBURY, R. S., DAHL, J. R., AND MONAHAN, W. G.: *The production of ^{13}N-labeled ammonia for medical use.* Radiochem. Radioanalyt. (letters) 8:317, 1971.

18. HOOP, B., JR., SMITH, T. W., BURNHAM, C. A., ET AL.: *Myocardial imaging with $^{13}NH_4{}^+$ and a multicrystal positron camera.* J. Nucl. Med. 14:181, 1973.

19. WALSH, W. F., HARPER, P. V., RESNEKOV, L., ET AL.: *Noninvasive evaluation of regional myocardial perfusion in 112 patients using a mobile scintillation camera and intravenous nitrogen-13 labeled ammonia.* Circulation 54:266, 1976.

20. BELLER, G. A., AND SMITH, T. W.: *Radionuclide techniques in the assessment of myocardial ischemia and infarction.* Circulation 53 (Suppl. I):123, 1976.

21. SMITH, T. W., BELLER, G. A., GOLD, H. K., ET AL.: *Three-dimensional myocardial imaging using a multicrystal positron camera.* Circulation 50:24, 1974.

22. TER-POGOSSIAN, M. D., HOFFMAN, E. J., WEISS, E. S., ET AL.: *Positron emission reconstruction tomography for the assessment of regional myocardial metabolism by the administration of substrates labeled with cyclotron produced radionuclides.* In *Cardiovascular Imaging and Image Processing—Theory and Practice.* Society of Photo-Optical Instrumentation Engineers, Palos Verdes Estates, Cal., 1976, pp. 277-282.

23. HARPER, P. V., SCHWARTZ, J., BECK, R. N., ET AL.: *Clinical myocardial imaging with nitrogen-13 ammonia.* Radiology 108:613, 1973.

24. MARTIN, N. D., ZARET, B. L., McGOWAN, R. L., ET AL.: *Rubidium-81: A new myocardial scanning agent.* Radiology 111:651, 1974.

25. BUDINGER, T. F., YANO, Y., AND HOOP, B.: *A comparison of $^{82}RB^+$ and $^{13}NH_3$ for myocardial positron scintigraphy.* J. Nucl. Med. 16:429, 1975.

26. BELLER, G. A., HOOP, B., PARKER, J. A., ET AL.: *Sequential myocardial imaging with rubidium-82, an ultra-short-lived radionuclide.* Circulation 51, 52(Suppl. II):110, 1975.

27. BELLER, G. A., ALTON, W. J., MOORE, R. H., ET AL.: *Detection of nitroglycerin-induced changes in regional myocardial perfusion during acute ischemia by serial imaging with $^{82}RB^+$.* Circulation 53, 54(Suppl. II): 216. 1976.

28. ALTON, W. J., BELLER, G. A., GOLD, H. K., ET AL.: *Three-dimensional myocardial imaging after intracoronary injection of ^{68}Ga-labeled albumin microspheres.* Circulation 53, 54(Suppl. II):81, 1976.

29. HNATOWICH, D. J.: *The labeling of tin-soaked albumin microspheres with gallium-68.* J. Nucl. Med. 17:57, 1975.

30. HOOP, B., LAUGHLIN, J. S., AND TILBURY, R. S.: *Cyclotrons in nuclear medicine.* In Hines, G. J., and Sorenson, J. A. (eds.): *Instrumentation in Nuclear Medicine,* Vol. 2. Academic Press, New York, 1974, pp. 407–457.

31. PARKER, J. A., BELLER, G. A., HOOP, B., ET AL.: *Assessment of regional myocardial blood flow and regional fractional oxygen extraction in dogs using ^{15}O-water and ^{15}O-hemoglobin.* Circ. Res. (in press).

32. SCHMIT-NOWARA, W. W., HOOP, B., JR., AND KAZEMI, H.: *Role of pulmonary blood flow in removal of gaseous isotopes from the lung.* J. Appl. Physiol. 35:655, 1972.

33. TAMER, D. M., WATSON, D. D., KENNY, P. J., ET AL.: *Noninvasive detection and quantitation of left-to-right shunts in children using oxygen-15 labeled carbon dioxide.* Circulation 56:621, 1977.

34. NICHOLS, A. B., COCHAVI, S., AND MOORE, R. H., ET AL.: *Detection of experimental pulmonary emboli in dogs by sequential positron imaging after inhalation of ^{15}O-carbon dioxide.* Circ. Res. 42:53, 1978.

35. NICHOLS, A. B., COCHAVI, S., HALES, C. A., ET AL.: *Scintigraphic detection of pulmonary emboli by serial positron imaging of inhaled ^{15}O-carbon dioxide.* N. Engl. J. Med. 297:279, 1978.

Measurement of Coronary Blood Flow Using Xenon-133*

Paul J. Cannon, M.D.

The evaluation of the coronary circulation and cardiac function prior to consideration of coronary artery bypass surgery involves selective coronary arteriography, measurements of intracardiac pressures and measurements of global and segmental ventricular function by ventriculography. In recent years, several radionuclide procedures have been introduced to assess myocardial perfusion in patients with coronary artery disease. These include the relatively noninvasive assessment of regional myocardial perfusion with labeled potassium analogues such as thallium-201 both at rest and in association with ECG testing during exercise, and the evaluation of left ventricular ejection fraction and wall motion at rest and during exercise by radionuclide angiocardiography.[1-4] Invasive radionuclide procedures which are performed at the time of cardiac catheterization include qualitative estimations of the distribution of myocardial perfusion by injection of radiolabeled albumin macroaggregates into the right and left coronary arteries with subsequent visual assessment of particle distribution in the myocardium from precordial scintiscans, and quantitative measurements of mean left ventricular (LV) and regional myocardial blood flow per unit mass of tissue with xenon-133 and a scintillation camera.[5-7]

The primary purpose of studies of regional myocardial blood flow with ^{133}Xe and a scintillation camera is to determine the physiologic significance of coronary lesions in terms of their effects on myocardial capillary blood flow, both at rest and during a stress which normally increases coronary artery flow. By coronary arteriography alone it is impossible to assess myocardial blood flow distal to stenotic lesions or the relationship between coronary blood flow and the metabolic demands of the myocardium for oxygen and other nutrients.

PHYSIOLOGIC CONSIDERATIONS

Each of the coronary arteries supplies a discrete area of myocardium. Because of the high extraction of oxygen from coronary blood by the myocardium under basal conditions, increases in the metabolic requirements of heart muscle (e.g., increases induced by tachycardia or exercise) are normally met by proportional increases in coronary blood flow with only minor increases in myocardial O_2 extraction.[8] If the increase in blood flow is sufficient to meet increased O_2 demands, there is no ischemia. However, with coronary artery disease, blood flow through an obstructed artery may be insufficient to meet the requirements of

---

*This work was supported by Grant HL 14148 and by SCOR Grant HL 21006 from the N.H.L.B.I. of the U.S. Public Health Service.

myocardium for oxygen, and ischemia of the myocardium may ensue with angina pectoris, ST-T wave alterations on the electrocardiogram and abnormalities of ventricular function. Measurements of regional coronary blood flow in relation to measurements of ventricular performance may facilitate the identification of myocardial ischemia in patients with coronary artery disease.

Experimental studies in dogs during progressive coronary artery constriction have indicated that resting coronary artery blood flow does not fall until the arterial diameter has been narrowed by 90 percent or more.[9] As the diameter of a coronary artery is narrowed, vasodilation occurs in more distal coronary arterioles and maintains myocardial capillary perfusion relatively constant at rest (so-called autoregulation of blood flow).[10] However, there is a limit to autoregulation. As obstruction of a larger coronary vessel progresses, the reserve ability of the coronary arterioles to dilate diminishes, and coronary blood flow becomes increasingly dependent on the pressure difference across the obstruction.

Resting myocardial perfusion may be normal with coronary obstructions less than 90 percent; nevertheless, coronary flow responses to stimuli that induce hyperemia are reduced progressively with obstructions that are greater than 50 percent of the lumen diameter.[9] Hyperemic responses are abolished with lesions greater than 90 percent.[9] As a consequence, capillary blood flow in myocardium distal to such a lesion may not increase normally or may fall during tachycardia or exercise.

When myocardial blood flow is significantly reduced, there are alterations of global and segmental ventricular performance.[11] In viewing coronary arteriograms and ventriculograms, it is frequently difficult to assess whether a region of the ventricular wall distal to a coronary obstruction that exhibits abnormal function does so because the myocardium is ischemic (such ischemic tissue might benefit from a revascularization procedure) or because it is composed of fibrous scar from previous infarction (which would not benefit from operation). Similarly, it is difficult to evaluate at arteriography the functional significance of collateral vessels which supply heart regions distal to significant coronary lesions.[12] In some of these situations, measurements of regional myocardial perfusion may be helpful in the analysis.

THEORETICAL CONSIDERATIONS

Measurement of myocardial blood flow with ^{133}Xe and a scintillation camera is an adaptation of the Kety method, whereby nutrient capillary blood flow per unit volume of tissue is estimated by making measurements of the saturation or desaturation of the tissue with an inert gas.[13,14] Inert gases that have been used for this purpose include the following: nitrous oxide, hydrogen, helium, argon, ^{133}Xe, and krypton-85.[7]

The basic approach is to inject radioactive xenon into a coronary artery. The inert gas diffuses instantaneously into the tissue supplied by the artery and then washes out as a function of nonradioactive myocardial capillary blood flow. Multiple ^{133}Xe myocardial washout curves are recorded from different regions of the heart using a scintillation camera. By means of a computer, rate constants of regional tracer clearance from the heart are calculated from the data recorded by each of the multiple crystals (multicrystal scintillation camera system) or in each area of interest over the heart (single-crystal scintillation camera system). Myocardial capillary blood flow rates per unit mass of tissue are calculated by the Kety formula using an assumed partition coefficient for ^{133}Xe between blood and myocardium.[15] The pattern of regional blood flow rates so obtained is then superimposed upon a tracing of the patient's coronary arteriogram, made during the same study, in order to localize regions of the heart and to relate myocardial perfusion abnormalities to the coronary vascular anatomy.

It is important to realize that several assumptions are involved in any measurement of tissue blood flow by the Kety formula:

196

1. The coronary arterial inflow and venous outflow are equal and unchanging during the course of the measurement.

2. The indicator used to make the measurement is homogeneously distributed throughout the tissue under study.

3. There must be continuous diffusion equilibrium of the indicator between capillary blood and the tissue according to its partition coefficient (i.e., there must be no arteriovenous shunting or diffusional limitation of tracer passage from blood into the tissue).

4. The indicator must not be metabolized and must be removed only by venous blood without recirculation.

5. In studies which the desaturation of the tissue is measured by sampling the venous outflow, the vein in which the sampling catheter is placed must contain all of the blood leaving the tissue under study and not include venous outflow from other regions.

The theoretical basis and some of the experimental validation of the Kety approach have been summarized in several review articles.[7, 13, 14, 16, 17] The Kety application of the Fick principle may be briefly expressed as

$$\frac{dQ}{dt} = F(Ca - Cv) \qquad 1$$

where Q is the amount of radioactive inert gas in the myocardium, Ca and Cv are the concentrations of gas in the arterial and venous blood respectively, t is time, and F is blood flow. After an intra-arterial bolus injection of a radioactive gas which does not recirculate, the arterial concentration Ca becomes zero. The venous concentration, Cv, can be expressed as $Q/V\lambda$, where V is the myocardial volume of distribution of the gas and λ the partition coefficient describing the relative solubility of the gas in blood and myocardium. Thus, Equation 1 becomes

$$\frac{dQ}{dt} = \frac{FQ}{V\lambda} \qquad 2$$

This can be rearranged as

$$\frac{dQ}{Q} = -\left(\frac{F}{V\lambda}\right) dt \qquad 3$$

Solving this equation yields:

$$Q_t = Q_o e^{-(F/V\lambda)t} \qquad 4$$

where Q_t and Q_o equal the amounts of radioactive gas at times t and o respectively. Therefore,

$$\frac{F}{V\lambda} = k \qquad 5$$

which is the exponential rate constant of tracer washout from the myocardium that can be determined experimentally.

Therefore,

$$\frac{F}{V} = k \times \lambda \qquad 6$$

However, V can be expressed as w/ρ, where w is the weight of the tissue in grams and ρ is the specific gravity (1.05).

Since the weight of the tissue is not known, blood flow is expressed by convention in terms of 100 g of tissue; thus, 100 is substituted for w and flow is expressed as milliliters per 100 g myocardium per minute:

$$F = \frac{k \times \lambda}{\rho} \times 100 \qquad\qquad 7$$

RADIONUCLIDE

^{133}Xe is a radioisotope of an inert gas which is chemically inert and physiologically inactive at the concentrations used for blood flow measurements. Because of its lipid solubility, it diffuses rapidly into myocardial cells after injection into an artery. Pulmonary excretion is rapid; 95 percent of an intraarterial dose of ^{133}Xe is eliminated in one passage through the lungs.[18] Thus, recirculation of indicator is not a significant problem.

^{133}Xe is produced in a reactor. It decays with a physical half-life of 5.3 days to ^{133}Ce by emission of β particles of 347 keV, gamma rays of 81 keV (37 percent), and x-rays of 31 keV.[19] An intravenous dose of 40 mCi of ^{133}Xe dissolved in saline yields a gonadal dose of approximately 10 mrads, a level considered safe for diagnostic studies.[20] The 81-keV gamma radiation emitted by ^{133}Xe is sufficiently energetic to be detected externally and is sufficiently low in energy to be easily collimated. Some disadvantages of ^{133}Xe, however, are its expensiveness, low count rates observed with thick collimators due to low energy and poor depth response, differing solubilities in muscle and in fat, and the production in tissue of Compton scatter radiation which is sufficiently energetic that it is detected in the primary photo peak of the scintillation camera. A new radionuclide, ^{127}Xe, which is not yet commercially available, has energy charactersitics which suggest it may be more useful than ^{133}Xe for studies of the myocardial circulation.[21]

SCINTILLATION CAMERA AND COLLIMATORS

The development of scintillation cameras and high speed digital computers has facilitated attempts to estimate capillary perfusion in multiple areas of the heart by precordial monitoring of regional myocardial tracer washout curves after intracoronary injection of ^{133}Xe. The feasibility of this approach was demonstrated by Cannon, Haft, and Johnson in 1969 when they visualized regional perfusion and areas of experimental infarction in dogs using ^{133}Xe and an image intensifier scintillation camera.[22] In subsequent studies in which quantitative estimates of regional myocardial perfusion were made in animals, normal subjects, and patients with coronary artery disease, Cannon, Dell, and Dwyer used ^{133}Xe and a multicrystal scintillation camera.[23,24] Other investigators have performed similar studies using single-crystal scintillation cameras.[25-32]

An extensive discussion of the relative merits of different scintillation camera systems exceeds the scope of this chapter.[33] However, it is important to reiterate that an intrinsic limitation of current scintillation cameras is that they can make only two-dimensional representations of tracer distribution in a three-dimensional organ. Each of the multiple scintillation crystals in the multicrystal camera, or each region of interest (subdivisions of a large scintillation crystal) in the single-crystal camera, monitors the washout of ^{133}Xe from all of the tissue within the solid angle viewed by that crystal (or region) and its collimator. For this reason it is currently not possible to distinguish epicardial from endocardial blood flow in any one myocardial region, or to distinguish relative flow rates when two cardiac walls supplied by one coronary artery overlie one another within the solid angle viewed by the detector.

Several different collimator systems have been used by investigators in studies of regional myocardial blood flow. Those working with multicrystal cameras have used a 1.5-in. multichannel parallel-hole collimator. Each hole had a radius of view that was 6 mm at 3 cm distance (corresponding roughly to the anterior wall of the heart) and 12 mm at 8 to 9 cm distance.[24] Groups using single-camera systems have used high resolution or high sensitivity parallel-hole collimators, or one of the pinhole type.[25-32]

With increasing distance of the object viewed (tracer in the myocardium) from the face of the collimator, there will be overlap of the field of view of adjacent detectors or regions of interest.[24] If there is much overlap, a significant percentage of counts recorded by a primary detector (or region of interest) positioned over one area of the heart may arise from an adjoining area with a different blood flow rate. The slope of a ^{133}Xe washout curve recorded from one area of myocardium will be in error by an amount that is proportional to the relative percentage of overlap times the difference in blood flow rates between the myocardial region viewed by the primary detector and the area viewed by adjacent detectors.

Other technical considerations include the effects of distance upon counting efficiency and of Compton scatter radiation on the measurements. Radiation from the posterior myocardium is detected by a precordial detector with less efficiency than in the anterior myocardium. Calculations using the absorption coefficient for ^{133}Xe and the response characteristics of a 1½-in. collimator and a multicrystal system indicated that the count rate in a tissue equivalent (H_2O) at 1-in. distance was 63 percent of the count rate at the collimator face; it was 16 percent at 4-in. distance. Studies of Compton scatter were performed with a 1-cm bar source of ^{133}Xe in saline positioned 1 cm from a 1.5-in. collimator with water as the scattering medium and a 75 to 250-keV window.[34] Crystals adjacent to the crystal directly over the source recorded an average of 6 percent of the counts recorded by the primary crystal.

TECHNIQUE

The technique for measuring regional myocardial blood flow with ^{133}Xe and a scintillation camera which is described here was developed and used at Columbia Presbyterian Medical Center in New York City in studies of more than 400 patients.[35] Cardiac catheterization and coronary arteriography are performed only for clinical indications. Written informed consent is obtained from the patients for the studies which are approved by both the Human Investigation and Joint Radioisotope Committees of the institution.

The patients are brought to the laboratory in the postabsorptive state after premedication with pentobarbital. A left heart (and frequently a right heart) cardiac catheterization is carried out; pressures in the aorta, left ventricle, pulmonary artery, right ventricle, and right atrium are recorded. Coronary arteriography is performed by a modified Judkins' technique.[36] Images of contrast material in the coronary vessels are recorded by cine and/or serial cut films. In most patients after the measurements of regional myocardial blood flow are completed, single or biplane left ventriculograms are also obtained. Contraction sequences of the left ventricle are analyzed and left ventricular end-diastolic volume and ejection fraction are calculated by an adaptation of the method of Dodge and coworkers.[37]

^{133}Xe (10 to 20 mCi) dissolved in 1 to 2 ml of isotonic saline is injected as a bolus into the main right or left coronary artery. The gamma radiation emitted by the radioisotope as it arrives in the heart, diffuses into myocardial cells and is washed out by nonradioactive coronary blood, is collimated by the 1.5-in. multichannel collimator and recorded by the scintillation camera for 4 to 7 minutes. Scintillations produced by incident photons in each crystal are converted to an electrical signal, conditioned (anticoincidence circuitry and pulse-height analysis) and recorded onto magnetic tape as counts per second (cps). ^{133}Xe arrival and washout from the myocardium are also visualized by replaying the tape on the oscilloscope. [133]Xe exhaled by the patient is transferred to a trap or to the outside air by a vacuum exhaust system, held close to the patient's face.

Data Processing

The data contained on magnetic tape are processed further by a digital computer (IBM 360/91) at the Columbia University Computer Center. Crystals which recorded myocardial ^{133}Xe washout are distinguished from crystals which record ^{133}Xe excretion from the lungs by a computer print-out of the peak counts per second recorded by each crystal and the number of seconds after injection of the tracer that the peak counts per second occurred. Crystals over the heart record higher peaks earlier, while those monitoring pulmonary excretion record lower peaks later.[24] Since ^{133}Xe is distributed as a function of total blood flow to each region, maps of the initial peak count distribution are also produced to assess spatial heterogeneity of initial blood flow distribution by the injected artery.

Using the method of least squares, the computer fits a monoexponential equation to the first 40 data points after the peak counts per second recorded by each crystal and calculates the slope (k) of this portion of each myocardial ^{133}Xe washout curve. A printout is produced containing the multiple slopes or clearance constants of ^{133}Xe washout from the myocardium which are measured by crystals overlying the heart. The myocardial blood flow rates in the different regions along with the standard deviation of each measurements are calculated by the Kety formula, $F = k \times \lambda/\rho$, where F is the myocardial blood flow rate in ml/100 g/min, k is the rate constant of ^{133}Xe clearance from the myocardium determined experimentally, λ is the blood/myocardium partition coefficient for ^{133}Xe obtained by Conn in normal dogs (0.72), and ρ is the specific gravity of the tissue (1.05).[15]

The measurements of local myocardial perfusion are related to lesions apparent on the coronary arteriograms by superimposing the printout of local myocardial blood flow rates upon an appropriately magnified tracing of the patient's coronary arteriogram. Correct alignment of the printout of local flow rates and the left anterior oblique coronary arteriogram is achieved by use of the radiopaque-radioactive markers which were recorded during the study. Local myocardial blood flow rates which were measured by crystals overlying specific cardiac regions (e.g., right ventricle, left ventricle, left anterior descending or circumflex areas) are averaged. In addition, mean perfusion rates proximal and distal to discrete coronary lesions are also calculated by averaging the flow rates from the appropriate crystals on the myocardial perfusion pattern. An analysis of the regional distribution of the initial peak counts per second recorded by all of the crystals overlying the myocardium after ^{133}Xe injection is also performed in relation to the coronary arteriogram and to regions of abnormal ventricular wall motion on the ventriculogram.

Rationale for Form of Data Analysis

The primary data obtained in the studies of regional myocardial perfusion with ^{133}Xe and a scintillation camera are the mathematical expressions of the clearance rates of the radionuclide from different areas of the myocardium calculated from the regional ^{133}Xe washout curves. For analysis of the washout curves, most investigators have performed monoexponential analysis of the intial portion of the multiple ^{133}Xe washout curves. There are three reasons for this approach.

1. Over a wide range of coronary blood flow rates in normal dogs, four groups have found close correlations between directly measured (flowmeter) coronary artery flow rates/gram tissue and the flow rates/gram tissue calculated by monoexponential analysis of the initial segment of single labeled inert gas myocardial washout curves.[38-41] Poorer correlations were obtained when height area or two compartment analysis was applied to the data.[40,41]

200

2. After ^{133}Xe leaves the coronary sinus, some of the radionuclide accumulates in the retrocardiac lungs; arrival of tracer in the lungs behind the heart distorts the tail of the myocardial washout curves.[24,42]

3. Indirect evidence suggests that later components of precordial ^{133}Xe curves may be affected by recirculation of tracer or movement of the radionuclide into nonmuscular cardiac tissue such as scar or pericardial fat.[40,41]

In order to express the primary data (clearance constants of tracer washout) in terms of blood flow by the Kety formula, the calculations involve use of an assumed constant (the partition coefficient, λ). The λ most widely used is 0.72 which was obtained from static studies of normal dog myocardium by Conn.[15] Some investigators also correct the λ for hematocrit, a correction which is important in anemic patients but is of minor importance in those without anemia.[43]

VALIDATION STUDIES

The four studies which reported that the measurements of mean LV myocardial blood flow/unit mass made by monoexponential analysis of single ^{133}Xe washout curves were highly correlated with LV flow/unit mass calculated from measurements of coronary artery flow and LV weight were performed in normal animals with a wide range of coronary blood flow.[38-41] When similar measurements were performed in dogs with heterogeneity of local myocardial perfusion rates induced by coronary artery constriction, however, the calculations from single ^{133}Xe or ^{85}Kr washout curves significantly overestimated mean LV perfusion.[17]

Because a scintillation camera can simultaneously make spatially discrete measurements of ^{133}Xe clearance in normal and ischemic myocardium, preliminary experiments in dogs have been performed in our laboratory comparing mean LV and regional myocardial blood flow rates obtained with ^{133}Xe and with radioactive microspheres under conditions of varying heterogeneity of blood flow induced by partial coronary constriction and infusions of isoproterenol.[44] In these experiments on animals with marked spatial and transmural heterogeneity of myocardial blood flow, measurements of mean LV myocardial blood flow/unit mass by ^{133}Xe and by radioactive microspheres were highly correlated (r = 0.95, slope not different from unity). Regional flow rates were also highly correlated; however, ^{133}Xe overestimated microsphere flow in ischemic tissue distal to the coronary lesion slightly. In experiments when transmural blood flow in one heart region was altered over an endocardial to epicardial flow ratio of 0.63 to 1.05, the ^{133}Xe measurements did not differ significantly from the mean transmural blood flow measured by microspheres. These data indicate that neither recirculation of tracer nor diffusion of ^{133}Xe into fat spuriously reduces measurements of mean LV or transmural blood flow/unit mass. In ischemic regions, a modest overestimation flow may occur with the ^{133}Xe technique, possibly because a bolus injection was performed together with monoexponential analysis of the data obtained from a region with marked heterogeneity of flow. Collectively, the data support the utility of the measurement technique for studies of patients with coronary artery disease.

MYOCARDIAL PERFUSION PATTERNS

Figure 1 shows the myocardial perfusion pattern obtained after intracoronary injection of ^{133}Xe in a patient with a normal left coronary arteriogram. A small but significant inhomogeneity of local myocardial blood flow rates in the left ventricle (LV) is apparent. The average coefficient of variation of local flow rates in a series of patients with normal coronary arteriograms was 15.8 percent.[45] That this represented true inhomogeneity of blood flow was indicated by the fact that the between-crystal variance of flows significantly exceeded the variance of the individual flow measurements. Other workers, in studies of dogs

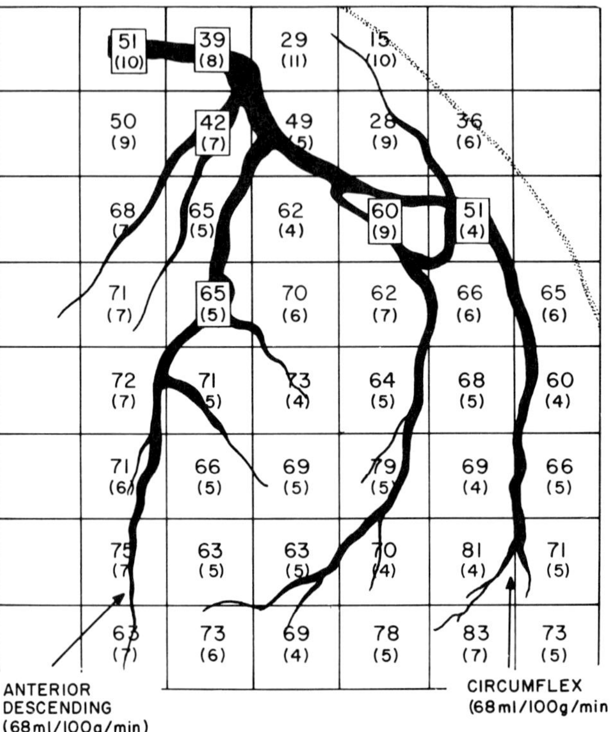

Figure 1. The myocardial perfusion pattern from a study of a patient with a normal left coronary arteriogram.[24] The number in each box is the local flow rate measured by an individual collimated scintillation detector. The number in parenthesis is the standard deviation of that measurement. Mean LV and regional blood flow (LAD, Circ) are calculated by averaging the local flow rates in the appropriate areas of the perfusion pattern.

with radioactive microspheres and of patients after direct intramyocardial injection of ^{85}Kr, have found a comparable degree of topographical heterogeneity of perfusion in hearts with normal coronary arteriograms.[46,47]

When ^{133}Xe is injected into the right coronary artery, right ventricular regions of the heart can be identified on the myocardial perfusion pattern (Fig. 2) from landmarks apparent on the coronary angiograms (i.e., the septal "blush" and the location of the A-V node artery and right marginal artery). In 17 patients with normal coronary arteriograms, mean left ventricular myocardial blood flow rates averaged 64 ml/100 g/min; mean right ventricular perfusion rates averaged 47 ml/100 g/min; and right atrial perfusion rates were 34 ml/100 g/min.[24] Similar differences in LV and RV myocardial capillary blood flow/unit mass of tissue have been found in normal dogs in studies using labeled ions or radioactive microspheres.[46]

MYOCARDIAL BLOOD FLOW IN PATIENTS WITH NORMAL CORONARY ARTERIOGRAMS

Studies of the determinants of myocardial blood flow in patients with normal coronary arteriograms have been performed in this laboratory. In 12 patients with normal coronary arteriograms, normal intracardiac pressures and normal left ventriculograms, mean LV

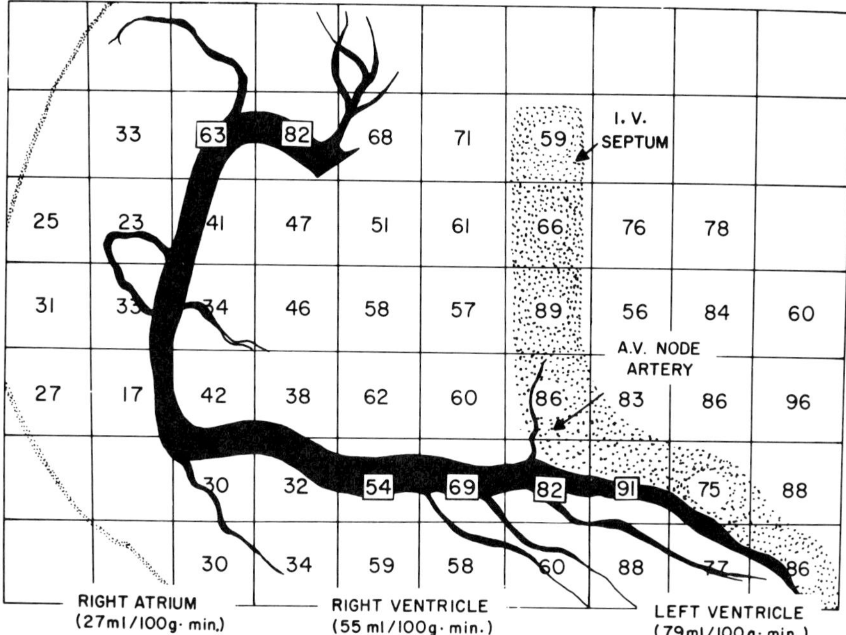

Figure 2. The myocardial perfusion pattern from a study of a patient with an arteriographically normal dominant right coronary arteriogram.[24] LV perfusion/unit mass exceeded that of the RV.

perfusion averaged 61 ml/100 g/min. In a smaller number of patients with hypertension, the average mean LV perfusion rates were not significantly different from the normals.[48] However, mean LV myocardial blood flow was significantly lower than normal in patients with congestive or hypertrophic cardiomyopathy and in a selected series of patients with severe isolated aortic stenosis.[49,50] Although LV myocardial blood flow/unit mass of tissue was reduced in the latter three groups with LV hypertrophy, the total LV perfusion calculated as the product of flow/unit mass × LV mass estimated from the ventriculograms was higher than in the normals.[49,50]

In experimental animals, coronary blood flow is directly related to the level of myocardial oxygen consumption.[8] Other studies have indicated that the three major determinants of myocardial oxygen consumption are heart rate, peak ventricular wall stress, and contractility of the heart muscle.[51] In the study of cardiomyopathy by Weiss and associates, multiple regression analysis of the data obtained in the normals and those with hypertrophic or congestive cardiomyopathy indicated that there was a significant direct relationship between the mean LV myocardial blood flow rate measured with ^{133}Xe and heart rate, peak wall stress, and a ventriculographic index of cardiac performance, the mean velocity of circumferential fiber shortening.[49] This suggested that in these patients with normal coronary arteries but different patterns of ventricular performance, the resting myocardial capillary blood flow rate is determined to a great extent by three of the major determinants of myocardial oxygen consumption: heart rate, wall stress, and contractility.

The importance of ventricular wall stress as a determinant of mean LV myocardial blood flow in man is suggested by the data in Figure 3. In four groups of patients with normal coronary arteriograms, mean LV myocardial blood flow/100 g/min/beat was linearly related to peak wall stress.[50] The slopes and intercepts of the regressions were almost identical for the two groups of patients with normal values for mean Vcf (normals, hypertrophic cardiomyopathy). The lower y intercepts for the patients with congestive cardiomyopathy and

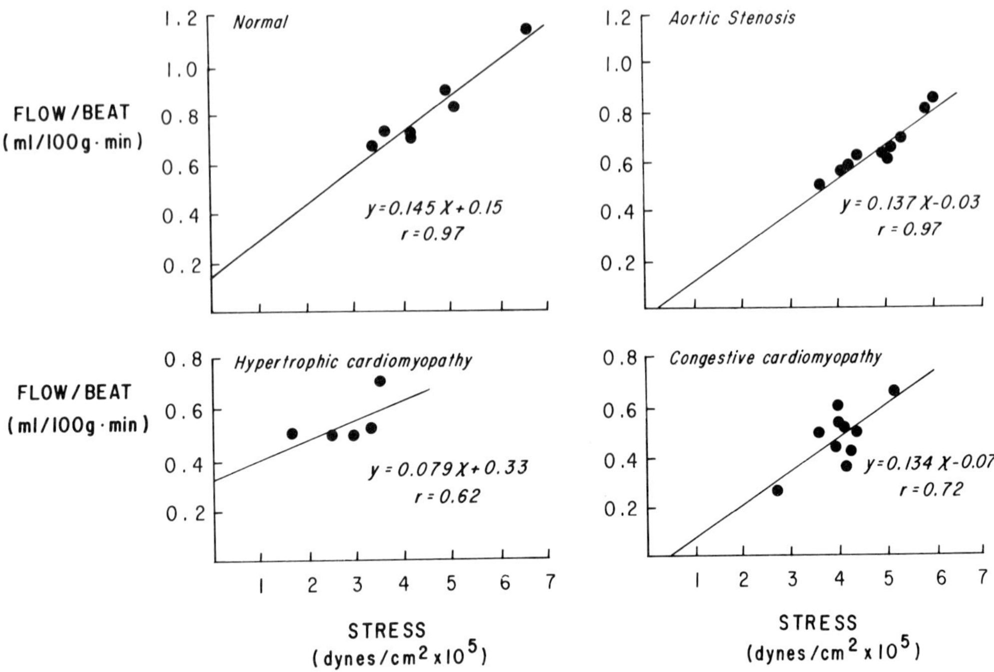

Figure 3. LV myocardial blood/100 g per heart beat is plotted against wall stress in four groups of patients.

with severe aortic stenosis may have been related to reduced ventricular contractility. Such data in patients without coronary artery disease implies that the effect of specific coronary artery lesions upon regional myocardial blood flow must be assessed against basal relationships between ventricular performance, myocardial oxygen consumption, and capillary blood flow existing in a given patient.

MEAN LV MYOCARDIAL BLOOD FLOW IN PATIENTS WITH CORONARY ARTERY DISEASE

Figure 4 shows the mean LV myocardial blood flow rates/unit mass of tissue measured in a large series of patients with radiographically documented coronary artery disease.[48] In those with less than 50 percent coronary obstructions, the average mean LV perfusion was not significantly different from the control patients. Similar observations have been reported by Holman and coworkers[28] and by Scheibel and associates.[30]

In patients with greater than 50 percent lesions of the LAD, average mean LV myocardial blood flow was also not significantly different from normal.[48] Reasons for this were:

1. The mean resting perfusion distal to less than 90 percent lesions was not selectively reduced.

2. A small area of reduced perfusion was counterbalanced by larger areas with higher blood flow rates.

In patients with radiographically significant double and triple vessel coronary disease (>50% lesions), mean LV myocardial blood flow/unit mass was significantly lower than in the patients with normal arteriograms and ventriculograms (Fig. 4). The average values in the various groups were: LAD + right disease, 47 ± 11 ml/100 g/min; LAD + Circ disease, 53±10 ml/100 g/min; and LAD + right + Circ disease, 45±9 ml/100 g/min.[48] These observations, which confirm a previous study using ^{133}Xe in a small number of patients,[45]

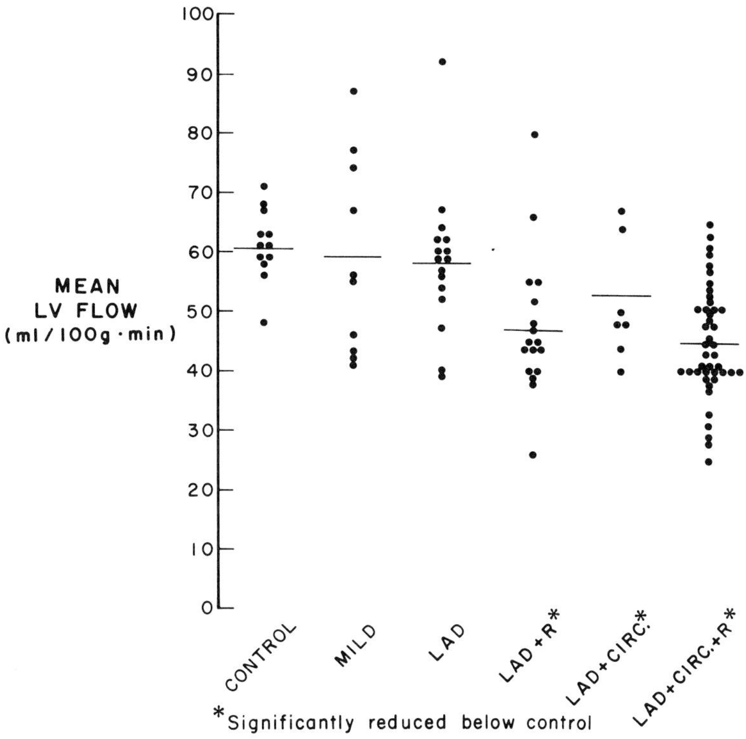

Figure 4. The mean left ventricular myocardial blood flow rates/unit mass of tissue are plotted for each patient; patients are grouped by the degree of atherosclerosis apparent on the coronary arteriograms (mild, <50% lesions). Average mean LV perfusion was significantly reduced (*) below the control group in patients with more than 50 percent lesions of two or three vessels.[48]

are different from results in patients with coronary disease obtained earlier by several groups who measured coronary flow with N_2O or measured single labeled inert gas washout curves disease.[39,52,53] Depressed mean LV perfusion in patients with diffuse coronary artery disease has also been observed by Klocke and coworkers using the H_2 technique with coronary sinus sampling.[54] Why average LV perfusion is reduced at rest in patients who are angina-free and do not manifest ischemic ECG changes is not clear. Mean myocardial O_2 extraction at rest in such patients has been reported to be normal.[55] A large number of the patients did not have historical, ECG or ventriculographic evidence of prior myocardial infarction or of left ventricular hypertrophy.[48] The data suggest the possibility that abnormal ventricular function in patients with diffuse coronary disease may lower the myocardial oxygen consumption so that a lower level of resting capillary perfusion is adequate to meet reduced metabolic needs of the diseased ventricle.

REGIONAL MYOCARDIAL BLOOD FLOW IN PATIENTS WITH CORONARY ARTERY DISEASE

The coefficient of variation of the local myocardial blood flow rates in the left ventricles was significantly greater in patients with radiographically significant coronary artery disease than in patients with normal arteriograms (23% versus 15%),[45] These data confirm other measurements obtained by direct injections of ^{85}Kr or from analysis of coronary sinus H_2 which indicated that there is a greater nonuniformity of myocardial perfusion in the presence of coronary artery stenosis.[47,56]

The detection of selective reductions of regional myocardial blood flow in patients with coronary lesions is complicated by the fact that many patients with coronary atherosclerosis exhibit diffuse depression of myocardial blood flow. Regional perfusion is usually reduced at rest, however, if the stenosis is 90 to 100 percent. In patients with residual transmural anterior myocardial infarctions, greater than 90 percent lesions of the left anterior descending artery (LAD), and abnormal anterior wall motion, Dwyer and associates found that myocardial perfusion in the LAD region, 44 ml/100 g/min, was significantly less than the LAD flow of 62 ml/100 g/min found in control patients without infarction who had normal coronary arteriograms.[57] In patients with residual inferior infarction and greater than 90 percent right coronary lesions, there was no evidence of tracer washout from the inferior LV after right coronary injection and there was a reduced RV perfusion rate.[57]

The data in Figure 5 are from studies of patients with radiographically significant LAD or LAD plus right coronary artery disease.[48,58] Selective reductions of regional myocardial blood flow/unit mass were observed distal to 90 to 100 percent lesions. In patients with 50 to 90 percent lesions there was a greater variety of blood flow rates; some values were normal and others were reduced. Klocke and coworkers recently measured myocardial H_2 washout curves by sampling blood from the LAD drainage in the great cardiac vein, and they also found resting reductions in LAD flow/weight at rest in patients with 90 to 100 percent LAD lesions and variable resting flow values in those with lesser degrees of stenosis.[17] In the study of patients with LAD and LAD plus right coronary disease from our laboratory, quantitative measurement of segmental ventricular wall motion was not performed. Engel and associates recently reported preliminary data which suggested that for a

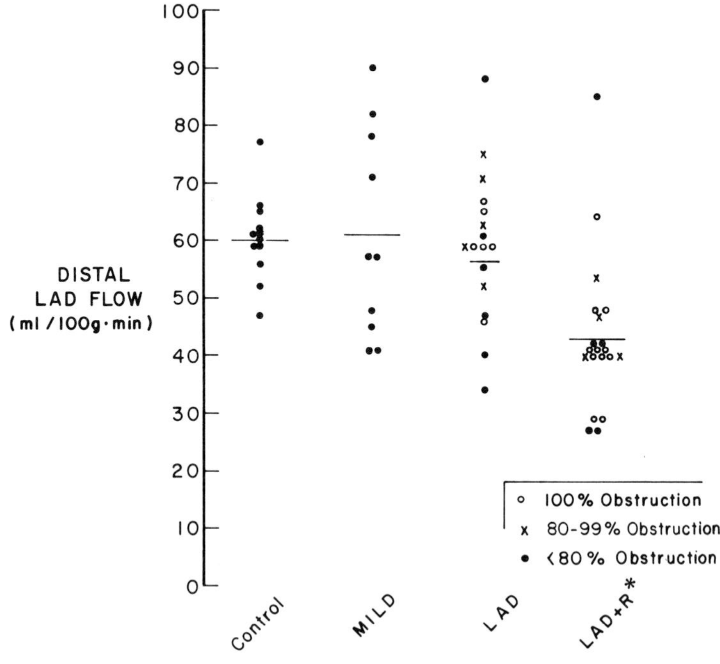

Figure 5. The regional myocardial blood flow distal to the LAD lesion is plotted for a large series of patients with normal coronary arteriograms, mild disease (<50% lesions), greater than 50 percent lesions of the LAD, and greater than 50 percent lesions of the LAD with associated right coronary lesions.

given degree of coronary stenosis from 50 to 100 percent, regional perfusion distal to the lesion was inversely related to the degree of wall motion (i.e., lower flow values were found with hypokinesis or akinesis).[60]

The presence of myocardial scar or aneurysm in regions of abnormal wall motion may be suspected from analysis of the initial distribution of peak ^{133}Xe counts after intracoronary injection (i.e., the initial tracer activity is reduced in the area of infarction).[35] These results are analogous to findings by groups that perform myocardial scintiscans of initial myocardial distribution of labeled albumin microspheres after intracoronary injection[5,6] or the distribution of ^{201}Tl in the myocardium after intravenous injection of the tracer.[2]

MYOCARDIAL BLOOD FLOW RESPONSES TO INTERVENTIONS IN CORONARY ARTERY DISEASE

Studies of mean LV and regional myocardial blood flow responses to interventions that increase myocardial oxygen consumption or induce hyperemia have been performed in order to detect abnormalities of coronary flow reserve in patients with atherosclerosis. Schmidt and coworkers increased heart rate gradually by atrial pacing to a steady-state rate of 150/min or to the onset of mild chest pain and made measurements of myocardial blood flow.[61] In 24 patients with normal coronary arteriograms or less than 50 percent coronary lesions, mean LV myocardial blood flow rose 24 ml/100 g/min during pacing; no patient developed angina or ST-T abnormalities. In 24 patients with greater than 50 percent obstructions, the rise in an index of myocardial oxygen consumption (the product of heart rate and systolic BP) during pacing was not significantly different from that of the group with no or minimal coronary disease. Nevertheless, 20 of the 24 patients with greater than 50 percent lesions developed angina or ischemic ECG changes. Mean LV perfusion rose only 16 ml/100 g/min in those patients with greater than 50 percent lesions (p < 0.01). Regional myocardial blood flow distal to the major coronary artery lesion rose significantly less than in the remainder of the ventricle during pacing. The data supported the hypothesis that angina pectoris occurs in some patients when coronary lesions restrict the increment in myocardial blood flow required to meet the increased metabolic needs of heart muscle of O_2 and other nutrients.

Other investigators have used different techniques to increase myocardial oxygen consumption. Maseri and associates induced rapid tachycardia abruptly in patients with greater than 90 percent LAD obstructions and observed a decreased rate of ^{133}Xe clearance distal to the lesions during the pacing-induced angina.[62] Latson and coworkers infused isoproterenol intravenously in patients during cardiac catheterization. These workers found that the percent increase in regional perfusion distal to coronary stenosis greater than 50 percent was inversely related to the degree to which the arterial diameter was narrowed by disease.[63] Several investigators have measured myocardial blood flow responses after intracoronary injection of agents which dilate the coronary vessels. Scheibel and coworkers found that the increment in regional myocardial perfusion induced by intracoronary papavarine was less in regions supplied by diseased arteries than in regions supplied by arteries that were angiographically normal.[30] Holman and associates found that the increased regional myocardial ^{133}Xe clearance induced by intracoronary injection of contrast material was significantly reduced in patients with coronary artery disease.[64]

COLLATERAL BLOOD FLOW

Collateral blood flow in man can also be assessed using ^{133}Xe and a scintillation camera. If sufficient isotope is delivered to a myocardial region supplied by collateral vessels so that washout curves with good counting statistics are obtained, the rate constants of local isotope clearance reflect the level of capillary perfusion regardless of whether the blood originated from the injected or other coronary arteries. Several studies have suggested that in some patients, collaterals can maintain normal resting myocardial blood flow and contribute to myocardial blood flow reserve during times of stress.[61,65,66]

The technique is also helpful in the postoperative assessment of coronary artery bypass operations. After injection of ^{133}Xe into the graft, the number of crystals (or the size of the region of interest) which record washout curves indicates the size of the area receiving diffusible nutrients from the graft, while the local perfusion rates reflect the adequacy of myocardial capillary blood flow both from the graft and from the diseased coronary arteries. Data by Korbuly and coworkers suggest that bypass grafts increase myocardial blood flow in the grafted areas.[67]

EFFECTS OF DRUGS ON CORONARY BLOOD FLOW

Other investigators have studied the effects of a variety of cardiac drugs using intracoronary injection of ^{133}Xe to measure regional myocardial blood flow. Dirschinger and associates found sodium nitroprusside and isosorbide dinitrate decreased blood flow in areas perfused by nonstenosed coronary arteries and increased blood flow in hypoperfused subregions with reversible asynergy.[68] Lichtlen and associates found that nitroglycerin reduced perfusion both in normal and poststenotic areas, whereas dipyridamole increased flow in these regions.[32] Cohn and coworkers also studied nitroglycerin using ^{133}Xe; they obtained suggestive evidence that regional flow reductions were smaller or that flow increased distal to lesions supplied by collaterals from nonstenosed vessels after administration of the drug.[69]

CONCLUSION

The studies reviewed in this chapter illustrate some of the existing uses for the measurement of regional myocardial blood flow with a scintillation camera after intracoronary injection of [133]Xe. The method is still undergoing modifications and improvements in different centers. It is expected that as new tracers such as [127]Xe are developed and more efficient and faster scintillation cameras are produced, some of the current limitations of the technique will be reduced. It appears likely that the significance of coronary obstructions will be more readily assessed when the measurements of blood flow distal to coronary obstructions are performed in multiple projections, not only at rest, but also during interventions which either increase myocardial oxygen consumption or induce coronary vasodilation.

ACKNOWLEDGMENTS

The author wishes to acknowledge the participation of Drs. M. Weiss, D. Blood, L. Johnson, D. Schmidt, R. Dell, E. Dwyer, R. Sciacca, and W. Casarella in the investigations summarized in this article, and the participation of Ms. Phyllis Wolf in the preparation of the manuscript.

REFERENCES

1. ZARET, B. L., STRAUSS, H. W., MARTIN, N. D., ET AL.: *Non-invasive regional myocardial perfusion with radioactive potassium: Study of patients at rest, with exercise and during angina pectoris.* N. Engl. J. Med. 288:809, 1973.

2. STRAUSS, H. W., HARRISON, K., LANGAN, J. K., ET AL.: *Thallium-201 for myocardial imaging. Relation of thallium-201 to regional myocardial perfusion.* Circulation 51:641, 1975.

3. MARSHALL, R. C., BERGER, H. J., COSTIN, J. C., ET AL.: *Assessment of cardiac performance with quanti-*

tative radionuclide angiocardiography:sequential left ventricular ejection fraction, normalized left ventricular ejection rate, and regional wall motion. Circulation 56:820, 1977.

4. BORER, J. S., BACHRACH, S. L., GREEN, M. V., ET AL.: Realtime radionuclide cineangiography in the noninvasive evaluation of global and regional left ventricular function at rest and during exercise in patients with coronary artery disease. N. Engl. J. Med. 296:839, 1977.

5. ASHBURN, W. L., BRAUNWALD, E., SIMON, A. L., ET AL.: Myocardial perfusion imaging with radioactive labeled particles injected directly into the coronary circulation of patients with coronary artery disease. Circulation 44:851, 1971.

6. JANSEN, C., JUDKINS, M. P., GRAMES, G. M., ET AL.: Myocardial perfusion color scintigraphy with MAA. Radiology 109:369, 1973.

7. CANNON, P. J., WEISS, M. B., AND SCIACCA, R. R.: Myocardial blood flow in coronary artery disease: studies at rest and during stress with inert gas washout techniques. Prog. Cardiovasc. Dis. 20:95, 1977.

8. KATZ, L. N., AND FEINBERG, H.: The relation of cardiac effort to myocardial oxygen consumption and coronary flow. Circ. Res. 6:656, 1958.

9. GOULD, K. L., LIPSCOMB, K., AND HAMILTON, G. W.: Physiological basis for assessing critical coronary stenosis. Am. J. Cardiol. 33:87, 1974.

10. MOSHER, P., ROSS, J. JR., McFATE, P. A., ET AL.: Control of coronary blood flow by an autoregulatory mechanism. Circ. Res. 14:250, 1964.

11. PARKER, J. O., CHIONG, M. A., WEST, R. O., ET AL.: Sequential alterations in myocardial lactate metabolism, S-T segments and left ventricular function during angina induced by atrial pacing. Circulation 40:113, 1969.

12. McGREGOR, M.: The coronary collateral circulation: a significant compensatory mechanism or a functionless quirk of nature. Circulation 52:529, 1975.

13. KETY, S. S.: Theory and applications of exchange of inert gas at lungs and tissues. Pharmacol. Rev. 3:1, 1951.

14. KETY, S. S.: Blood tissue exchange methods. Theory of blood tissue exchange and its application to measurement of blood flow. Methods Med. Res. 8:223, 1960.

15. CONN, H. L., JR.: Equilibrium distribution of radioxenon in tissue: xenon-hemoglobin association curve. J. Appl. Physiol. 16:1065, 1961.

16. ZIERLER, J. L.: Equations for measuring blood flow by external monitoring of radioisotopes. Circ. Res. 16:309, 1965.

17. KLOCKE, F. J.: Coronary blood flow in man. Prog. Cardiovasc. Dis. 2:117, 1977.

18. CHIDSEY, C. A., III, FRITTS, H. H., JR., HARDEWIG, A., ET AL.: Fate of radioactive krypton (Kr^{85}) introduced intravenously in man. J. Appl. Physiol. 14:63, 1959.

19. LEDERER, C. M., HOLLANDER, J. M., AND PERLMAN, I.: Table of Isotopes, ed. 6. John Wiley and Sons, New York, 1967.

20. LASSEN, N. A.: Assessment of tissue radiation dose in clinical use of radioactive inert gases with examples of absorbed dose from 3-H, 85-Kr, 133-Xe. Minerva Nucl. 8:211, 1964.

21. ATKINS, H. L., SUSSKIND, H., KLOPPER, J. F., ET AL.: A clinical comparison of Xe-127 and Xe-133 for ventilation studies. J. Nucl. Med. 18:653, 1977.

22. CANNON, P. J., HAFT, J. I., AND JOHNSON, P. M.: Visual assessment of regional myocardial perfusion using radioactive xenon and scintillation photography. Circulation 40:277, 1969.

23. CANNON, P. J., DELL, R. B., AND DWYER, E. M., JR.: Regional myocardial perfusion in man. J. Clin. Invest. 49:16a, 1970.

24. CANNON, P. J., DELL, R. B., AND DWYER, E. M., JR.: Measurement of regional myocardial perfusion in man with ^{133}xenon and a scintillation camera. J. Clin. Invest. 51:964, 1972.

25. MASERI, A., MANCINI, P., L'ABBATE, A., ET AL.: Method for regional dynamic study of myocardial blood flow in man. J. Nucl. Biol. Med. 15:54, 1971.

26. STOKELY, E. M., NARDIZZI, L. R., PARKEY, R. W., ET AL.: Regional myocardial perfusion data with spatial and temporal quantitation. J. Nucl. Med. 14:669, 1973.

27. BONTE, F. J., PARKEY, R. W., STOKELY, E. M., ET AL.: Radionuclide determination of myocardial blood flow. Semin. Nucl. Med. 3:153, 1973.

28. HOLMAN, B. L., ADAMS, D. F., JEWITT, D., ET AL.: Measuring regional myocardial blood flow with ^{133}Xe and the Anger camera. Radiology 112:99, 1974.

29. KORHOLA, O.: Myocardial scintigraphy and estimation of regional blood flow with xenon-133. Acta Radiol. Suppl. 337:7, 1974.

30. SCHEIBEL, R. L., MOORE, R., KORBULY, D., ET AL.: Regional myocardial blood flow measurements in the evaluation of patients with coronary artery disease. Radiology 115:379, 1975.

31. RUDOLPH, W., FLECK, E., DIRSCHINGER, J., ET AL.: *Regional myocardial blood flow determined by the xenon-133 washout technique with respect to coronary artery stenosis and wall motion abnormalities.* Herz Kardiovaskulare Erkankungen 2:16, 1977.

32. LICHTLEN, P. R., ENGEL, H. J., AND HUNDESHAGEN, H.: *Regional myocardial blood flow in normal and post-stenotic areas after nitroglycerin, beta blockade (atenolol), coronary dilatation (dipyridamole) and calcium antagonism (nifidipine).* Herz Kardiovaskuläre Erkrankungen 2:81, 1977.

33. JONES, R. H., LIDOFSKY, L., BUDINGER, T., ET AL.: *Instrumentation.* In Pierson, R. M., Jr., Kriss, J. P., Jones, R. H., MacIntyre, W. J., (eds.): *Nuclear Cardiography.* John Wiley and Sons, New York, 1975.

34. CANNON, P. J., SCIACCA, R. R., BRUST, J.C.M., ET AL.: *Measurement of regional cerebral blood flow with 133-xenon and a multiple-crystal scintillation camera.* Stroke 5:371, 1974.

35. CANNON, P. J., SCIACCA, R. R., FOWLER, D. L., ET AL.: *Measurement of regional myocardial blood flow in man: Description and critique of the method using xenon-133 and a scintillation camera.* Am. J. Cardiol. 36:783, 1975.

36. JUDKINS, M. P.: *Selective coronary arteriography. I. A percutaneous femoral technique.* Radiology 89:815, 1967.

37. DODGE, H. T., SANDLER, H., BAXLEY, W. A., ET AL.: *Usefulness and limitations of radiographic methods for determining left ventricular volume.* Am. J. Cardiol. 18:10, 1966.

38. HERD, J. A., HOLLENBERG, M., THORBURN, G. D., ET AL.: *Myocardial blood flow determined with krypton[85] in unanesthetized dogs.* Am. J. Physiol. 203:122, 1962.

39. ROSS, R. S., UEDA, J., LICHTLEN, P. R., ET AL.: *Measurement of myocardial blood flow in animals and man by selective injection of radioactive inert gas into the coronary arteries.* Circ. Res. 15:28, 1964.

40. BASSINGTHWAIGHTE, J. B., STRANDELL, T., AND DONALD, D. E.: *Estimation of coronary blood flow by washout of diffusible indicators.* Circ. Res. 23:259, 1968.

41. SHAW, D. J., PITT, A., AND FRIESINGER, G. C.: *Autoradiographic study of the 133-xenon disappearance method for measurement of myocardial blood flow.* Cardiovasc. Res. 6:268, 1972.

42. HORWITZ, L. D., CURRY, G. C., PARKEY, R. W., ET AL.: *Differentiation of physiologically significant coronary artery lesions by coronary blood flow measurements during isoproterenol infusion.* Circulation 49:55, 1974.

43. CARLIN, K., AND CHIEN, S.: *Effect of hematocrit on the washout of xenon and iodoantipyrine from dog myocardium.* Circ. Res. 40:505, 1977.

44. SCIACCA, R. R., WEISS, M. B., BLOOD, D. K., ET AL.: *Comparison of regional myocardial blood flow measurements with 133-xenon and radioactive microspheres in dogs with coronary artery constrictions.* Submitted for publication.

45. CANNON, P. J., DELL, R. B., AND DWYER, E. M., JR.: *Regional myocardial perfusion rates in patients with coronary artery disease.* J. Clin. Invest. 51:978, 1972.

46. DOMENECH, R. J., HOFFMAN, J. I. E., NOBLE, N. I. M., ET AL.: *Total and regional coronary blood flow measured by radioactive microspheres in conscious and anesthetized dogs.* Circ. Res. 25:581, 1969.

47. SULLIVAN, J. M., TAYLOR, W. J., ELLIOTT, W. C., ET AL.: *Regional myocardial blood flow.* J. Clin. Invest. 46:1402, 1967.

48. CANNON, P. J., SCHMIDT, D. H., WEISS, M. B., ET AL.: *The relationship between regional myocardial perfusion at rest and arteriographic lesions in patients with coronary atherosclerosis.* J. Clin. Invest. 56:1442, 1975.

49. WEISS, M. B., ELLIS, K., SCIACCA, R. R., ET AL.: *Myocardial blood flow in congestive and hypertrophic cardiomyopathy: Relationship to peak wall stress and mean velocity of circumferential fiber shortening.* Circulation 54:484, 1976.

50. JOHNSON, L. L., SCIACCA, R. R., ELLIS, K., ET AL.: *Reduced left ventricular myocardial blood flow per unit mass in aortic stenosis.* Circulation 57:582, 1978.

51. BRAUNWALD, E.: *Control of myocardial oxygen consumption: physiologic and clinical considerations.* Am. J. Cardiol. 27:416, 1971.

52. ROWE, G. G., THOMSEN, J. H., STENLUND, R. R., ET AL.: *A study of hemodynamics and coronary blood flow in man with coronary artery disease.* Circulation 39:139, 1969.

53. HOLMBERG, S., PAULIN, S. PREROVSKY, I., ET AL.: *Coronary blood flow in man and its relation to the coronary arteriogram.* Am. J. Cardiol. 19:486, 1967.

54. KLOCKE, F. J., BUNNELL, I. L., GREENE, D. G., ET AL.: *Average coronary blood flow per unit weight of left ventricle in patients with and without coronary artery disease.* Circulation 50:547, 1974.

55. MESSER, J. V., AND NEIL, W. A.: *The oxygen supply of the human heart.* Am. J. Cardiol. 9:384, 1962.

56. KLOCKE, F. J., AND WITTENBERG, S. M.: *Heterogeneity of coronary blood flow in human coronary artery disease and experimental myocardial infarction.* Am. J. Cardiol. 24:782, 1969.

57. DWYER, E. M., JR., DELL, R. B., AND CANNON, P. J.: *Regional myocardial blood flow in patients with residual anterior and inferior transmural infarction.* Circulation 48:924, 1973.

58. CANNON, P. J., WEISS, M. B., AND CASARELLA, W. J.: *Studies of regional myocardial blood flow: results in patients with left anterior descending coronary artery disease.* Semin. Nucl. Med. 6:279, 1976.

59. ENGEL, H. J., LICHTLEN, P. R., AND HUNDESHAGEN, H.: *Effects of coronary obstructions and segmental left ventricular dysfunction on regional myocardial blood flow.* Circulation 57 (Suppl. III):10, 1977.

60. ENGEL, H. J., LICHTLEN, P., AND HUNDESHAGEN, H.: *Assessment of regional myocardial perfusion: comparison of the precordial xenon clearance technique and perfusion scintigraphy.* Herz Kardiovaskuläre Erkrankungen 2:130, 1977.

61. SCHMIDT, D. H., WEISS, M. B., CASARELLA, W. J., ET AL.: *Regional myocardial perfusion during atrial pacing in patients with coronary artery disease.* Circulation 53:807, 1976.

62. MASERI, A., L'ABBATE, A., PESOLA, A., ET AL.: *Regional myocardial perfusion in patients with atherosclerotic coronary artery disease, at rest and during angina pectoris induced by tachycardia.* Circulation 55:423, 1977.

63. LATSON, T., GRAVES, B., AND HORWITZ, L.: *Detection of physiologically significant coronary artery disease by regional myocardial blood flow measurements during isoproterenol infusion.* Circulation 57 (Suppl. III):34, 1977.

64. HOLMAN, B. L., COHN, P. F., ADAMS, D. F., ET AL.: *Regional myocardial blood flow during hyperemia induced by contrast agent in patients with coronary artery disease.* Am. J. Cardiol. 38:416, 1976.

65. FRICK, M. H., VALLE, M., KORHOLA, O., ET AL.: *Analysis of coronary collaterals in ischemic heart disease by angiography during pacing induced ischemia.* Br. Heart J. 38:186, 1976.

66. FRICK, M. H., KIORHOLA, O., VALLE, M., ET AL.: *Radiologically detected collaterals and regional myocardial flow responses to ischemia in ischemic heart disease.* Ann. Clin. Res. 8:241, 1976.

67. KORBULY, D. E., FORMANEK, A., GYPSER, G. ET AL.: *Regional myocardial blood flow measurements before and after coronary bypass surgery.* Circulation 52:38, 1975.

68. DIRSCHINGER, J., FLECK, E., REDL, A., ET AL.: *Effects of sodium nitroprusside and isosorbide dinitrate on regional myocardial blood flow in patients with coronary artery disease and left ventricular asynergy.* Herz Kardiovaskuläre Erkrankungen 2:71, 1977.

69. COHN, P., MADDOX, D., HOLMAN, B. L., ET AL.: *Effect of sublingually administered nitroglycerin on regional myocardial blood flow in patients with coronary artery disease.* Am. J. Cardiol. 39:672, 1977.

Radioactive Particle Imaging of Coronary Blood Flow Distribution

James H. Caldwell, M.D., James L. Ritchie, M.D., and Glen W. Hamilton, M.D.

The use of radionuclides for myocardial imaging began in the early 1960s. Carr and associates[1] reported on rubidium-86 as an intravenous agent that behaves similarly to potassium ion in the ischemic myocardium. Subsequently, cesium-31,[2] radioiodinated fatty acid,[3] potassium-43,[4] rubidium-81,[5] and thallium-201 have been used intravenously as myocardial imaging agents. The clinical usefulness of these agents is hindered by their relatively poor imaging qualities and uptake in contiguous organs such as liver and lungs, leading to images which are often indistinct due to radioactivity in structures surrounding the heart. Direct intracoronary injection of radioactive particles into the coronary circulation eliminates the problem of uptake by adjacent organs but requires cardiac catheterization. Radioactive macroaggregated albumin (MAA) particles were first used by Quinn and coworkers[6] to visualize the myocardium in dogs. Following balloon occlusion of the ascending aorta,[131]I-labeled MAA was injected into the aorta. Using a rectilinear scanner, they demonstrated image defects that corresponded to areas of hypoperfusion secondary to coronary artery ligation.

In 1970 Endo and colleagues[7] reported on the direct intracoronary injection of ^{131}I-labeled MAA particles in humans. No adverse effects were seen on the ECG or hemodynamic parameters in eight dogs and four patients.

Because large doses of intracoronary particles had been used previously as a method for producing cardiogenic shock,[8,9] there was initial hesitation in using the particle technique for studying the coronary circulation in humans. Between 1970 and 1972, Weller,[10] Poe,[11] and Schelbert[12] examined carefully the effect of MAA and human albumin microspheres (HAM) injected into the canine coronary circulation on coronary blood flow, contractile force, the ECG, cardiac enzymes, heart rate, systemic blood pressure, and histologic sections of the heart. Based on these studies, it was shown that the procedure was safe if the particles were small (50 μ or less in diameter) and the number of particles injected was less than 300,000. Within a short time, studies on more than 400 patients were reported without adverse effects.[13,14] During the last five years, over 4000 studies have been performed without morbidity or mortality, thus establishing the safety of the procedure.[15,16]

PHYSIOLOGIC BASIS OF PARTICLE IMAGING

Particle Distribution

Measurement of blood flow to various organs by radionuclide techniques has been available for many years and was reviewed in depth by Wagner and coworkers.[17] The basic principle of perfusion imaging with radioactive particles is that small particles when injected into the afferent blood supply of an organ are trapped in the capillary or precapillary beds in a distribution which is proportional to regional blood flow. If the particles are tagged with radionuclides, the distribution of radioactivity is directly proportional to blood flow distribution at the moment of injection.

Blood flow distribution can then be measured by in vitro scintillation counting of tissue samples or by scintigraphic imaging over the organ in question. When applied to the intracoronary injection of radioactive particles, activity in a particular region of the heart reflects relative blood flow to that region provided adequate mixing of the particles occurred before branching of the coronary artery. For studies employing tissue counting, particle size should be between 8 and 15 μ in order to accurately record the ratio of epicardial to endocardial flow. For external imaging, particle size is not critical within the limits of safety as epicardial-endocardial flow cannot be distinguished.

Coronary Physiology

In the normal heart, blood flow to the left ventricular myocardium is relatively uniform—approximately 100 ml/100 g/min. Flow to the right ventricle and atrial muscle is slightly less on a ml/g basis. There is also some heterogeneity of flow when analyzed by in vitro tissue sample counting techniques; however, this is unimportant for external imaging purposes. Maximal flow during stress (the coronary reserve) can reach 4 to 5 times the basal levels and is achieved through a decrease in coronary resistance. When a coronary stenosis is present, distal coronary perfusion *at rest* is maintained through the mechanism of coronary reserve, i.e., diminished coronary resistance, until the stenosis is 80 percent or greater of the vessel diameter. Thus, a stenosis in the range of 40 to 80 percent becomes hemodynamically significant only during times of *increased flow demand*. This concept has been critically demonstrated by Gould and associates in awake animal models[18] (Fig. 1). The concept of coronary reserve is important in understanding how radioactive particles can be used to assess regional coronary blood flow at rest and during maximal stress (or coronary vasodilatation) in relation to the degree of coronary stenosis.

Angiographic contrast material, such as Renagrafin-76® and Hypaque-75®, is a potent coronary vasodilator which can increase coronary blood flow by 3 to 4 times normal.[18] The exact mechanism of this vasodilation is uncertain, although it probably relates to hyperosmolarity of the agent. When contrast material is injected into a coronary system with various patterns of coronary stenosis, flow increases normally in vessels with no or minor stenosis. In vessels with significant stenosis, the flow increase is diminished due to decreased coronary reserve. This results in flow maldistribution, with the vessel with the greatest stenosis having the most severe deficit in flow. Radioactive particles injected during the period of contrast-induced vasodilatation will likewise be maldistributed, with the vessel with the most severe stenosis receiving the least amount of radioactive particles. Since particle distribution in the coronary circulation is rapid (1 to 2 seconds), the pattern of particle distribution is related to the physiologic conditions at the moment of injection even though counting or imaging may be carried out at a later time. This principle can be used to assess coronary stenosis.[19] An isotope with one energy is attached to MAA or HAM and injected into the coronary artery under basal conditions. These particles will be normally distributed in spite of significant coronary stenoses. A second injection of particles with another radioisotope label is made immediately (6 to 10 seconds) following intracoronary contrast agent; these particles will be maldistributed due to diminished coronary reserve in vessels with significant stenosis, and the coronary bed with the most impaired hyperemic response will appear relatively hypoperfused on the scintigraphic image.

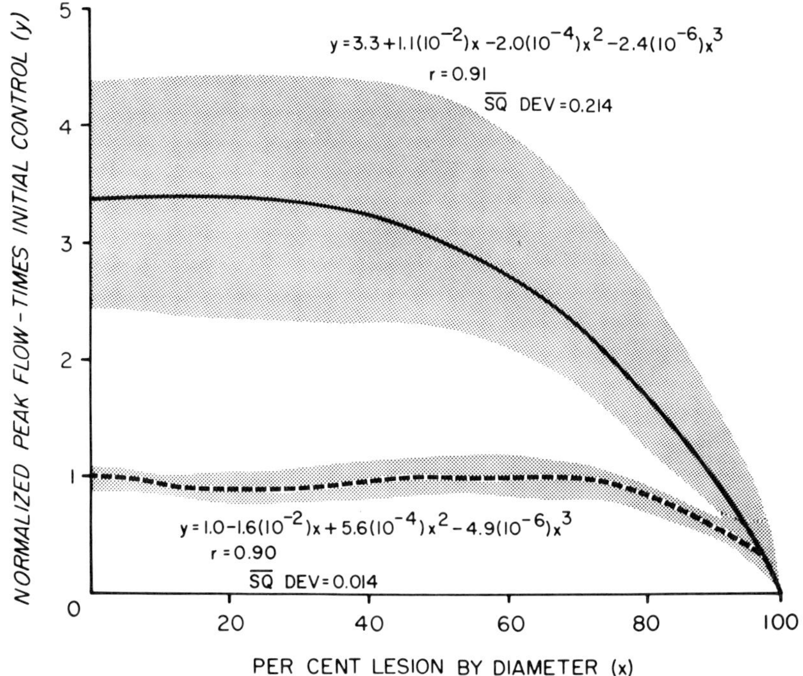

Figure 1. Coronary blood flow at rest (bottom) and during maximal contrast induced hyperemia (top) in the awake dog. The shaded areas represent data points from studies in eight dogs with coronary stenosis ranging from 0 to 100 percent. Coronary flow was measured by electromagnetic flow probe. Resting flow is maintained until the stenosis exceeds 80 percent. The hyperemic response becomes impaired when the stenosis exceeds 50 percent, and there is no vasodilator capacity remaining once the stenosis is in the range of 90 percent. (From Gould et al.,[18] with permission of the American Journal of Cardiology)

Many patients with coronary artery disease demonstrate particle maldistribution (regional perfusion defects) at rest. This is most commonly due to regions of myocardial fibrosis due to previous infarction. Occasional patients with preinfarction angina or rest angina will also have regional perfusion defects at rest, presumably due to diminished regional blood flow at the time of injection.

TECHNICAL CONSIDERATIONS

The safety of intracoronary particle injections depends upon proper preparation and strict quality control of particle size and number. Either MAA particles or HAM between 20 and 50 μ in diameter may be used satisfactorily and can be labeled with either ^{99m}Tc or ^{113m}In. Normally, 1 to 3 mCi of either isotope are labeled to 50,000 or fewer particles. These two isotopes have good imaging characteristics, short half-lives, and are readily available. Iodine-131 has also been used but is less desirable because it has a low photon yield and the larger number of particles required decrease the safety factor.

If only a resting study is to be performed, an injection of one isotope can be made in the left coronary artery and the second isotope injected into the right coronary. To perform a resting study, 1.5 mCi of ^{113m}In-labeled particles may be injected into the right coronary artery and 1.5 to 3.0 mCi of ^{99m}Tc particles are injected into the left coronary. Or, if one has only one isotope for the study, one-third can be injected into the right coronary and two-thirds into the left coronary artery. It is difficult, however, to assure that the activity

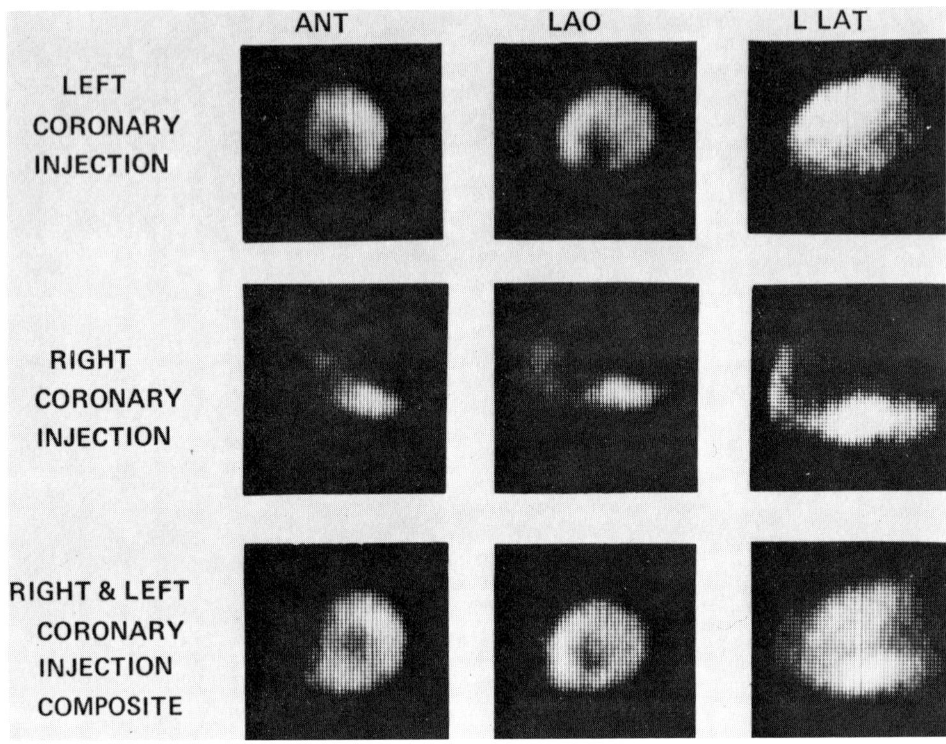

Figure 2. The normal myocardial image following injection of labeled MAA into both the right and left coronary arteries. Different radioactive labels were used for each artery. The inferior defect in the image of the left coronary artery represents the myocardium supplied by the right coronary artery. The lower row demonstrates the composite image after computer addition of the right and left coronary injection images. (From Ritchie et al.,[16] in Radiology, with permission)

per gram of muscle mass will be the same for both ventricles, in which case, greater activity in the region of one artery could lead to an erroneous interpretation.

For a rest-hyperemic study, only one coronary artery can be studied. The rest-hyperemic study is performed by injecting one isotope at rest (as described above). Five to ten milliliters of contrast agent are then injected as for a coronary arteriogram and followed in 6 to 10 seconds by injection of the second isotope.

Following catheterization, the patient is taken to the nuclear medicine laboratory where scintigraphic images are obtained in the anterior, 45 degrees LAO, and left lateral views. Additional LAO views at 30 and 60 degrees are often helpful.

Imaging should be performed using a gamma camera with dual spectroscopy, if available, so that images of both isotopes can be collected simultaneously. Collimation will depend upon which isotopes are used and should be selected to give the highest resolution possible. Collection of data from both isotopes in a dedicated computer system is useful for image processing and display (Fig. 2). If such a system is not available, adequate studies can be obtained by imaging each isotope serially and photographically superimposing the images. Great care must be taken, however, to prevent patient motion between the images of the two isotopes in each view.

216

INTERPRETING THE MYOCARDIAL IMAGE

Rest Image

The coronary beds are conceptually best visualized in the LAO projection (Fig. 3). In this view, the area supplied by the left anterior descending coronary artery lies to the left, that supplied by the circumflex on the right, and that by the right coronary is present as a smaller inferior wedge. The normal image in each of the projections is displayed in Figure 4. The distribution of isotope for each coronary bed is demonstrated as well as the composite image.

The distribution of the isotope with the right coronary injection produces a somewhat variable image due to the wide variation in muscle mass but, in general, assumes a "ball-with-tail" appearance, with the ball representing the inferior wall of the left ventricle (Fig. 4). With the left coronary injection, the image resembles a sphere with diminished activity inferiorly in the LAO projection. Abnormalities of coronary perfusion appear as areas of decreased activity (Fig. 5) and are usually seen in two or more projections. Figure 6 schematically shows the common location of defects and how they appear in each projection. Representative examples are presented in Figure 7.

Hyperemic Image

On the basis of the previous discussion of the physiology of hyperemia and coronary stenosis, four image patterns are seen. In the normal pattern, there will be no difference between isotope distribution at rest and during the hyperemic study. In patients with old

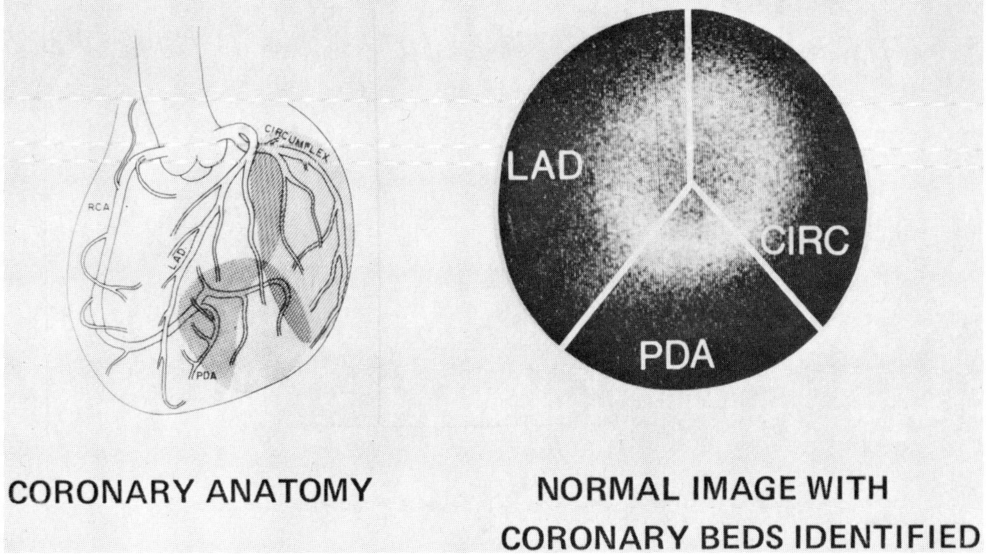

Figure 3. The coronary artery distribution and myocardial image in the LAO projection. The least overlap of the three major coronary beds is provided in this view, as shown on the left. The region of the myocardial image perfused by each of the major coronary arteries is shown on the right. (From Ritchie et al.,[16] in Radiology, with permission)

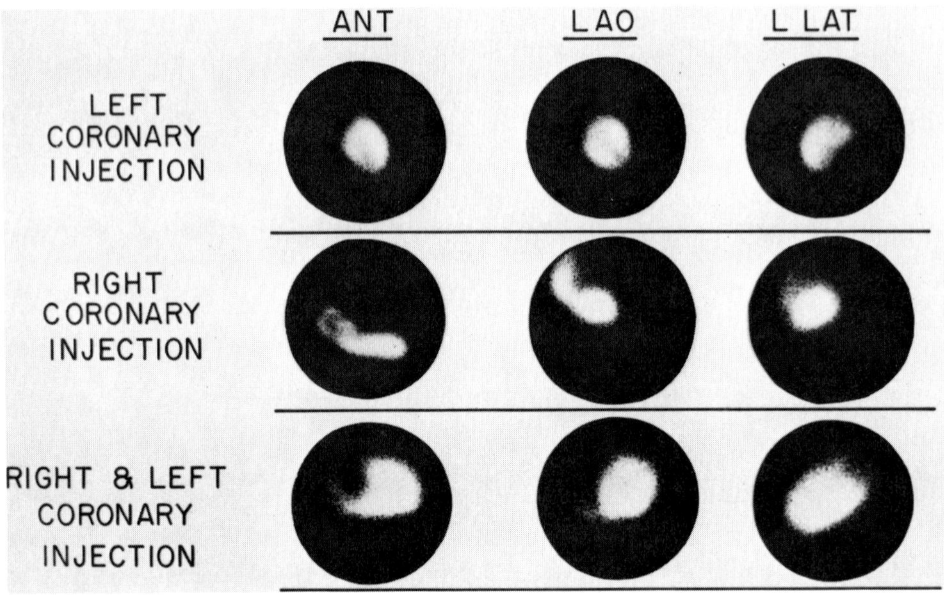

Figure 4. The normal myocardial image following injection of the right and left coronary arteries. Unlike Figure 2, there has not been any computer processing. In the image of the left coronary injection, the defect representing the right coronary artery is not as well seen. Also, the central cavity is not as evident in the unprocessed image. (From Hamilton, G. W., et al., Am. Heart J. 89:708, 1975, with permission)

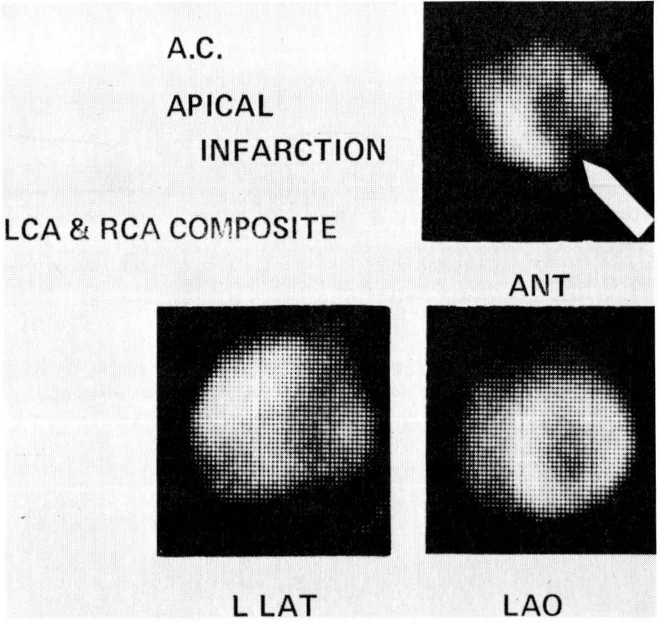

Figure 5. Example of abnormal perfusion. The patient has had a previous apical myocardial infarction. This is represented by the area of decreased activity at the apex in the anterior projection. (From Ritchie et al.,[16] in Radiology, with permission)

218

DEFECT	RAO	LAO	LLAT	
APICAL				Tip of image amputated in RAO and L. Lateral. prominent central cavity on LAO. The latter also occurs with cardiac dilitation.
ANTERIOR SEPTAL				Often difficult to detect on ANT or RAO. Easily seen on LAO & L LAT
INFERIOR				Often hard to visualize. Small inferior defect in all views.
POSTERIOR LATERAL				Often difficult to detect on ANT or RAO. Easily seen on LAO & L LAT

Figure 6. Schematic representation of the major defects as seen in each of the standard projections. The right anterior oblique projection is shown but the anterior view is comparable. Most defects are present in at least two views. (From Ritchie et al., in Radiology, with permission)

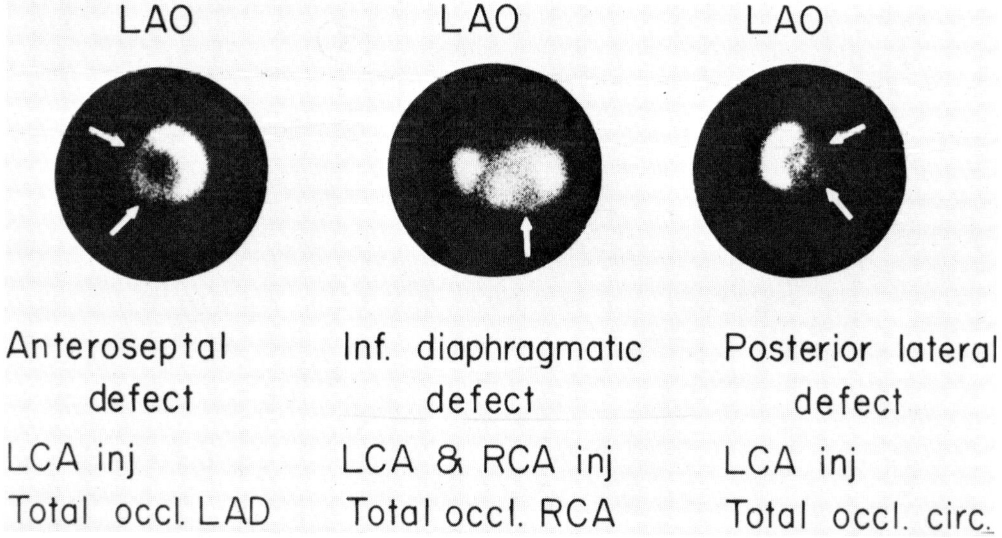

Figure 7. Abnormal myocardial images representing three of the major defects seen in patients with myocardial infarction. Abbreviations: occl. = occluded; inf. = inferior; inj. = injection; LCA = left coronary artery; RCA = right coronary artery; LAD = left anterior descending coronary artery; circ = circumflex artery. (From Hamilton, G. W., et al., Am. Heart J. 89:708, 1975, with permission)

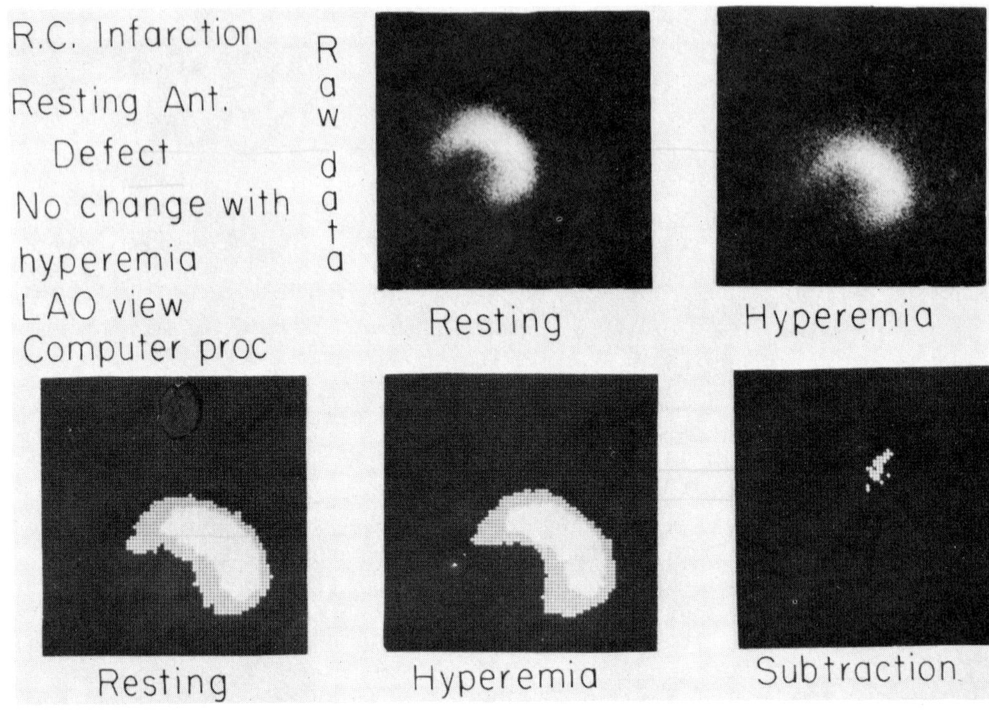

Figure 8. Abnormal myocardial image in the anteroseptal region at rest in the LAO projection. During hyperemia, no change is seen in the defect. The patient had a total occlusion of the left anterior descending coronary artery with a previous anterior myocardial infarction. (From Ritchie et al.,[19] with permission of the American Journal of Cardiology)

myocardial infarction, a resting defect is present that is unchanged following hyperemia, except that the defect may be slightly more apparent because of the relatively greater activity in the surrounding tissue during hyperemia (Fig. 8). Regions of myocardium with normal resting blood flow but abnormal hyperemic response due to coronary stenosis, demonstrate a normal resting image with a defect during contrast hyperemia (Fig. 9). Lastly, regions of previous infarction with an image defect at rest may show an increase in the size of the defect during hyperemia due to impairment of coronary reserve in the preinfarction zone.

CLINICAL CORRELATIONS AND APPLICATIONS

In our series of 127 patients[20] studied for suspected or known coronary artery disease, 56 (44%) had a resting defect. Of these, 53 (95%) had evidence of previous myocardial infarction by ECG, ventriculography, biopsy, or direct inspection at time of surgery. Only two patients with ECG evidence of infarct had normal scans. However, three of six patients with preinfarction angina had defects on the rest image with no other evidence of infarction. In this setting, it seems likely that transmural resting coronary flow was diminished and that hypoperfusion without infarction was identified. Maseri and co-workers[22] have reported similar findings with ^{201}Tl injected at the time of pain, with subsequent complete resolution with relief of pain.

The resting image is not closely related to coronary anatomy. Patients with normal coronary anatomy will routinely have normal resting studies. The presence of a resting defect is predictive of significant coronary artery disease except in the rare cases of infarction with normal coronary arteries. The converse is not true, and only a fraction

220

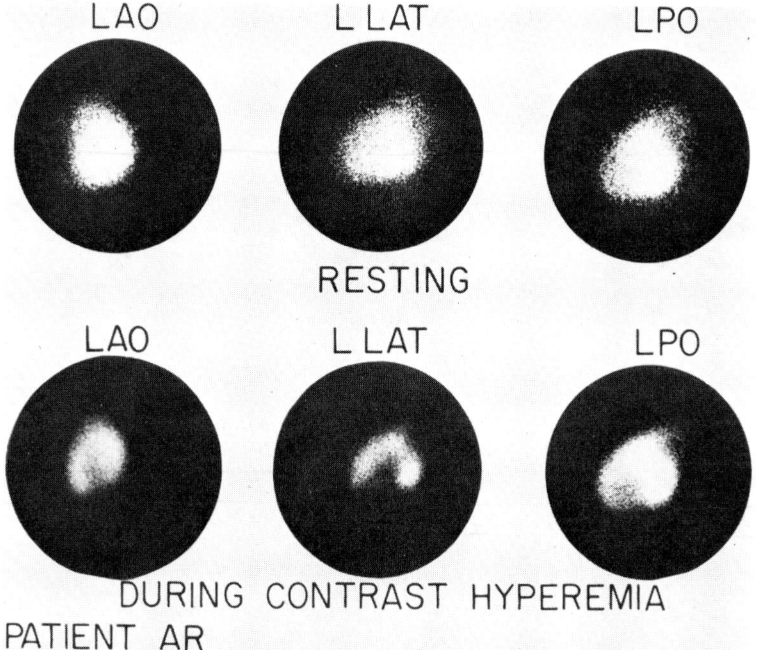

Figure 9. Rest-hyperemic study. The resting image is normal in spite of a severe proximal RCA stenosis. Following contrast induced hyperemia, a large inferior defect appears in all three images but is most evident in the LAO and left lateral views.

(⅓ to ½) of patients with coronary artery disease will have a resting perfusion defect.[20]

When rest-hyperemic studies were performed in 49 patients, all patients with insignificant coronary artery disease (no stenosis greater than 50%) or none at all had normal images.[19] Of 39 patients with significant coronary artery disease 23 had a resting defect. Seventy-five percent of those with a stenosis greater than 50 percent had a new or additional defects with hyperemia. In four patients with severe stenosis in both the left anterior descending and circumflex coronary arteries, the rest-hyperemic images were identical, presumably due to nearly balanced disease in both arterial beds which diminished coronary reserve equally in both regions. Thus, while some coronary stenoses were not detected, a resting defect and/or a defect with hyperemia was invariably associated with a significant coronary stenosis.

Kirk and associates[15] used the resting distribution of labeled human albumin microspheres to predict improved ventricular function following coronary artery bypass surgery. Patients were grouped according to the presence or absence of segmental contraction abnormalities by contrast ventriculogram and by regional myocardial perfusion. In patients with localized contraction abnormalities, but normal regional myocardial perfusion, 97 percent demonstrated improved contraction following bypass surgery. In those with localized ventricular contraction abnormalities and mild to moderate regional myocardial perfusion defects,* 65 percent improved following surgery. Only 20 percent of patients with abnormal ventriculograms and severe perfusion defects* showed postoperative improvement. The myocardial perfusion image was thus useful in predicting the changes in ventricular function following surgery.

---

*Percent of normal perfusion was defined as percent of activity in a normal region using 10 percent level color-coded analysis. Mild to moderate defects showed activity of from 40 to 70 percent of normal, and severe defects less than 40 percent of normal activity.

COMPARISON WITH OTHER RADIONUCLIDE METHODS

Thallium-201 is currently the most commonly used radionuclide for noninvasive myocardial imaging, replacing [43]K and [81]Rb, because of improved imaging qualities. Rest-exercise studies are probably less sensitive than that of rest-hyperemia intracoronary particle studies. In our laboratory, intracoronary MAA detected 95 percent of patients with known coronary disease, whereas ^{201}Tl rest-exercise studies only identified 79 percent of similar patients.[19,21] While these studies were not performed in an identical patient group, simultaneous studies using both techniques showed a 10 percent greater detection of resting defects with particles as compared to ^{201}Tl imaging.[21] The use of both particles and ^{201}Tl studies would be redundant in many instances. For example, an inferior imaging defect at rest with ^{201}Tl would also be present with a resting particle study. Similarly, an exercise defect demonstrated with thallium would also be present on a hyperemic particle study. However, due to the reduced sensitivity of ^{201}Tl, some patients with normal rest-exercise thallium will demonstrate abnormalities with intracoronary particle studies. At this time, the rest-hyperemic particle method is most useful when the exercise thallium study is normal and the coronary arteriogram demonstrates a stenosis of questionable significance. In this situation, an image defect with hyperemia would indicate a hemodynamically significant stenosis.

The major limitations of the particle technique are the requirements for cardiac catheterization and that rest-hyperemia studies are limited to either the left or right coronary systems. Both intravenous agents and direct particle injections occasionally fail to detect patients with balanced coronary stenosis because blood flow is equally depressed in both beds. This is less common with intravenous agents because the entire myocardium is visualized.

SUMMARY

The intracoronary injection of radionuclide-labeled macroaggregated albumin particles has been shown to be safe and to provide excellent myocardial imaging. When performed in the resting state, it is more sensitive than either the ECG or ventriculogram in identifying areas of scarring and fibrosis due to previous myocardial infarction. Combined with contrast induced coronary hyperemia, intracoronary MAA particles are useful in identifying stenoses of moderate severity that do not alter resting coronary blood flow. Particle studies are likely more sensitive than the rest-exercise [210]Tl myocardial scan in identifying patients with significant coronary artery disease. Resting perfusion abnormalities are more closely related to improved or unimproved left ventricular function after surgery than are the ECG or resting left ventriculogram. While the long-range clinical utility of the particle technique is uncertain, its most useful application currently is identification of significant stenoses undetected by ^{201}Tl and of questionable significance by arteriography.

REFERENCES

1. CARR, E. A., JR., BEIERWALTES, W. H., WEGST, A. V., ET AL.: *Myocardial scanning with rubidium-86.* J. Nucl. Med. 3:76, 1962.

2. CARR, E. A., JR., GLEASON, G., SHAW, J., ET AL.: *The direct diagnosis of myocardial infarction by photoscanning after administration of cesium-131.* Am. Heart J. 68:627, 1964.

3. EVANS, J. R., GUNTON, R. W., BAKER, R. G., ET AL.: *Use of radioiodinated fatty acid for photoscans of the heart.* Circ. Res. 16:1, 1965.

4. ZARET, B. L., STRAUSS, H. W., MARTIN, N. D., ET AL.: *Noninvasive regional myocardial perfusion with radioactive potassium. Study of patients at rest, exercise, and during angina pectoris,* N. Engl. J. Med. 288:809, 1973.

5. MARTIN, N. D., ZARET, B. L., McGOWAN, R. L., ET AL.: *Rubidium-81: A new myocardial scanning agent. Noninvasive regional myocardial perfusion scans at rest and exercise and comparison with potassium-43.* Radiology 111:651, 1974.

6. QUINN, J. L., SENATTO, M., AND KEZDI, P.: *Coronary artery bed photoscanning using radioiodine albumin macroaggregates.* J. Nucl. Med. 7:107, 1966.

7. ENDO, M., YAMAZAKI, T., KONNO, S., ET AL.: *The direct diagnosis of human myocardial ischemia using ^{131}I-MAA via the selective coronary catheter.* Am. Heart J. 80:498, 1970.

8. AGRESS, C. M., ROSENBERG, M. J., JACOBS, M. J., ET AL.: *Protracted shock in the closed-chest dog following coronary embolization with graded microspheres.* Am. J. Physiol. 179:536, 1952.

9. GUZMAN, S. V., SWENSON, E., AND MITCHELL, R.: *Mechanism of cardiogenic shock.* Circ. Res. 10:746, 1962.

10. WELLER, D. A., ADOLPH, R. J., WELLMAN, H. N., ET AL.: *Myocardial perfusion scintigraphy after intracoronary injection of ^{99m}Tc-labeled human albumin microspheres.* Circulation 46:963, 1972.

11. POE, N.: *The effects of coronary arterial injection of radio-albumin macroaggregates on coronary hemodynamics and myocardial function.* J. Nucl. Med. 12:724, 1971.

12. SCHELBERT, H. R., ASHBURN, W. L., COVELL, J. W., ET AL.: *Feasibility and hazards of the intracoronary injection of radioactive serum albumin macroaggregates for external myocardial perfusion imaging.* Invest. Radiol. 6:379, 1971.

13. ASHBURN, W. L., BRAUNWALD, E., SIMON, A. L., ET AL.: *Myocardial perfusion imaging with radioactive-labeled particles injected directly into the coronary circulation of patients with coronary artery disease.* Circulation 44:851, 1971.

14. JANSEN, C., JUDKINS, M. P., GRAMES, G. M., ET AL.: *Myocardial perfusion scintigraphy with MAA.* Radiology 109:369, 1973.

15. KIRK, G. A., ADAMS, R., JANSEN, C., ET AL.: *Particulate myocardial perfusion scintigraphy: Its clinical usefulness in evaluation of coronary artery disease.* Semin. Nucl. Med. 7:67, 1977.

16. RITCHIE, J. L., HAMILTON, G. W., WILLIAMS, D. L., ET AL.: *Myocardial imaging with radionuclide-labeled particles.* Radiology 121:131, 1976.

17. WAGNER, H. N., RHODES, B. A., SASAKI, Y., ET AL.: *Studies of the circulation with radioactive microspheres.* Invest. Radiol. 4:374, 1969.

18. GOULD, K. L., LIPSCOMB, K., AND HAMILTON, G. W.: *Physiologic basis for assessing critical coronary stenosis.* Am. J. Cardiol. 33:87, 1974.

19. RITCHIE, J. L., HAMILTON, G. W., GOULD, K. L., ET AL.: *Myocardial imaging with indium-113m and technetium-99m-macroaggregated albumin: New procedure for identification of stress induced regional ischemia.* Am. J. Cardiol. 35:380, 1975.

20. HAMILTON, G. W.: *Myocardial imaging with radioactive particles.* In Pitt, B., and Strauss, H. W. (eds.): *Cardiovascular Nuclear Medicine,* ed. 2. C. V. Mosby Co., in press.

21. RITCHIE, J. L., HAMILTON, G. W., WILLIAMS, D. L., ET AL.: *Myocardial imaging with ^{201}thallium: Correlation with intracoronary macroaggregated albumin imaging.* Circulation 52:231, 1975.

22. MASERI, A., PARODI, O., SENERI, S., ET AL.: *Transient transmural reduction of myocardial blood flow, demonstrated by thallium-201 scintigraphy, as a cause of variant angina.* Circulation 54:280, 1976.

Radionuclide Studies in Patients with Congenital Heart Disease*

Robert H. Jones, M.D., Peter M. Scholz, M.D., and Page A. W. Anderson, M.D.

The management of patients with congenital heart disease would be facilitated by diagnostic procedures which impose less risk and discomfort than cardiac catheterization. Serial documentation of changes in hemodynamics by these procedures might also aid the selection and timing of medical and surgical therapy. Radionuclides have been used for almost 30 years to assess congenital heart disorders.[1] However, recent improvements in instrumentation and data processing have greatly increased the accuracy and clinical usefulness of these procedures.[2] Techniques for imaging myocardial infarction and for measuring regional myocardial blood flow have been used to evaluate children with congenital heart defects. However, indications for these procedures occur infrequently and studies in children do not differ substantially from principles outlined for adults in chapters on these topics. Therefore, the only procedure for assessment of congenital heart disease described in this chapter utilizes data recorded during the passage of a tracer bolus through the central circulation.

Diagnostic procedures for congenital heart disease must be applicable for studying very young children. Several limitations and difficulties which do not occur as commonly in adult studies may be encountered with radionuclide angiocardiography in children. Spatial resolution available from detecting instruments is severely stressed by the small heart-size of children. The low radiopharmaceutical dose which may be administered limits the statistical accuracy of counts which are recorded during the brief transit of tracer through the heart. Also, rapid heart rate and brisk transit of blood through the central circulation, typical of children, further limit the number of counts recorded during the initial transit of tracer. Moreover, many congenital heart disorders are associated with intracardiac shunting of blood, and complex pathways of blood flow are often difficult to completely define with invasive procedures and prove nearly impossible to define by less invasive approaches. Often only the most prominent hemodynamic abnormality can be recognized by radionuclide studies. Despite these potential limitations, radionuclide angiocardiography appears to hold great promise for evaluating children with congenital heart disorders. The purposes of this review are to summarize applications of radionuclide angiocardiography that have been documented to be of clinical benefit to patients with congenital heart disease and to suggest areas which appear promising for future development.

---

*Supported in part by NIH Grant HL20677-02.

TECHNIQUE OF STUDY

Instrumentation

Initial radionuclide studies of congenital heart disorders utilized single probe detectors which were first Geiger Müller tubes and subsequently scintillation counters.[3,4] Studies using these simple devices were best performed concurrently with cardiac catheterization which permitted tracer injection into selected cardiac chambers with external recording over the heart, lung, and head.[5,6] Single probe studies produced useful data but were excessively complex for general use. Development of the gamma camera provided a tool for dynamic imaging of the movement of radioactivity through the heart.[7] Interfacing a computer with the gamma camera permitted quantitative data to be obtained from detector regions selected to correspond to individual cardiac chambers.[8] The computerized gamma camera is currently the instrument of choice for radionuclide studies of congenital heart disorders, but further advances in nuclear cardiology may ultimately lead to the reinstitution of more simple instruments with specific applications.

A high count rate capability is the single most critical characteristic of gamma camera performance for radionuclide angiocardiography in children. Collimator design limits the sensitivity of most single-crystal gamma cameras to under five thousand counts per second per millicurie of Tc-99m pertechnetate. For this reason, all studies illustrated in this chapter were obtained using the Baird-Atomic System Seventy-Seven multicrystal gamma camera which can achieve forty thousand counts per second per millicurie of Tc-99m pertechnetate.

Radiopharmaceuticals

Radiopharmaceutical characteristics which would be optimal for radionuclide angiocardiography may be easily defined. An ideal tracer should remain in the intravascular space during at least one pass through the central circulation and thereafter disappear with a short biologic or physical half-life. The energy of photons emitted should be optimal for available instruments and the tracer should be readily available and inexpensive. The tracer which now most nearly approaches these needs is ^{99m}Tc-pertechnetate; this radiopharmaceutical is used almost exclusively for radionuclide angiocardiography because of general availability, low cost, gamma energy of 140 keV, and physical half-life of six hours. Occasionally, a radiopharmaceutical which remains within the vascular space longer than ^{99m}Tc-pertechnetate may be desired to perform gated cardiac blood pool imaging or to obtain a blood volume essential for cardiac output determinations. Human serum albumin or red blood cells labeled with ^{99m}Tc-pertechnetate have been shown to remain within the vascular space for a time sufficient to perform these studies. A minimal dose of 2 mCi is recommended, and a dose of 0.3 mCi/kg provides counts necessary for an adequate study in older children. Table 1 summarizes total body irradiation to children resulting from administration of ^{99m}Tc as pertechnetate or as human serum albumin. The relatively small radiation dose imposed by radionuclide angiocardiography compares favorably with a total body irradiation exceeding 1 rad which may result from cardiac catheterization.

An alternate radiopharmaceutical which holds promise for study of congenital heart disorders is ^{15}O in the form of $C^{15}O_2$.[9] This accelerator-produced agent has the disadvantage of a 511-keV energy which exceeds the range optimal for gamma cameras but has the advantage of a physical half-life of 2 minutes which permits sequential studies and provides a low radiation dose (Table 1). Moreover, $C^{15}O_2$ may be administered by inhalation and the carbonic anhydrase reaction rapidly transfers the ^{15}O onto $H_2^{15}O$ within the pulmonary capillary. Administration of the tracer during the initial phase of

Table 1. Radiation dose from radionuclide angiocardiography

	Newborn	1 Year	5 Years	10 Years	15 Years	Adult
Tc-99m Pertechnetate						
Administered activity (mCi)	2.1	4.5	6.5	9.1	16	30
Critical organ (G.I.) dose (rads)	4.0	3.0	3.0	3.0	3.8	6.0
Whole body dose (rads)	.29	.23	.22	.20	.25	.36
Tc-99m Human Serum Albumin						
Administered activity (mCi)	2.1	4.5	6.5	9.1	16	30
Critical organ (blood) dose (rads)	1.7	1.1	1.0	.85	.96	1.4
Whole body dose (rads)	.38	.28	.26	.23	.29	.45
$C^{15}O_2$ Inhalation						
Administered dose (mCi)	2.1	7.3	12	20	30	30
Critical organ (heart) dose (rads)	.5	.69	.69	.72	.82	.74
Whole body dose (rads)	.06	.06	.07	.07	.08	.08

inspiration results in a rapid entry of $H_2^{15}O$ into the pulmonary capillary with discrete return of the bolus to the left heart. The atraumatic method of administration is important in children. Also, the normal absence of tracer in the right heart following inhalation may provide more accurate shunt detection and localization than intravenously injected radiopharmaceuticals, particularly in patients with bidirectional shunting. [10]

Short-lived radiopharmaceuticals such as [191m]Ir with a physical half-life of five seconds are being developed for radionuclide angiocardiography in children. [11] More experience will be required before these agents can be recommended for clinical application. However, further developments in instrumentation and radiopharmaceuticals promise to greatly enhance the information available from radionuclide angiocardiography in patients with congenital heart disease.

Data Acquisition

Older children and adults accept the minimally traumatic radionuclide procedure readily. Young children may require light sedation to overcome anxiety associated with venipuncture. A single venipuncture can be used to obtain blood samples for other necessary diagnostic studies in addition to the radionuclide study, thereby imposing no additional pain than would otherwise be required. Cooperation in remaining quiet during study may be improved in children by establishing the injection site prior to transport to the place of study. Although satisfactory studies may be accomplished with injection into very peripheral veins of the scalp, hands, or feet, a more central venous injection more predictably produces a compact tracer bolus. An antecubital vein is an acceptable injection site but the external jugular vein is the site which most consistently provides good injections.

Selection of the patient position for study involves the considerations typical of contrast angiography. A single view which provides the maximum data in most patients is a straight anterior projection, but a left anterior oblique projection may be occasionally useful to separate right and left cardiac chambers. Data recording is begun after patient positioning and the tracer is introduced rapidly using a flush of nonradioactive fluid. A recording interval of 50 msec is the longest framing rate which will permit later definition of changes occurring within individual cardiac cycles, and studies performed in patients with very rapid heart rates do not require recording intervals more brief than 10 msec. The initial transit of tracer through the heart requires 10 seconds and a total of 1 minute provides sufficient recording time for the entire study.

Data Manipulation

Corrections for differences in uniformity over the field of view of the detector and for counting losses associated with instrument dead time are mandatory in certain instruments and less critical, but desirable in other instruments. The next step in data processing is generation of images which reflect tracer distribution at sequential intervals during its passage through the central circulation. These may be displayed as a group of serial images or viewed as a cine study. The anatomic configuration of the heart is often apparent on images, but blood flow through the central circulation is best characterized quantitatively by data representing count changes within cardiac chambers. Serial images permit definition of the boundaries of detector areas which correspond to individual cardiac chambers as regions of interest for curve generation. In addition to images of count intensity, images or numeric definition of the time of maximum count or mean transit time over the detector surface may aid selection of regions of interest.[12] Even careful selection of regions of interest does not produce curves which reflect counts within only the defined chamber. Each curve from defined cardiac chambers includes counts scattered into the region of interest and counts included within the region of interest because of anatomic overlap of adjacent cardiac chambers. Complex curves representing data recorded from adjacent cardiac regions at subsequent time intervals may be separated by exponential extrapolation or by more complex mathematical techniques such as application of the gamma variate function. Thus, appreciation of all hemodynamic information available in radionuclide angiocardiography requires both inspection of images to appreciate the spatial orientation of counts and use of curves to quantitate the magnitude of defect.

THE NORMAL RADIONUCLIDE ANGIOCARDIOGRAM IN CHILDREN

Serial radionuclide images following tracer administration to a 4-year-old child depict blood flow through a normal heart (Fig. 1). Definition of regions of interest provides high time-resolution indicator dilution curves from individual cardiac chambers (Fig. 2). Data recorded over the lung return exponentially to baseline prior to systemic recirculation because of the anatomic isolation of this region. Biphasic curves from right and left heart chambers include counts from adjacent chambers. In the anterior projection, the pulmonary artery overlies the left atrium and a biphasic curve is recorded from this region. However, these curves may be separated by extrapolation for area determinations or transit time calculations. The high time-resolution recording of data retains count changes occurring within individual cardiac contractions which are particularly apparent in data recorded from the left ventricle.

The mean transit time defines the time when the average tracer element appears in the region of interest. The mean transit time for any curve is calculated by the sum of the product of counts and time for each interval divided by the sum of counts. Because the tracer mixes progressively with the blood volume within the central circulation, the mean transit time occurs at increasingly longer intervals after the time of maximum count in sequential chambers. However, the difference in mean transit time between the outflow and inflow of each cardiac chamber represents the time required for blood passage through that chamber. For example, the pulmonary transit time is obtained by subtracting the mean transit time of the curve recorded over the pulmonary outflow tract from the mean transit time of the curve recorded over the left atrium. This transit time does not differ significantly from the transit time of the curve recorded over the lung

228

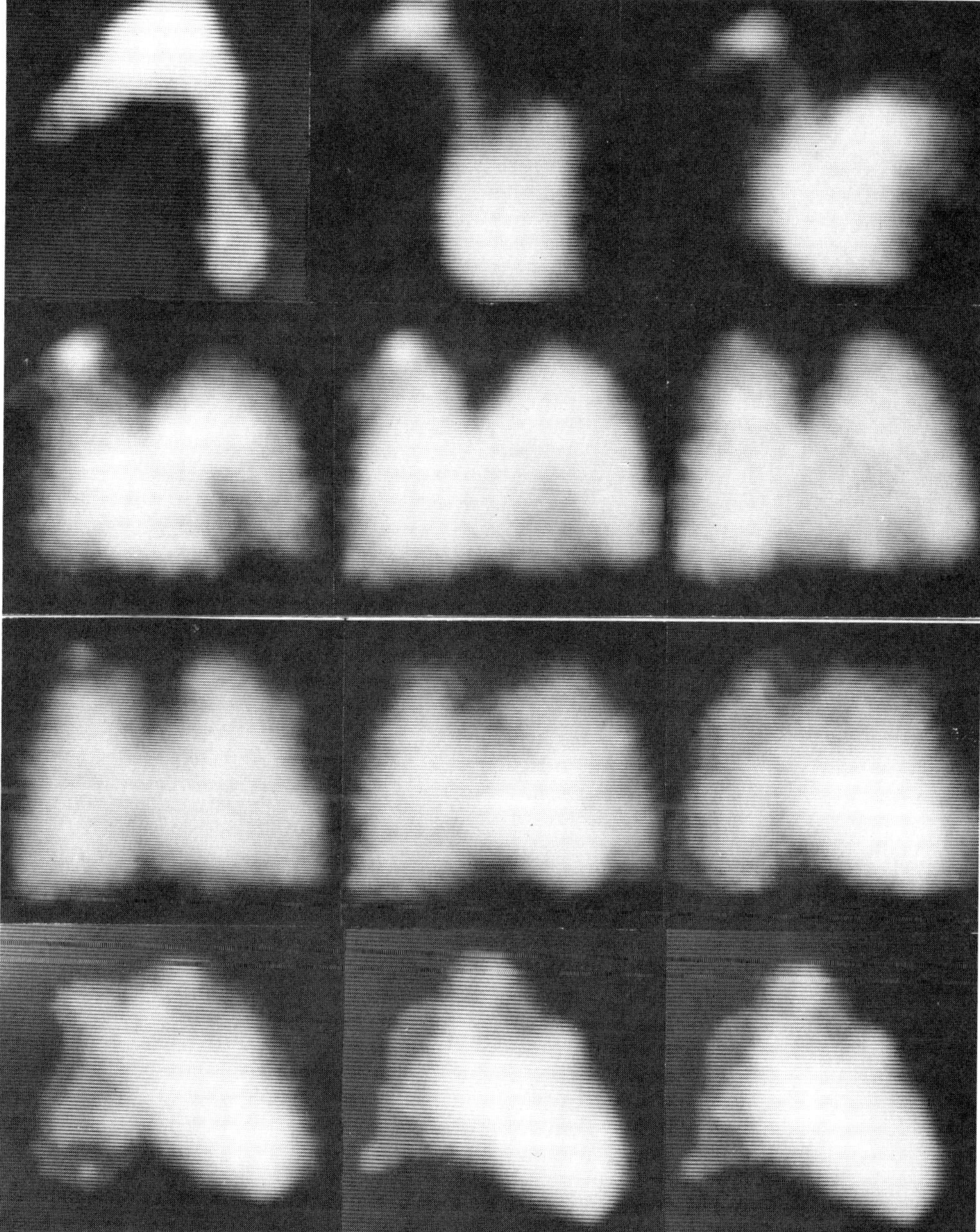

Figure 1. These images were obtained at 0.5-sec intervals following injection of 5 mCi ^{99m}Tc-pertechnetate in a 4-year-old child without cardiovascular disease.

if the appearance time in the lung is regarded as zero time. However, definition of the appearance time in a chamber often proves difficult; therefore, differences between individual chamber mean transit times provide the most reproducible measurements. The normal pulmonary transit time in adults is 6.3 $\pm$ 1.5 seconds. The pulmonary transit time in normal children decreases with age (Fig. 3). If tracer passing through a volume

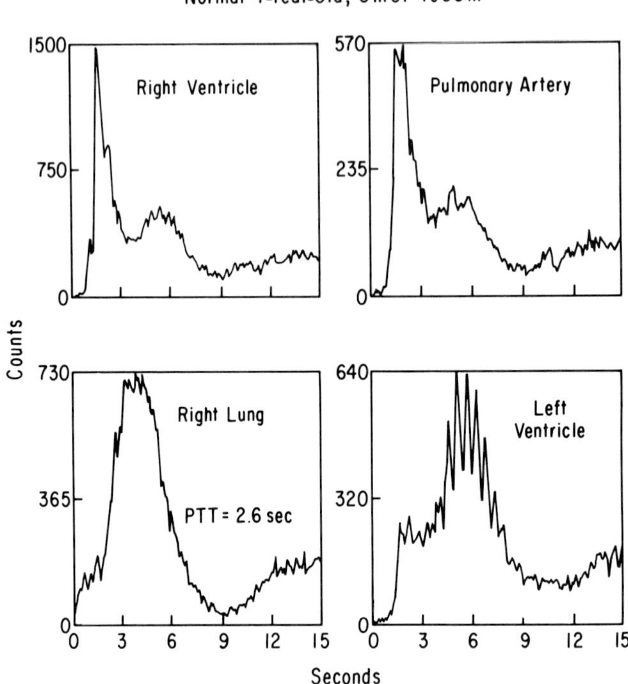

Figure 2. These indicator-dilution curves were obtained in a normal 4-year-old child (see Fig. 1) by definition of regions of interest which correspond to individual cardiac chambers.

of fluid mixes evenly, the mean transit time is directly related to static volume of the chamber and inversely related to the flow rate. Therefore, definition of the pulmonary transit time and the cardiac output permits calculation of the pulmonary blood volume.

Recording data at a rapid rate documents changes in counts in individual cardiac chambers within each contraction. Actual cardiac contractions can be visualized on cine images and appreciated as fluctuations in counts on curves recorded from right and left ventricles. The low statistical quality of data imposed by the count rate limitations of

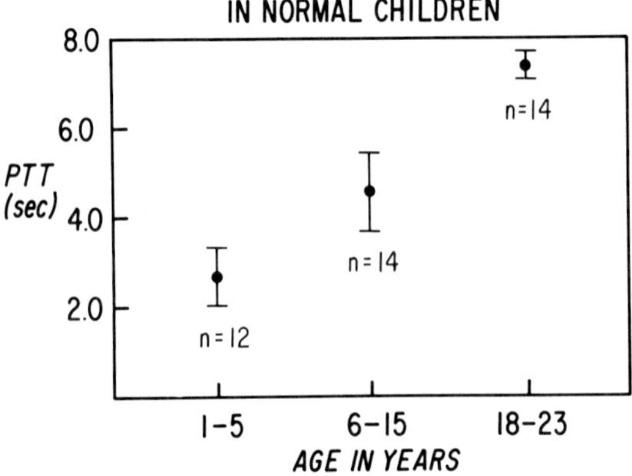

Figure 3. Pulmonary transit times obtained in 26 normal children and 14 young adults document the decrease in transit time in younger children.

available instruments limits the usefulness of visualization of wall motion or determination of volume changes within cardiac chambers during single contractions. However, since at least five individual cardiac contractions are commonly seen during one passage of tracer bolus through the heart, data from each of these contractions can be added at each corresponding interval during the heart beat to form a single representative cardiac cycle. The resulting images viewed either in a static or cine format permit evaluation of wall motion of cardiac chambers. Curves of count changes within the ventricles during the representative cardiac cycle reflect volume changes within the chamber (Fig. 4). The left ventricular ejection fraction may be calculated from the data by subtraction of background counts which do not arise from the blood pool of the chamber of interest. A variety of techniques have been devised by different investigators to perform the background correction, and the resulting determinations of left ventricular ejection fraction correlate closely with simultaneous measurements by contrast angiography in adults.[13] The validity of the technique has not been documented in children and may be less accurate primarily because of the lower counting rate. However, determinations of left ventricular ejection fraction using a standard approach in children of different age groups show a similar ejection fraction in each group (Fig. 5). The small standard deviation of the measurement suggests reproducibility of the technique.

Radionuclide angiography is a simple method to document normal hemodynamics in children. Normal children with functional cardiac murmurs may be regarded as less than normal because of parental or teacher anxiety. For this reason, objective documentation of normal hemodynamics by radionuclide angiocardiography often proves useful even

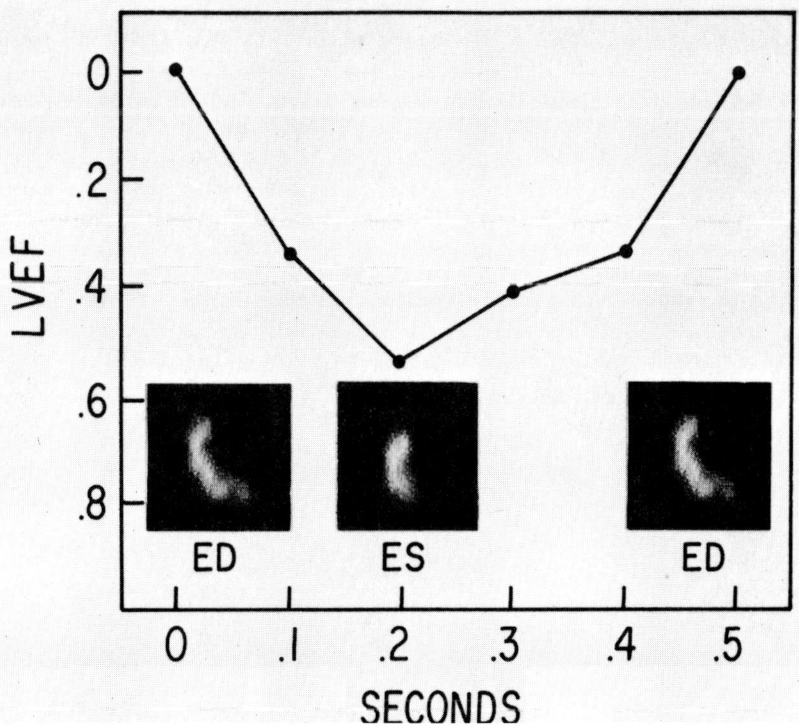

Figure 4. This left ventricular volume curve was obtained by summing data recorded during individual cardiac contractions during the initial transit of tracer through the heart. Images of the left ventricle at end-diastole and end-systole demonstrate normal wall motion.

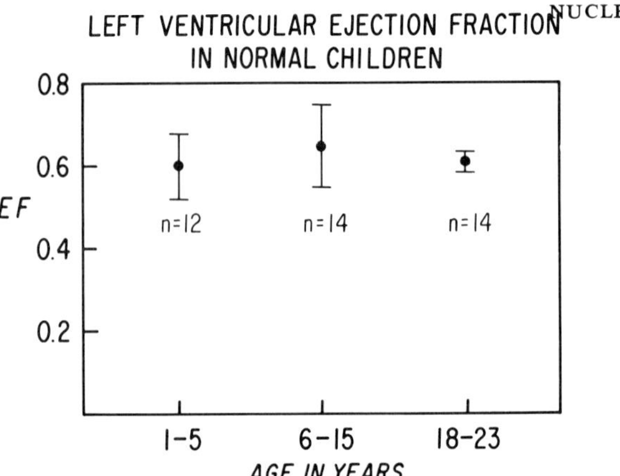

Figure 5. The left ventricular ejection fraction measured in 26 normal children and 14 young adults showed no significant change with age.

in situations where the probability of an intracardiac shunt is considered quite remote. Also, the procedure proves useful in assessing hemodynamics in patients following operative correction of the congenital defect. Moreover, radionuclide angiocardiography can be performed during exercise to document the functional reserve of the heart. For example, a rest and exercise radionuclide angiocardiogram was performed in a 7-year-old boy who had ligation of an anomalous coronary artery in infancy. Normal left ventricular wall motion at rest and an increase in wall motion and ejection fraction with exercise documented both a normal resting function and an adequate functional reserve, suggesting the safety of physical activity (Fig. 6).

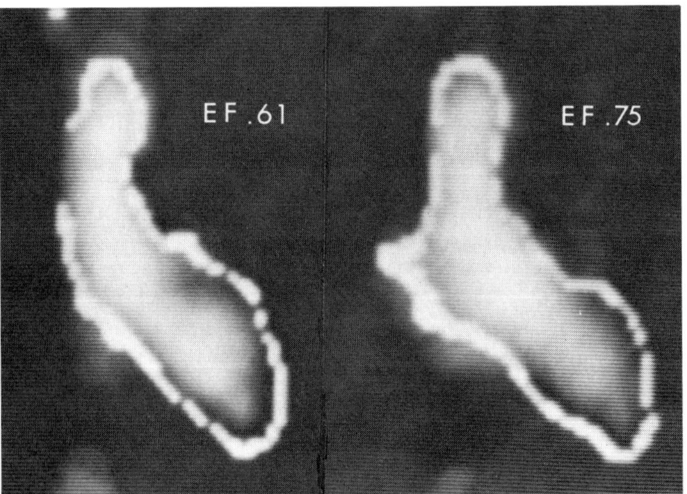

Figure 6. These images of the end-diastolic border of the left ventricle placed about the end-systolic counts were obtained at rest (left) and exercise (right) in a 7-year-old child who had ligation of an anomalous coronary artery in infancy. Normal wall motion is present at rest and is increased by exercise suggesting a functional reserve of the left ventricle.

CONGENITAL CARDIAC DISORDERS

Abnormal Blood Flow Patterns

The major value of radionuclide angiocardiography in some patients may reside in the simple anatomic definition of the path of blood flow through the central circulation. An example is an 18-year-old man with an atrial septal defect in whom a preoperative injection made in the left arm revealed a persistent left superior vena cava (Fig. 7). The postoperative study obtained with a right arm injection showed a normal right superior vena cava and a normal blood flow pattern. Definition of the presence of the persistent left superior vena cava aided the choice of the operative incision in this patient. Imaging of blood flow also proved useful in a 13-year-old child with tunnel aortic stenosis treated by a conduit inserted from the apex of the left ventricle to the abdominal aorta (Fig. 8). The postoperative radionuclide angiocardiogram confirms patency of the conduit, and the quantity of tracer flowing through the conduit suggests more blood passes through the conduit than is ejected through the stenotic subvalvular aortic tunnel. This procedure may be repeated serially to confirm continued patency of the conduit in this patient.

Right-to-Left Shunts

Cyanosis and a cardiac murmur commonly suggest the diagnosis of heart disease in older children. However, cyanosis in the newborn which occurs frequently with primary lung disorders may prove difficult to differentiate from cyanotic heart disease. Moreover, development of a technique which might serially quantitate the magnitude of right-to-left shunting could prove useful in selecting the most appropriate time for surgical correction. Right-to-left intracardiac shunting may frequently be recognized by images of blood flow showing direct transit of tracer between adjacent cardiac chambers (Fig. 9). The orientation of the cardiac septa relative to the body surface is unpredictable, and occasionally excessive rotation of the heart will result in images suggestive of right-to-left shunting in patients with only right heart enlargement.

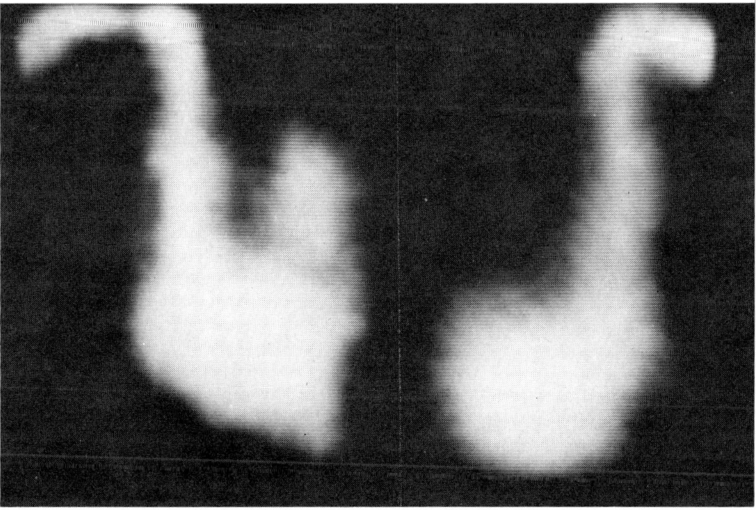

Figure 7. Radionuclide injection into the right arm (left) and left arm (right) of an 18-year-old patient with an atrial septal defect document the presence of a persistent left superior vena cava.

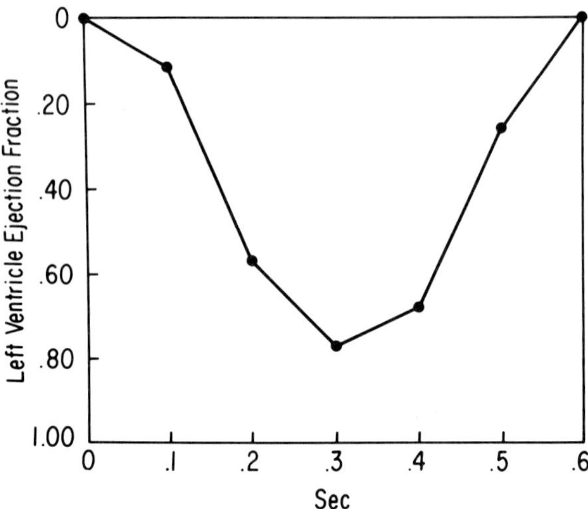

Figure 8. These images and curve of ejection fraction during a representative cardiac cycle were obtained in a 13-year-old patient following insertion of a conduit from the apex of the left ventricle to the abdominal aorta for treatment of tunnel aortic stenosis. Good left ventricular ejection is noted from the curve and by the near obliteration of the left ventricular cavity apparent on the image. Moreover, the images suggest that most of the tracer is ejected through the conduit and a lesser amount enters the ascending aorta during systole.

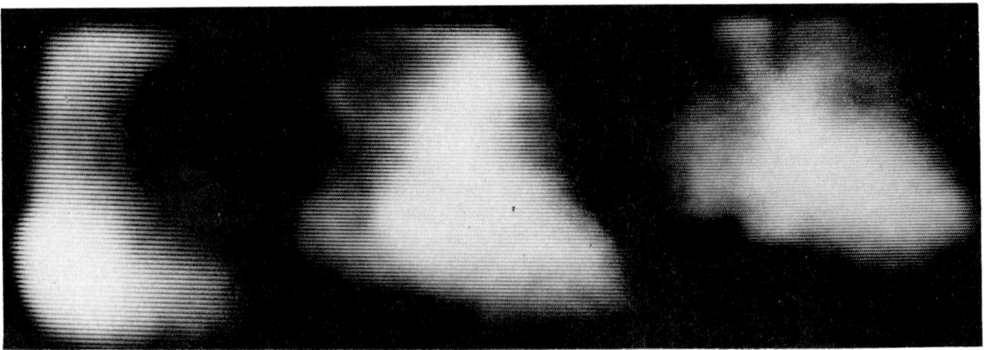

Figure 9. These sequential 0.5-sec images were obtained following injection of 5 mCi of ^{99m}Tc-pertechnetate in a 4-year-old child with pulmonary stenosis and a ventricular septal defect. The appearance of radioactivity in the left heart soon after right heart appearance document the presence of a right-to-left shunt.

Large amounts of right-to-left shunting may be recognized visually, but small amounts of shunting are apparent only when curves are generated from the left heart and aorta. Right-to-left shunting may be absolutely confirmed by the early appearance of tracer in the aorta soon after tracer appearance in the right heart. Selection of an isolated portion of the aortic outflow such as the carotid arteries is imperative since counts from the region of the ascending aorta may include counts from the pulmonary artery. The radionuclide curve in patients with right-to-left shunting has a typical configuration similar to a dye indicator curve in these patients. A mathematical approach used for calculating right-to-left shunts from dye dilution data would appear applicable to radionuclide data. Although the amount of right-to-left shunting appears related to the magnitude of curve abnormality, no report has yet confirmed the accuracy of quantitation of right-to-left intracardiac shunting using this approach.

Despite many research reports, radionuclide angiocardiography has not been routinely used for clinical evaluation of patients with cyanotic heart disease.[2] Complex blood flow patterns often limit interpretation of the data and most patients with cardiac disorders sufficiently severe to result in cyanosis merit early cardiac catheterization for full anatomic characterization of the disorder. However, the recent improvement in techniques for study of the newborn promises to expand application of these methods of evaluating these seriously ill infants. The usefulness of radionuclide angiocardiography for detection and quantitation of right-to-left shunting holds great promise, but will require further controlled experience prior to widespread clinical application.

Left-to-Right Shunts

Left-to-right intracardiac shunts greater than 30 percent of the systemic blood flow can be consistently recognized on serial images by a more brisk than normal transit of tracer through the lungs and by reappearance of tracer in right heart chambers distal to the site of shunt. Moreover, the most proximal site of left-to-right shunting can be accurately located in infants as small as 4.0 kg. With further improvement in data manipulation, the localization of more distal sites of left-to-right shunting will be possible in patients with more than one site of shunting. Shunts of smaller magnitude return less tracer to the right heart and therefore prove more difficult to recognize by images alone. Curves over the right and left heart chambers may show typical alterations with tracer entering the right heart chambers prior to transit through the lungs and again shortly following left heart appearance. However, this typical curve configuration may also be produced by including counts from the left cardiac chambers within regions of interest designated as right atrium and right ventricle. Therefore, data recorded over the lung at a site remote from the heart are necessary for quantitation of shunt flow.

The curve configuration typical of left-to-right shunting shows an initial transit of tracer through the lungs followed by an early reappearance of tracer returned by the shunt flow which interrupts the exponential decline of counts and may actually increase counts sufficiently to cause a second curve peak. More commonly, the recirculated counts blend with the initial counts and the multiple rapid recirculations cause the curve to break from an exponential decline and remain relatively flat with a high background. All approaches for quantitation of left-to-right shunts using these curves require separation of the first passage of tracer from the subsequent transit of tracer through the lungs. One method used an exponential extrapolation of the decline of counts to an arbitrary end-point and compared the area of first transit to the area reflecting the recirculated blood (Fig. 10).[14] In 26 patients with atrial and ventricular septal defects, this technique proved accurate when compared to Fick determination of shunts with a correlation coefficient of .96 and with recognition of shunts as small as 10 percent of the cardiac output (Fig. 11). An alternate approach of curve separation utilizing the gamma variate function to separate the complex curves also provided accurate quantitation of left-to-right shunts.[15]

An alternate approach to left-to-right shunt calculation uses $C^{15}O_2$ inhalation which introduces a discrete tracer bolus into the left heart.[16] Left-to-right shunting is recognized by return of a portion of this tracer to the lungs again by blood shunted through the pulmonary circulation which interrupts the initial exponential decline in lung counts. This approach which has been documented to be accurate for detection and quantitation of left-to-right shunts has the potential advantage of providing accurate quantitation of left-to-right shunting in patients who have bidirectional shunting.

Radionuclide angiocardiographic data to be used for shunt detection demands a discrete bolus injection since a biphasic injection will produce a lung curve which resembles a left-to-right shunt pattern. Moreover, introduction of a large amount of systemic blood

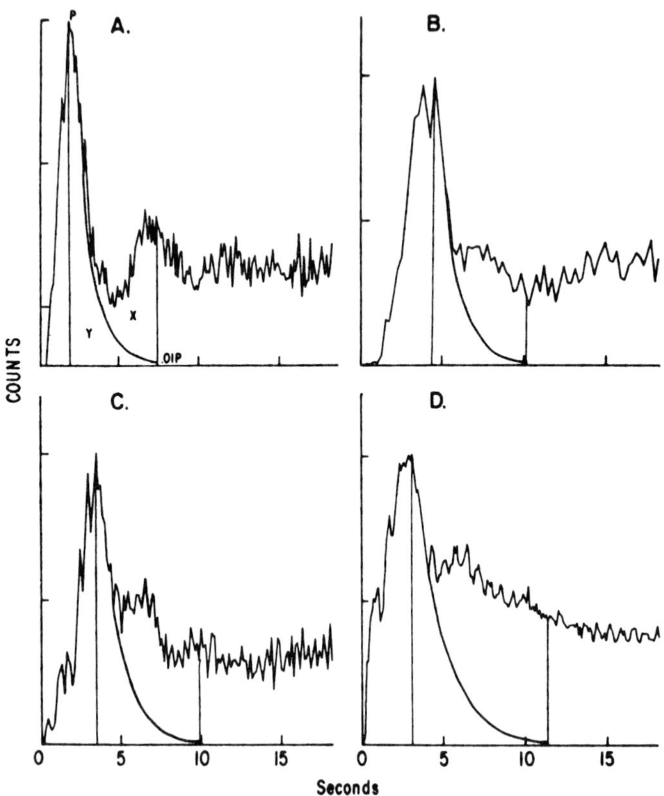

Figure 10. These curves of lung counts were obtained from (A) normal subject, (B) patient with ventricular septal defect and 26 percent left-to-right shunt, (C) patient with partial anomalous pulmonary venous drainage and 46 percent left-to-right shunt, and (D) ventricular septal defect and 70 percent left-to-right shunt. Ratios of areas representing the first and recirculated transit of tracer were determined for shunt calculation.

into the lung field of view, such as may occur with disorders which increase bronchial or intercostal artery blood flow, may result in data erroneously suggestive of left-to-right shunting. Massive mitral regurgitation may cause an abnormal lung curve, but tracer transit through the lungs is prolonged in this disorder in contrast to the more brisk transit associated with left-to-right shunting.

Objective documentation of the presence and magnitude of left-to-right intracardiac shunts represents an important application of noninvasive radionuclide studies. This simple outpatient procedure which objectively documents the presence or absence of a left-to-right shunt adds to the clinical evaluation of the patient without the risk and discomfort of cardiac catetherization. Functional cardiac murmurs are common in children and often raise the consideration of congenital heart disease. Even when the diagnosis of cardiac pathology appears highly unlikely in many of these patients, the objective documentation of the normal blood flow may provide worthwhile information to reassure the patient and his family. Moreover, many intracardiac defects associated with left-to-right shunting are not repaired surgically when the diagnosis is first recognized, and radionuclide angiocardiography can document the magnitude of shunting. In addition, postoperative studies may provide documentation of complete closure of septal defects (Fig. 12).

236

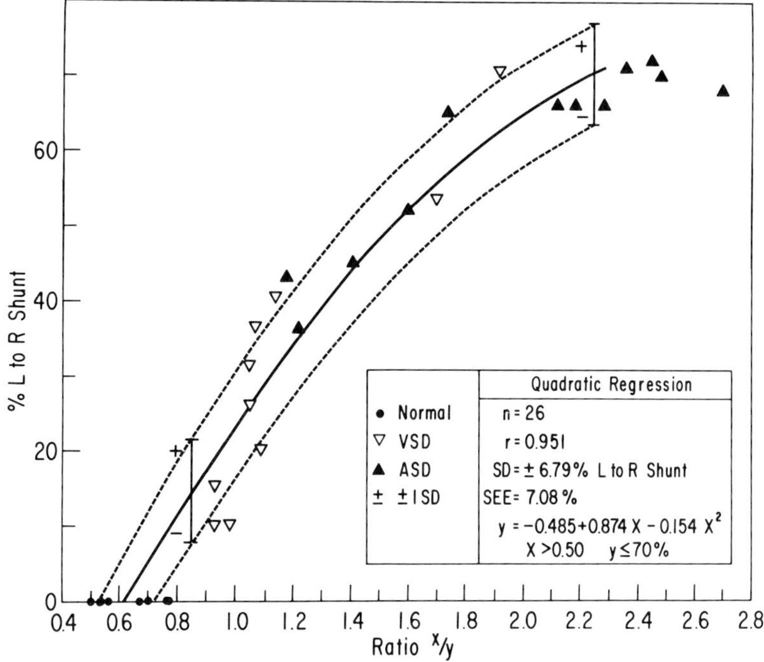

Figure 11. Determination of the ratio of areas representing recirculated tracer and the initial transit of tracer through the lungs permits accurate quantitation of left-to-right shunt flow as documented by oximetry data.

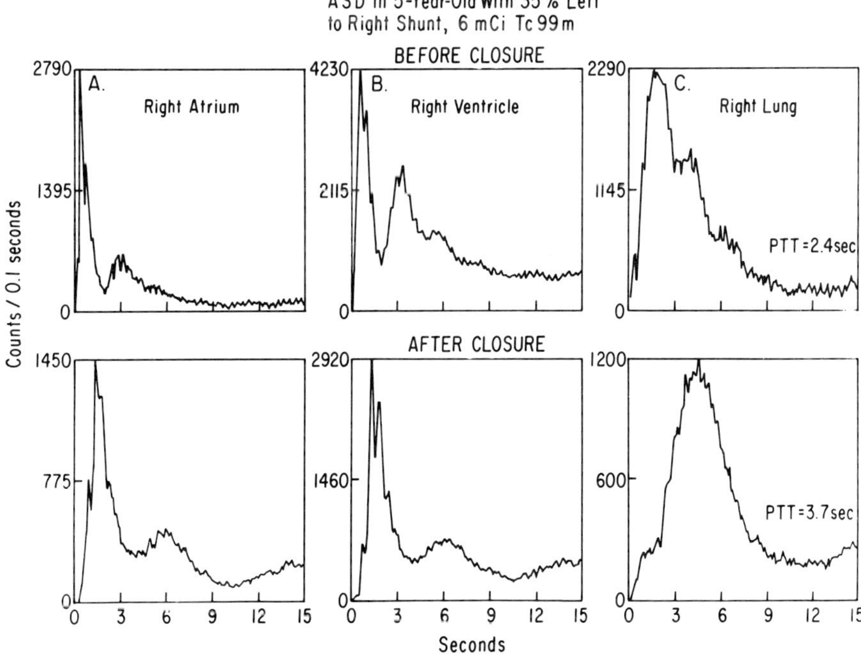

Figure 12. These indicator-dilution curves were obtained by radionuclide angiocardiography in a 5-year-old child before and after closure of an atrial septal defect. Curves from the right atrium, right ventricle, and right lung all demonstrate early recirculation of tracer confirming the left-to-right shunt. The postoperative blood flow pattern is normal.

CONCLUSIONS

Evaluation of the patient with congenital heart disease involves identification of anatomic abnormalities and delineation of their hemodynamic significance. Although cardiac disorders are usually definitively assessed by cardiac catheterization, this diagnostic procedure involves a significant risk in the infant or the severely ill patient. Development of reliable noninvasive techniques which provide similar information with less patient risk and discomfort would represent an important advance in the management of cardiac disorders. Radionuclides are ideally suited to characterize blood flow noninvasively. Radionuclide angiocardiography may be used to define abnormal blood flow patterns and to quantitate intracardiac shunting. Moreover, ventricular function may be assessed by these simple techniques. Further developments in instrumentation and data processing promise to greatly expand the usefulness of radionuclide angiocardiography in evaluation of patients with congenital heart disorders.

REFERENCES

1. PRINZMETAL, M., CORDAY, E., SPRITZLER, R. H., ET AL.: *Radiocardiography and its clinical applications.* JAMA 130:617, 1949.

2. JONES, R. H., AND ANDERSON, P. A.: *Congenital heart disease: imaging and analytic methods.* In Pierson et al. (eds.): *Quantitative Nuclear Cardiography.* John Wiley and Sons, Inc., New York, 1975.

3. GREENSPAN, R. H., LESTER, R. G., MARVIN, J. F., ET AL.: *Some clinical aspects of isotope circulation studies.* Radiology 73:345, 1959.

4. SHAPIRO, W., AND SHARPE, A. R.: *Precordial isotope-dilution curves in congenital heart disease, a simple method for the detection of intracardiac shunts.* Am. Heart J. 60:607, 1960.

5. SPACH, M. S., CANENT, R. V., BOINEAU, J. P., ET AL.: *Radioisotope-dilution curves as an adjunct to cardiac catheterization. I. Left-to-right shunts.* Am. J. Cardiol. 16:165, 1965.

6. FLAHERTY, J. T., CANENT, R. V., BOINEAU, J. P., ET AL.: *Use of externally recorded radioisotope dilution curves for quantitation of left-to-right shunts.* Am. J. Cardiol. 20:341, 1967.

7. GRAHAN, T. P., GOODRICH, J. K., ROBINSON, A. E., ET AL.: *Scintiangiocardiography in children. Rapid sequence visualization of the heart and great vessels after intravenous injection of radionuclide.* Am. J. Cardiol. 25:387, 1970.

8. JONES, R. H., SABISTON, D. C., BATES, B. B., ET AL.: *Quantitative radionuclide angiocardiography for determination of chamber to chamber cardiac transit times.* Am. J. Cardiol. 30:855, 1972.

9. WATSON, D. D., KENNY, P. J., GELBAND, H., ET AL.: *A noninvasive technique for the study of cardiac hemodynamics utilizing $C^{15}O_2$ inhalation.* Radiology 119:615, 1976.

10. TAMER, D. M., WATSON, D. D., KENNY, P. J., ET AL.: *Noninvasive detection and quantification of left-to-right shunts in children using oxygen-15 labeled carbon dioxide.* Circulation 56:626, 1977.

11. TREVES, S., AND COLLINS-NAKAI, R. I.: *Radioactive tracers in congenital heart disease.* Am. J. Cardiol. 38:711, 1976.

12. JONES, R. H., AND SCHOLZ, P. M.: *Data enhancement techniques for radionuclide cardiac studies.* Med. Radionuclide Imaging 2:255, 1977.

13. HOWE, W. R., JONES, R. H., AND SABISTON, D. C., JR.: *Radionuclide assessment of left-ventricular function following cardiac surgery.* Surg. Forum 27:253, 1976.

14. ANDERSON, P. A., JONES, R. H., AND SABISTON, D. C.: *Quantitation of left-to-right cardiac shunts with radionuclide angiography.* Circulation 49:512, 1974.

15. ASKENAZI, J., AHNBERG, D. S., KORNGOLD, E., ET AL.: *Quantitative radionuclide angiocardiography: detection and quantitation of left-to-right shunts.* Am. J. Cardiol. 37:382, 1976.

16. BOUCHER, C. A., AHLUWALIA, B., BLOCK, P. C., ET AL.: *Inhalation imaging with oxygen-15 labeled carbon dioxide for detection and quantitation of left-to-right shunts.* Circulation 54:632, 1977.

Computed Tomography: Imaging the Heart

William R. Gray, Jr., M.D.

Computed tomography (CT) of the head has been rapidly developed since Hounsfeld's initial presentation of a prototype in 1973,[1] and CT of the head has evolved as the primary diagnostic tool in neuroradiology. The natural outgrowth of the CT head scanner was the CT whole body scanner capable of producing transverse sections through any portion of the body. Even before whole body CT scanners were available, considerable interest developed in the potential use of CT in cardiac diagnosis. Early studies have demonstrated the vast potential of cardiac CT, and clinical usefulness is on the horizon.

SCANNER REQUIREMENTS FOR CARDIAC IMAGING

The basic operation of CT scanners involves axial rotation of an x-ray beam around the patient with detectors opposite the x-ray source which measure the attenuation of x-rays through the tissues. From the data collected in multiple projections a computer calculates the attenuation coefficient of each point of a matrix thus producing an image of the sections scanned. This image is stored and may be recorded photographically. CT has a highly sensitive capacity for resolving small differences in x-ray attenuation, a capability which is not possible with routine x-ray studies. Both CT scanners and reconstruction techniques are undergoing rapid development and the reader is referred to the bibliography for further review of these subjects.[2-6]

Whole body CT scanning is now available with a scan time of three seconds per slice. However, even this relatively short scan time is too long to image the heart without marked degradation of the reconstructed image due to cardiac motion. For high resolution reconstruction each point in the matrix must remain in a fixed position during the rotation of the scanner. When points of reference move during the time of scanning, image quality suffers greatly. To satisfy this requirement in early studies of CT of the heart, sacrificed animals with arrested hearts were used. In vivo studies demanded that the problem of cardiac motion be resolved by some other means. Three approaches have been suggested to enable studies of the beating heart: (1) decreasing the total data collection time per slice to milliseconds, (2) gating or synchronizing data collection to occur only during selected segments of the cardiac cycle, and (3) synchronizing the data with the cardiac cycle after data acquisition. The last of these three methods has undergone the most recent development and promises to be the most immediately useful technique for CT imaging of the heart.

DETECTION OF ACUTE MYOCARDIAL INFARCTION

Early investigations of CT imaging of the heart were performed by our group along with those of Adams, Ter-Pogossian, and Powell. Adams and coworkers[7] published the earliest CT images of the heart. Isolated canine hearts scanned in a CT head scanner demonstrated that the blood-myocardial wall interface was best delineated with either a substantial alteration in hematocrit or with the intravenous administration of iodinated contrast prior to sacrifice. On these scans, cardiac structures including papillary muscles and specific chambers were easily visualized. Also included in this series were hearts in which infarction had been produced two days prior to sacrifice. In the majority of hearts, the infarction was visualized as an area of significantly decreased attenuation. This decrease in attenuation was readily visualized on CT scan. The decreased attenuation was felt at least in part to be caused by the edema associated with acute infarction.

Ter-Pogossian and associates[8] utilized a body scanner to image both live and sacrificed dogs in which infarctions had been produced. The poor image quality produced in the living animal confirmed the need for means to reduce the degrading effects of cardiac motion. Scans of arrested hearts confirmed the finding of decreased attenuation associated with acute myocardial infarction. In animals with coronary artery occlusion with reflow and those with permanent coronary artery ligation, Powell and coworkers[9] reported decreased attentuation to be associated with acute myocardial ischemia in isolated hearts. They found good correlation between the amount of myocardial edema and the relative decrease in regional myocardial attenuation with CT scanning. They therefore propose CT as a potential means for sequential noninvasive quantitation of myocardial edema associated with ischemia.

Our early investigation of isolated canine hearts with 18 to 48-hour-old myocardial infarcts produced by permanent coronary artery occlusion also confirmed a decreased attenuation associated with acutely infarcted myocardium.[10] In addition, we sought to detect the histologically present calcification which occurs in the peripheral zones of infarcts 24 to 48 hours old. This calcification would have been expected to increase attentuation but was not seen by CT in our study. When adequate blood-myocardial wall interface was produced with the intravenous injection of iodinated contrast prior to sacrifice, measurements of ventricular wall thickness demonstrated a significant reduction in thickness in the infarcted myocardium. Early mild thinning of the infarcted wall appears to be related to dyskinesis resulting in fixed stretching of muscle cells during the evolution of infarction.

These early in vitro studies have demonstrated the potential of CT scanning to detect myocardial infarction. They show that in arrested hearts, cardiac anatomy is readily visualized, and the thinning associated with acute myocardial infarction is detectable. Acute myocardial infarction produces a decrease in attenuation coefficient relative to normal muscle, and this change can be at least partly attributed to injury-associated edema.

INFARCT SIZING BY COMPUTED TOMOGRAPHY

The logical progression after demonstrating the potential of CT to detect acute myocardial infarction was to evaluate its potential to quantitate infarct size. To study this potential we evaluated 20 canine hearts in which either anterior or posterior infarctions were produced by coronary artery ligation.[11] Each animal had been injected intravenously with iodinated contrast immediately prior to sacrifice which occurred 24 to 48 hours postinfarction. The hearts were removed and contiguous scans of 3-mm thickness were made from apex to base in the CT head scanner. Thus, arrested hearts were scanned in the optimal situation for identifying and quantitating acute myocardial infarction. To determine CT infarct volume the infarcts were outlined on prints from each 3-mm scan slice (Fig. 1), and the area was determined by sonic planimetry. The sum of the infarct areas was an estima-

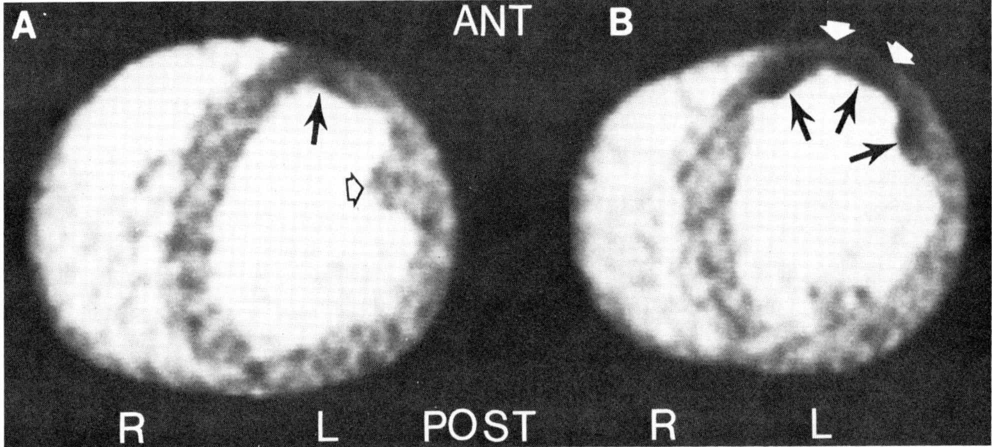

Figure 1. Two CT scans of an isolated arrested canine heart sacrificed immediately after intravenous injection of iodinated contrast. A, Both the right (R) and left (L) ventricles are easily identified as is the anterior papillary muscle (hollow black arrow). The solid black arrow points to a small area of infarction which becomes more evident on the second scan (B) which is 1.2 cm closer to the apex. The infarct (solid black arrows) is displayed as black representing decreased x-ray attenuation. This anterior infarct was 24 hours old and weighed 19 grams. The white arrows point to subepicardial sparing. A portion of the anterior papillary muscle is infarcted; the posterior papillary muscle is seen on this scan.

tion of total infarct volume, and the results were compared with gross infarct weight. There was correlation of infarct volume determined by CT with actual infarct weight. However, CT tended to underestimate infarct volume. This underestimation of infarct volume by CT has also been reported by Siemers and coworkers.[12] When contrast is given just prior to sacrifice the infarcted myocardium is easily differentiated as an area of decreased x-ray attenuation due to the absence of the high density contrast material. The underestimation of infarct volume seems logically to be related to collateral flow of contrast into the periphery of the infarcted tissue.

In this study, CT readily visualized all infarcts greater than one-half gram in size, and one of three infarcts weighing a half-gram or less was localized and accurately sized by this technique. Thus, CT potentially represents an extremely sensitive technique for infarct detection. Although there is underestimation of infarct size by this technique, correlation between infarct volume and actual infarct weight has been shown.

CURRENT AND POTENTIAL CAPABILITIES

Currently, multiple pathologic conditions have been demonstrated in nongated CT scans. These conditions include specific chamber enlargement, vascular calcification, calcified aneurysms which have not been seen on routine radiographs, and pericardial effusions. CT is not presently the diagnostic instrument of choice for recognition of these conditions; however, these current capabilities should of course not be overlooked.

The practical use of the potential to identify and quantitate areas of myocardial infarction primarily awaits the development of techniques to reduce the detrimental effect of cardiac motion on CT reconstruction. Harell and associates[13] were the first to report the successful use of cardiac computed tomography in a patient. Utilizing the postdata acquisition correlation of data with electrocardiographic cycle, they divided the cardiac cycle into seven equal segments. This "stop-action" imaging of the heart enabled improved anatomic visualization not possible in routine CT scanning. More sophisticated techniques must be developed, and as scan time decreases, high resolution CT studies of the heart will become reality.

The current diagnostic usefulness of CT imaging of the heart is relatively limited. However, the potential usefulness of this noninvasive technique for imaging the heart is tremendously exciting. With its very sensitive capacity to resolve small differences in x-ray attenuation, CT in combination with vascular contrast can easily distinguish ischemic from normally perfused tissues. In vitro studies have shown this technique to be extremely sensitive in infarct detection and to have great potential in sizing acute myocardial infarction. This imaging technique should allow accurate measurements of ventricular dimensions and be a noninvasive means to detect various regional myocardial abnormalities. Rapid development of CT imaging systems for cardiac work is anticipated and certainly will be the focus of much attention in the immediate future.

REFERENCES

1. HOUNSFELD, G. N.: *Computerized transverse axial scanning (tomography): Part 1. Description of the system.* Br. J. Radiol. 46:1016, 1973.

2. TER-POGOSSIAN, M. M., PHELPS, M. E., BROWNELL, G. L., ET AL.: *Reconstruction Tomography in Diagnostic Radiology and Nuclear Medicine.* University Park Press, Baltimore, 1977.

3. ALFIDI, R. J., MacINTYRE, W. J., AND HAAGA, J. R.: *The effects of biological motion on CT resolution.* Am. J. Roentgenol. 127:11, 1976.

4. BROOKS, R. A., AND DiCHIRO, G.: *Theory of image reconstruction in computed tomography.* Radiology 117:561, 1975.

5. WOOD, E. H.: *Cardiovascular and pulmonary dynamics by quantitative imaging.* Circ. Res. 38:131, 1976.

6. TER-POGOSSIAN, M. M.: *Computerized cranial tomography: equipment and physics.* Semin. Roentgenol. 12:13, 1977.

7. ADAMS, D. F., HESSEL, S. J., JUDY, P. F., ET AL.: *Computed tomography of the normal and infarcted myocardium.* Am. J. Roentogenol. 126:786, 1976.

8. TER-POGOSSIAN, M. M., WEISS, E. S., COLEMAN, R. E., ET AL.: *Computed tomography of the heart.* Am. J. Roentogenol. 127:79, 1976.

9. POWELL, W. J., JR., WITTENBERG, J., MATURI, R. A., ET AL.: *Detection of edema associated with myocardial ischemia by computerized tomography in isolated, arrested canine hearts.* Circulation 55:99, 1977.

10. GRAY, W. R., JR., PARKEY, R. W., BUJA, L. M., ET AL.: *Computed tomography:* in vivo *evaluation of myocardial infarction.* Radiology 122:511, 1977.

11. GRAY, W. R., JR., LEWIS, S. E., PARKEY, R. W., ET AL.: *Computed tomography for localization and sizing of acute anterior myocardial infarcts in dogs.* Circulation 56(Suppl. III):18, 1977.

12. SIEMERS, P. T., HIGGINS, C. B., AND SCHMIDT, W.: *Computed tomographic evaluation of myocardial infarction: Its changing image with time and contrast.* Presented to the Radiological Society of North America, December, 1977.

13. HARELL, G. S., GUTHANER, D. F., BRIEMAN, R. S., ET AL.: *Stop-action cardiac computed tomography.* Radiology 123:515, 1977.

Index